Unit Operations of Agricultural Processing

(SECOND REVISED EDITION)

K.M. Sahay
Sr. Scientist,
Central Institute of Agricultural Engineering,
Bhopal

and

K.K. Singh
Scientist,
Central Institute of Agricultural Engineering,
Bhopal

VIKAS® PUBLISHING HOUSE PVT LTD

VIKAS® PUBLISHING HOUSE PRIVATE LIMITED
E-28, Sector-8, Noida-201301 (UP), India
Phone: +91-120-4078900 • Fax: +91-120-4078999
Registered Office: A-27, 2nd Floor, Mohan Co-operative Industrial Estate, New Delhi-110044
E-mail: helpline@vikaspublishing.com • Website: www.vikaspublishing.com

SALES AND BRANCH OFFICES OF VIKAS® AND S CHAND AND COMPANY LIMITED

Chennai : Ph: 044-2363 2120, chennai@schandpublishing.com
Guwahati : Ph: 0361-2738 811, 2735 640, guwahati@schandpublishing.com
Hyderabad : Ph: 040-4018 6018, hyderabad@schandpublishing.com
Jalandhar : Ph: 0181-4645 630, jalandhar@schandpublishing.com
Kolkata : Ph: 033-2335 7458, 2335 3914, kolkata@schandpublishing.com
Lucknow : Ph: 0522-4003 633, lucknow@schandpublishing.com
Mumbai : Ph: 022-2500 0297, mumbai@schandpublishing.com
Patna : Ph: 0612-401 1400, patna@schandpublishing.com

Unit Operations of Agricultural Processing
ISBN: 978-81259-1142-5 Product Code: V3AGP64AGRI10ENAB01O

Second Revised Edition 2001
Reprinted several times between 2001 and 2023, 2024 (Thrice), 2025
Reprint 2026

PRINTED IN INDIA
By Shri Shyamlal Printing Press Private Limited, Plot 20/4, Site-IV, Industrial Area Sahibabad, Ghaziabad-201010 and Published by Vikas Publishing House Private Limited, E-28, Sector-8, Noida-201301

Dedicated to our Wives
Mrs. Renu Sahay
and
Mrs. Usha Singh

Preface to the Second Edition

The authors were deeply heartened by the acceptance of the first edition of *Unit Operations of Agricultural Processing* which came out in 1994. After a lapse of nearly five years, this basic textbook is now being revised. The purpose of this revision is to add some more basic theory and materials in storage, milling and food engineering chapters. Heat Exchangers which was not previously included in the chapter Elementary Food Engineering has now been added to it.

We hope that readers will find this book informative and useful. The authors would welcome valuable suggestions from readers to further improve the text of the book.

K.M. Sahay
K.K. Singh

Preface to the First Edition

Almost all the food, feed, fibre and fuel commodities go through a number of post harvest processing operations such as cleaning, grading, separation, drying, storage, milling, food engineering operation, packaging, transport and marketing before reaching to the consumers. Post harvest technology is in practice since dawn of human civilization. Agricultural processing has been defined as any processing activity that can be done on the farm or by local enterprise in which the farmer has an active interest. Post harvest processing is comparatively a new discipline and is directed to conservation of produce, process goods and value adding to make the material more readily useable and economically more remunerative. It has become very important in Indian context with the introduction of high yielding varieties and modern agricultural practices ushering Indian agriculture from subsistence farming to commercial enterprise. The importance of post harvest technology and associated equipment is increasing and will continue to increase as agricultural productivity increases and growers make strides to become producer-cum-processor from present status of mere producer. A large number of industries came in during the World War II. They are rice mills, flour mills, mills for animal feed, oil extraction plants, fruit processing etc.

As there is always a limit to land use and agricultural productivity, to meet the ever increasing requirements of food and other agro based products, avoid losses and for utilising by-products in economical way knowledge of appropriate post harvest processing is essential. To meet the need of trained man-power in the field of post harvest processing of agricultural commodities, transfer of knowledge of the basic unit operations of agricultural processing is necessary.

An attempt has been made in this book to describe and discuss some of the very basic unit operations of agricultural processing. The authors are hopeful that this book will serve as a good teaching aid in the field of post harvest technology.

K.M. Sahay
K.K. Singh

Preface to the First Edition

[illegible]

[illegible]

[illegible]

[illegible]

[illegible]

Contents

Chapter 1

Introduction

Agricultural processing may be defined as an activity which is performed to maintain or improve the quality or to change the form or characteristics of an agricultural product. Processing operations are undertaken to add value to agricultural materials after their production. The main purpose of agricultural processing is to minimise the qualitative and quantitative deterioration of the material after harvest.

Some of the agricultural processing operations performed on various crops (cereals, pulses and oilseeds) are cleaning, grading, sorting, drying, cooling, storage, milling, size reduction, expelling, mixing, blending, packaging, waste utilisation, seed treatment etc.

A processing job consists of a series of 'unit operation'. The sequence of processing operations may broadly be classified as (1) cleaning, grading and sorting, (2) drying and dehydration, (3) storage, (4) milling, (5) handling, packaging and transportation and (6) agricultural wastes and by-products utilisation. Some of the unit operations are performed in more than one processing job like cleaning, drying and material handling.

Apart from the above unit operations many basic information on seeds and products like their engineering properties, heat transfer, refrigeration, structure and composition of food grains are required to evolve or develop appropriate processing procedures for a particular product. Information on design of fans and blowers is also necessary.

Cleaning, grading and sorting : The crops after harvesting and threshing contain organic and inorganic impurities like straw, chaff, weed seeds, iron pieces, stones, mud etc. Grains with impurities fetch lower price. Presence of foreign matters increases the bulk hence increases cost of handling and transportation. Contaminated grains cause high losses during storage. It may cause wear and tear in processing machines. Hence cleaning of grains and other agricultural products is very important and should be performed before any further processing operation. Grading of cleaned seed is done to obtain good quality seed with higher germination and viability. Grading of fruits and vegetables is necessary to group them as per various commercial standards and usage.

Drying : Grains are generally harvested at higher moisture contents to reduce harvesting losses. This moisture content is unsafe for storage or milling. Such grains need immediate treatment to prevent quality deterioration due to mold development. Post harvest treatments like sealed storage, chemical or thermal treatment, cooling and drying are given to the grains for their short and long term preservation. Drying of grains is most widely used method for their preservation. Table 1.1 shows the moisture content of some of the grains for their harvest and storage. Improper drying causes increase in breakage of maize and soybean and decrease in milling quality of rice and wheat.

Storage : Bulk of the post harvest losses occur during storage. The storage losses of grains are due to the insects, fungus, rodents, birds, moisture etc.

Table 1.1

Moisture content of some of the cereals, pulses and oilseeds at harvest and for safe storage

Crop	*Optimum m.c. at harvest % w.b.*	*Storage m.c., % w.b.*	
		6–12 months	*Over 1 year*
Wheat	16–18	12	10
Paddy	20–22	12	10
Maize	20–23	12	10
Sorghum	20–22	12	10
Pigeonpea	—	10	8
Bengalgram	—	10	8
Soybean	—	10	8
Groundnut	10–12	8	7
Mustard	—	8	6

To minimise storage losses knowledge of scientific ways of storage is necessary. Information on design of modern sealable storage structures is essential. Maintenance of hygeinic conditions during storage is also necessary.

Milling : The food grains contain a large centrally located starchy endosperm which is also rich in protein. These seeds are covered with protective layers of hull and bran. The hull is coloured and may contain tannins and is largely indigestible. The grains also contain germ which is high in oil and is enzymatically active and under certain conditions may produce rancidity in the grain. Therefore the hull and germ are removed in milling operation. The starchy and proteinous endosperm is obtained by milling and in some cases the endosperm is further comminuted.

The pulses or food legumes which are rich in protein supply 15–30% of daily protein to Indian people. Pulses are mostly consumed in the form of dhal, the dehusked and split kernels. Dehusking and splitting of pulses are performed in milling operation.

Oil is extracted from the oil bearing materials either by solvent or by mechanical expeller. Juice from various products is also extracted by milling machines.

Material handling : The performance of any processing activity is considerably affected by the efficiency of movement of material from one unit operation to another. Optimisation of time and movement of material in a processing plant are necessary to minimise cost of operation and maximise profit. Knowledge of correct layout of handling equipment is desirable.

Waste and by-products utilisation : About 80% of the total energy consumed in the rural areas falls under cooking of food. It is also estimated that about 80% of energy required for cooking is obtained mainly from wood, crop residue and cowdung. Nearly 50% of the energy used in rural areas can be saved by using improved cooking stoves and furnaces. Cowdung and other bio-mass can be used for generation of bio-gas. Economic utilisation of food grain milling by-products can pave ways for modernisation of pulse and rice mills.

Structure and composition of food grains

The knowledge of grain structure and composition is desirable for selecting the specific operation to be performed for processing it into various products. Basically there is similarity in the anatomical structure of all cereal grains, however there is difference between them in details only. The structure of some of the important food grains are shown in Fig. 1.1. to 1.5.

Wheat, maize, sorghum and rye consist of pericarp and seed. The seed is comprised of seed coat, germ and endosperm. Paddy, barley and pulses have an additional outer layer of husk. The pericarp, seed coat, germ and endosperm are further subdivided in various layers and tissues or regions.

Wheat grains are 5–8 mm long and 3.7 to 4.5 mm in equivalent diameter. The shape is ovoid and rounded at both ends. At one end the germ is prominent whereas at the other end there is a tuft of fine hairs. The pericarp and outer endosperm (or aleurone) layer consisting of 3–15% of the kernel weight are removed during milling operation. The removed portion is called 'bran'. The germ constitutes 2–3% of the kernel weight and is extracted during milling operation.

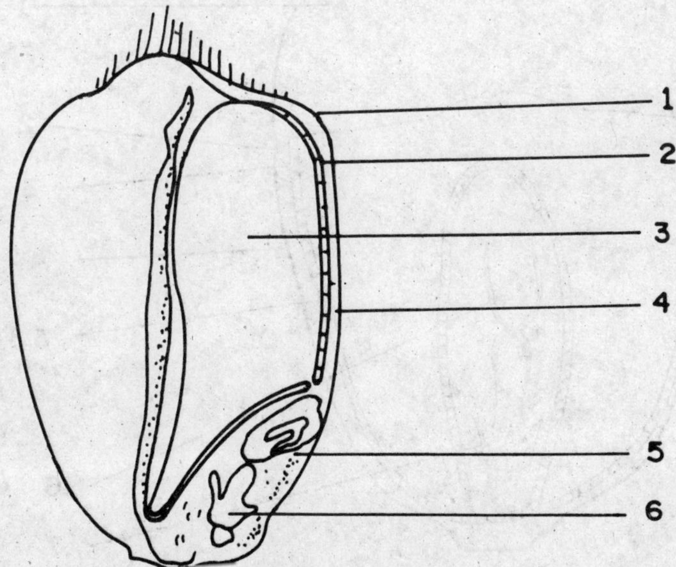

Fig. 1.1 : Structure of wheat grain
1. pericarp 2. aleurone layer 3. endosperm 4. seed coat 5. plumule 6. radicle

There is an indentation, called crease, along the ventral side of the wheat grain. The crease is an infolding of the aleurone and all covering layers. The presence of crease poses problem in separating endosperm from the enclosing layers during milling process.

Paddy grains are 5–10 mm long and 1.2 to 4.6 mm in equivalent diameter. The grains are covered with husk, flattened laterally with small point at the end. There is no crease in paddy grains. A paddy grain consists hull which is about 20% of weight of grain and brown rice kernel. The outer most six layers of cells of kernel form pericarp which is about 5 to 7% of the brown rice weight. These layers called bran are removed during milling to produce white rice. The layer next and below the pericarp constitutes seed coat. Below the seed coat are the layers of cells (upto six layers) called aleurone layer. This layer contains mainly protein. The seed coat together with aleurone layer form 4–6% of rice kernel. The entire mass of grain below the aleurone layer forms the endosperm. This is starchy portion of a rice grain. It consists of oblong rectangular cells and constitutes about 89 to 94% of the grain mass. At one end of the vertical side of the grain is embryo or germ, responsible for germination. It constitutes about 2 to 3% (by weight) of the rice grain.

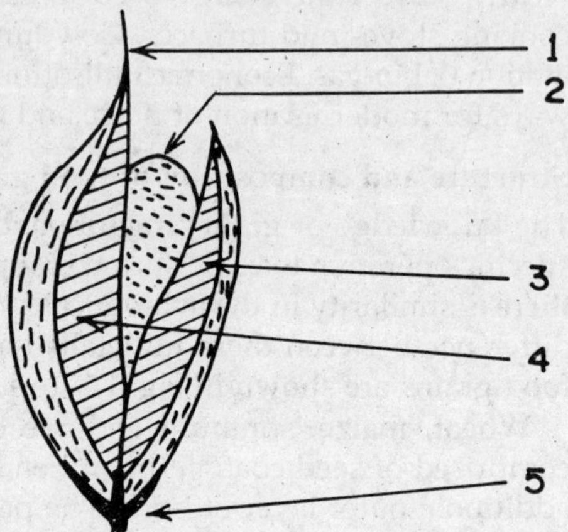

Fig. 1.2 : Structure of paddy grain
1. awn 2. rice kernel 3. palea 4. lemma 5. empty glume

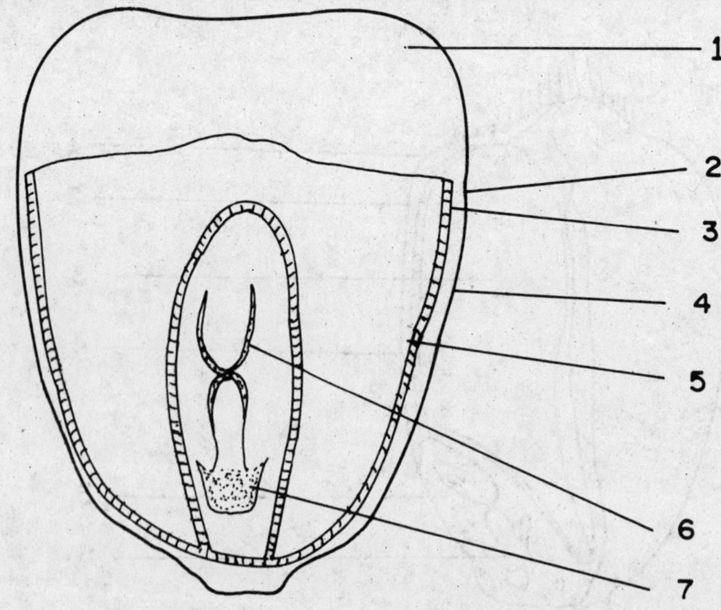

Fig. 1.3 : Structure of maize grain
1. husk 2. epidermis 3. mesocarp 4. seedcoat 5. aleurone layer 6. plumule 7. radicle

Maize grain is larger in size (8 to 17 mm long and 7.8 to 9.8 mm equivalent diameter) than other cereals. Its basal part is narrow and the apex is broad. A mature maize kernel is composed of four major parts. Pericarp (hull or bran), germ, endosperm and tip cap. The outermost layer of dead thick walled and elongated cells forms pericarp. Below this layer is a spongy layer of cells known as cross and tube cells. Next to this is seed coat or testa. Beneath the testa, there is a single thick layer known as aleurone layer. The layer constitutes about 3% of the kernel weight. The part of the hull overlying the germ is known as tip cap which constitutes about 0.83% of the kernel weight. This is the smallest fragment which connects the kernel to the cob. The tip cap is generally removed with the hull in milling operation. The germ of maize grain constitutes about 12% of the total kernel weight. It stores nutrients which are mobilised during germination. The mature endosperm of maize grain is composed of floury and horny regions. It constitutes about 82% of the kernel weight.

Sorghum grain is small and round in shape. It consists of an embryo (germ), a relatively large scutellum, an endosperm enclosed in a seed coat or testa, and a fruit coat or pericarp. Testa and pericarp are fused together.

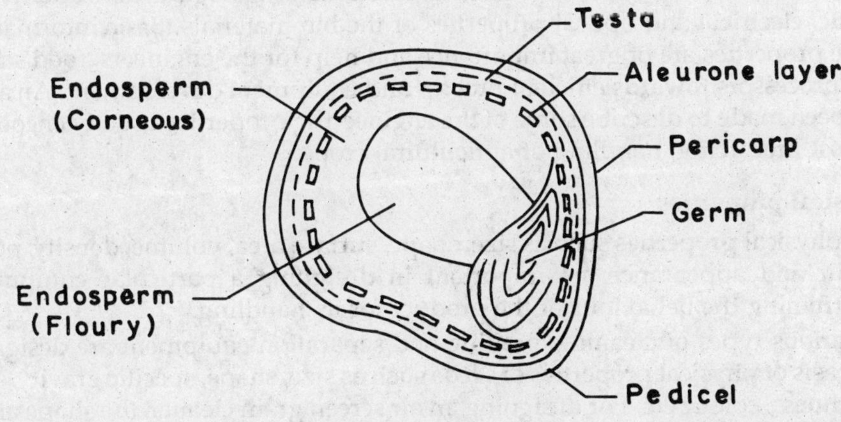

Fig. 1.4 : Structure of sorghum grain

There is much similarity in the structure of seeds of various pulses or food legumes. Pulses are dicot plants, however these differ from each other in colour, shape and size. The outermost layer of the seed is the testa or seed coat. The endosperm of a mature seed is in the form of a thin layer surrounding the cotyledons or embryo. There is a few external structure like hilum, microphyle and raphe. The hilum is an oval shaped scar in the middle where the seed is attached from the stalk. The microphyle is a small opening in the seed coat beside the hilum. The raphe is in a ridge from beside hilum and opposite to the microphyle.

Inside the seed coat the seed is composed of embyonic structure. The shoot of the seed consists of two cotyledons. The embyo consists of a radicle, a plumule or epicotyl, cotyledons and

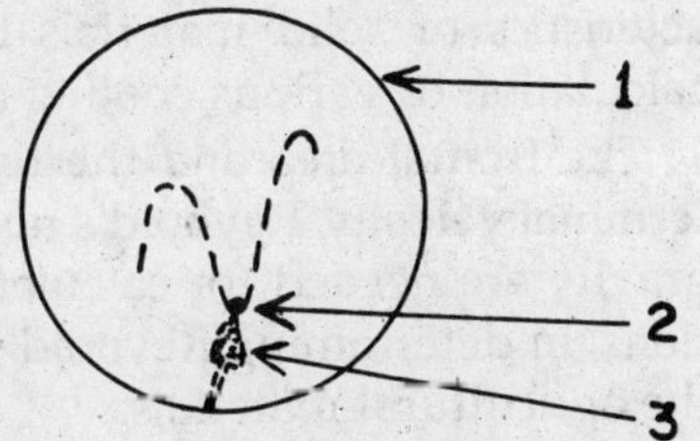

Fig. 1.5 : Structure of a pulse grain 1. seed coat 2. micropyle 3. hilum

hypocotyl which connects the radicle and the plumule.

Pulses contain 17–25% of protein by weight. The protein content of pulses is about double of the cereals and more than meat and egg. Pulses contain lipids and fats in low quantities and are good source of thiamin, niacin, calcium and iron. Though the ascorbic acid content of pulses is low in dry grains, but during their germination stage ascorbic acid is available. Some of the common food legumes are pigeonpea, greengram, blackgram, Bengalgram (chick-pea), lentil, field peas, horse gram etc.

Engineering properties of agricultural materials

Engineering properties are the properties which are useful and necessary in the design and operation of various equipment employed in the field of agricultural processing and also for design and development of other farm machinery. As mentioned earlier unit operations such as cleaning, grading, drying, dehydration, storage, milling, handling and transportation, thermal processing of foods are among the important operations in agricultural processing. In these operations while handling of grains and other commodities the properties which play an important role are physical, mechanical, frictional, rheological, aero and hydrodynamic, electrical and optical properties of the bio materials. Basic information on these properties are of great importance and help for the engineers, food scientists and processors towards efficient process and equipment development. An attempt has been made to describe some of the engineering properties usually encountered in post production handling of agricultural crops.

Physical properties

The physical properties such as size, shape, surface area, volume, density, porosity, colour and appearance are important in designing a particular equipment or determining the behaviour of the product for its handling.

Various types of cleaning, grading and separation equipment are designed on the basis of physical properties of seeds such as size, shape, specific gravity, surface roughness, colour etc. For designing an air screen grain cleaner, the shape and size of the grain determine the shape and size of screen openings, angle of inclination and vibration amplitude and frequency of screens. The density of the grains decides the size of screening surface. The frontal area and related diameters and density are essential for determination of terminal velocity of the grain. Terminal velocity is necessary to decide about the winnowing velocity of air blast for separation of lighter materials in air screen grain cleaners.

The shape of product is an important parameter which affects conveying characteristics of solid materials by air or water. The shape is also considered in calculation of various cooling and heating loads of food materials.

The frontal area and the related diameters are essential for determination of terminal velocity, Reynold's number and drag coefficient. The density and specific gravity are needed for calculating the thermal diffusivity in heat transfer operations, in determining Reynold's number, in pneumatic and hydraulic handling of the agricultural materials.

The surface characteristics, colour and appearance are exploited for selective separation and storage of fruits and vegetables.

Some of the important physical properties are described below.

Shape and size

The following parameters may be measured for describing the shape and size of the granular agricultural materials.

(i) *Roundness* : It is a measure of the sharpness of the solid material. The most widely accepted methods for determining the roundness of irregular particle are given below.

Roundness (Fig. 1.6a)

$$= \frac{\text{Largest projected area of the particle when it is in natural rest position, } A_p}{\text{Area of smallest circumscribing circle, } A_c}$$

$$\text{Roundness ratio (Fig.1.6b)} = \frac{\text{Radius of curvature, } r, \text{ of the sharpest corner}}{\text{Mean radius of the particle, } R}$$

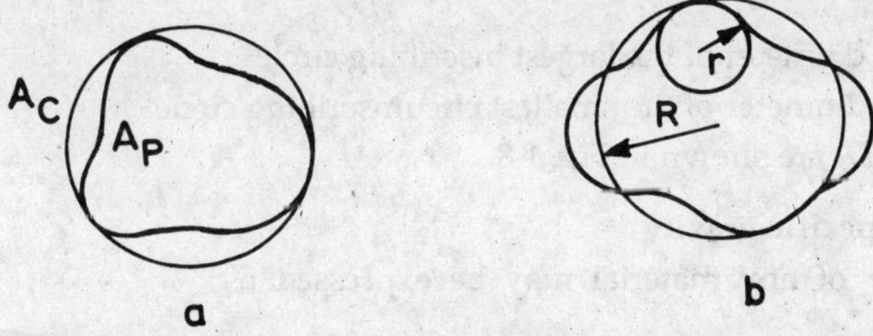

Fig. 1.6: Diagram for roundness and roundness ratio of a particle

(ii) *Sphericity* : Sphericity may be defined as the ratio of the diameter of a sphere of the same volume as that of the particle and the diameter of the smallest circumscribing sphere or generally the largest diameter of the particle. This parameter shows the shape character of the particle relative to the sphere having same volume. If D_e is the diameter of a sphere having same volume as that of the particle and D_c is the diameter of the smallest circumscribing sphere, then the sphericity can be expressed as under.

$$\text{Sphericity} = \frac{D_e}{D_c}$$

Considering the volume of the particle is equal to the volume of a ellipsoid having three axes with intercepts l, b and t and diameter of the circumscribed sphere is the largest intercept, l, of the ellipsoid as shown in Fig. 1.7., the degree of sphericity can be expressed as given below.

$$\text{Sphericity} = \sqrt[3]{\frac{\text{Volume of the particle}}{\text{Volume of circumscribed sphere}}}$$

$$= \left[\frac{\frac{\pi}{6} l\, b\, t}{\frac{\pi}{6} l^3}\right]^{1/3}$$

Fig. 1.7: Intercepts of an elongated grain

$$= \frac{(l\,b\,t)^{1/3}}{l}$$

$$= \frac{\text{geometric mean diameter}}{\text{major intercept}}$$

Where, l = largest intercept

b = largest intercept perpendicular to l

t = largest intercept perpendicular to l and b

It is not necessary that the three intercepts of the particle intersect each other at a common point.

The geometric mean diameter of the particle is also called as the "equivalent diameter".

The sphericity can also be defined as:

$$\text{Sphericity} = \frac{D_i}{D_c}$$

Where, D_i = diameter of the largest inscribing circle

D_c = diameter of the smallest circumscribing circle

The D_i and D_c are shown in Fig. 1.8.

Density and specific gravity

The density of any material may be expressed as below,

$$\text{Density} = \frac{\text{Weight of the material, } kg}{\text{Volume of the material, } m^3}$$

The density and specific gravity values of grains and other commodities are used in design of storage bins and silos, separation of desirable materials from impurities, cleaning and grading, evaluation of the grain maturity, texture and softness of fruits, quality evaluation of the products etc.

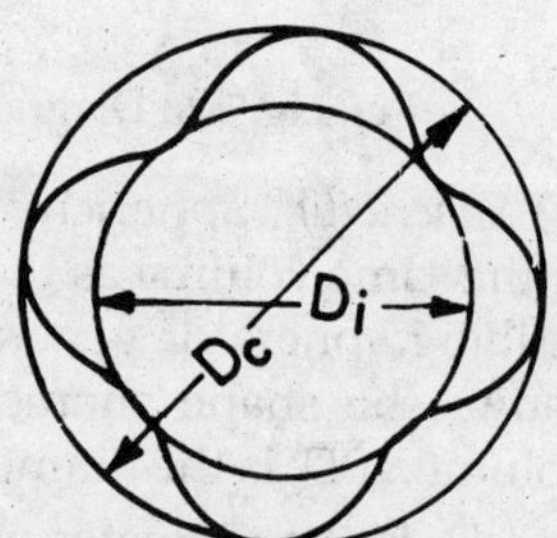

Fig. 1.8: Diagram of smallest circumscribing and largest inscribing circles of a particle

Some of the agricultural products having irregular shape, small size and void spaces pose certain problems in measurement of volume and density. The volume of such irregular shaped material is generally determined by displacement method which has been described here.

Determination of the volume by Platform scale : The platform scale is a simple technique which is commonly used for determination of volume of large materials like fruits and vegetables. The weight of material is determined by weighing on the scale in air, thereafter, the material is forced into the water with the help of a rod. The later reading of the scale while the material is submerged minus the weight of container and water is the actual weight of the displaced water. The volume, density and specific gravity of the material are estimated by following formulae.

$$\text{Volume, } m^3 = \frac{\text{weight of displaced water, } kg}{\text{weight density of water, } kg/m^3}$$

$$\text{Weight density, } kg/m^3 = \frac{\text{weight of material in air, } kg}{\text{volume of the material, } m^3}$$

$$\text{Specific gravity} = \frac{\text{weight in air} \times \text{specific gravity of water}}{\text{weight of displaced water}}$$

Determination of Specific gravity by Pycnometer Method : The specific gravity bottle or pycnometer and toluene ($C_6H_5CH_3$) are used for determination of specific gravity of granular agricultural materials. The following procedure is used for determination of the specific gravity.

(*i*) The weight of empty pycnometer and the weight of pycnometer filled with water at 20°C are taken.

(*ii*) The specific gravity of toluene is determined as the ratio of the weight of toluene in the bottle and weight of distilled water in the bottle at same temperature.

$$\text{Specific gravity of toluene} = \frac{\text{weight of toluene}}{\text{weight of water}}$$

(*iii*) Ten gram sample of grain is placed in the pycnometer and filled with toluene to cover the sample.

(*iv*) The air is completely exhausted from the bottle by a vacuum pump.

(*v*) When there is no air bubble remain in the bottle, the bottle is filled with toluene and the temperature is allowed to reach 20°C.

(*vi*) The bottle is weighted and the specific gravity of grain is calculated by following expressions,

Specific gravity of grain

$$= \frac{\text{specific gravity of toluene} \times \text{weight of grain}}{\text{weight of the toluene displaced by the grain}}$$

$$= \frac{\text{specific gravity of toluene} \times \text{weight of grain}}{[(\text{weight of sample}) - (\text{weight of toluene} + \text{sample} + \text{bottle}) - \text{weight of toluene} + \text{bottle})]}$$

Aero and Hydrodynamic properties

The aero and hydrodynamic properties such as terminal velocity of agricultural products are important and required for designing of air and water conveying systems and the separation equipment. The physical properties such as density, shape and size, drag coefficient etc. are required for calculating the terminal velocity of the material. For example, in pneumatic conveying and separation processes the material is lifted only when the air velocity is greater than its terminal velocity.

Drag coefficient

When a particle is immersed in a fluid current the forces acting on the particle has been illustrated in Fig. 1.9.

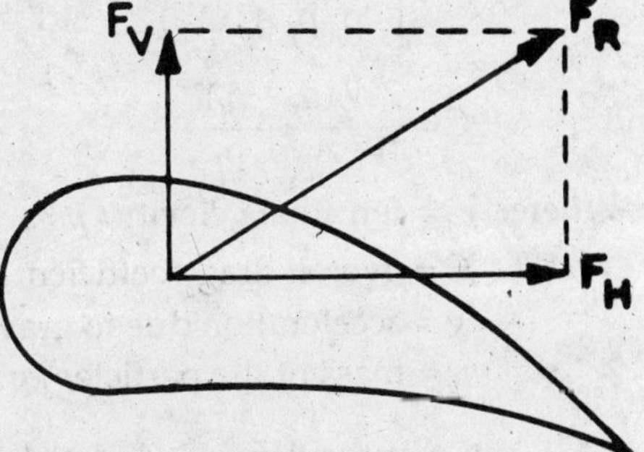

Fig. 1.9: Illustration of the forces acting on a body immersed in a fluid current

If F_V is the drag force and F_H is lift force, then the resultant force, F_R may be calculated by resolving the forces.

The F_V and F_H can be derived by dimensional analysis technique. It is assumed that the particle is smooth and having a projected area, A_p, moving through a fluid of mass density, ρ_f, viscosity η, and modulus of elasticity, E, with a velocity, V then;

$$F_V = f_1 (A_p, \rho_f, \eta, E, V) \quad ...1.1$$

and $$F_H = f_2 (A_p, \rho_f, \eta, E, V) \quad ...1.2$$

Using dimensional analysis technique, the drag and lift forces are:

$$F_V = C_V \, A_p \frac{\rho_f V^2}{2} \quad ...1.3$$

and $$F_H = C_H \, A_p \frac{\rho_f V^2}{2} \quad ...1.4$$

The C_V and C_H are drag and lift coefficient of the material respectively.

The resultant force F_R can be given as;

$$F_R = C \, A_p \frac{\rho_f V^2}{2} \quad ...1.5$$

Where, F_R = resistance drag force or weight of particle at terminal velocity, kg

C = Overall drag coefficient

ρ_f = mass density of fluid, $\frac{Kg\, s^2}{m^4}$

A_P = projected area of the particle normal to direction of motion, m^2

V = relative velocity between main body of fluid and material, m/s

Terminal velocity

The terminal velocity of a particle may be defined as equal to the air velocity at which a particle remains in suspended state in a vertical pipe. In the condition of free fall, the particle attains a constant terminal velocity, V_t, the net gravitational accelerating force, F_g, equals the resisting upward drag force, F_r.

If $V = V_t$, $F_g = F_r$

By substituting the values of F_g and F_r the terminal velocity can be expressed as;

$$m_p \, g \left[\frac{(\rho_p - \rho_f)}{\rho_p} \right] = 1/2 \, C \, A_p \, \rho_f \, V_t^2 \quad ...1.6$$

$$\therefore \quad V_t = \left[\frac{2 \, W (\rho_p - \rho_f)}{\rho_p \, \rho_f \, A_p \, C} \right]^{1/2} \quad ...1.7$$

and $$C = \frac{2W (\rho_p - \rho_f)}{V_t^2 \, A_p \, \rho_p \, \rho_f} \quad ...1.8$$

Where, V_t = terminal velocity, m/s

C = overall drag coefficient

g = acceleration due to gravity, m/s^2

m_p = mass of the particle, kg

ρ_p = mass density of particle, $\frac{kgs^2}{m^4}$

ρ_f = mass density of fluid, $\frac{kgs^2}{m^4}$

A_p = projected area of particle in perpendicular direction of motion, m^2
W = weight of particle, kg

In the steady state conditions, after attaining the terminal velocity, if the density of the particle is greater than the density of the fluid, the particle will move downward. If the density of the particle is lesser than the density of the fluid, the particle will rise upward. If the separation of mixture of wheat and foreign matters is to be achieved by air stream, the terminal velocities of each component of mixture decide the range of air velocity to be used for a definite extent of separation. In Table 2.4 the terminal velocity values of various components of wheat-chaff mixture to achieve 50% and above separation are given. Table 1.2 shows the terminal velocities of some of the agricultural grains and oilseed for airborne condition. In Table 1.3 the values of the terminal velocity and drag coefficient of groundnut and soybean are given.

The drag coefficient of the material and its resistance to air flow depend upon, (1) the bed thickness of the material, (2) type, shape and size of grain, (3) the air velocity and (4) orientation and packing of the material.

In the design of blowers for aeration and drying gadgets, the resistance exerted by the grain bed to air current blown through it must be known.

Table 1.2

Air velocity requirement for air borne of some of the agricultural materials

Grain	*Unit density, kg/m³*	*Terminal velocity, m/s*
Wheat	998–1238	9–11.5
Rye	1158–1218	8.5–10
Oats	738–968	8–9
Corn	1138–1198	34.9
Soybean	1029–1152	44.3

Table 1.3

Terminal velocity and drag coefficient for groundnut and soybean

Grain	*Terminal velocity m/s*		*Drag Coefficient*	
	Range	*Mean*	*Range*	*Mean*
Groundnut kernel (RS–1)	12.31–13.78	13.23	0.52–0.64	0.58
Soybean (Punjab–1)	12.30–13.92	13.40	0.38–0.62	0.47
Soybean (Lee)	13.30–14.55	14.17	0.33–0.51	0.41

Frictional properties

The frictional properties such as coefficient of friction and angle of repose are important in designing of storage bins, hoppers, chutes, pneumatic conveying system, screw conveyors, forage harvesters, threshers etc.

The rolling resistance or maximum angle of stability in rolling of round shape agricultural materials is useful in designing handling equipment e.g. conveying of fruits and vegetables by gravity flow.

In mechanical and pneumatic conveying systems, the material generally moves or slides in direct contact with the trough, casing and other components of the machine. Thus various parameters affect the power requirement to drive the machine. Among these parameters, the frictional losses is one of the factors which must be overcome by providing additional power to the machine. Hence, the knowledge of frictional properties of the agricultural materials is necessary, therefore, some of the important frictional properties of agricultural products have been described here.

Static friction : The friction may be defined as the frictional forces acting between surfaces of contact at rest with respect to each other.

Kinetic friction : It may be defined as the friction forces existing between the surfaces in relative motion.

If F is the force of friction, and W is the force normal to the surface of contact, then the coefficient of friction 'f' is given by the relationship

$$f = \frac{F}{W}$$

The coefficient of friction may also be given as the tangent of the angle of the inclined surface upon which the friction force tangential to the surface and the component of the weight normal to the surfaces are acting.

Rolling resistance : If a round or cylindrical shaped object rolls over a horizontal surface with force, F, and the deformation in surface occurs, there will be a resultant force, R, exerted by the surface on the body as shown in Fig. 1.10. If the moment of forces is taken about point of application of R and the accelerating force is neglected, then

$$\sum M_b = F \, . \, a - W \, . \, c = 0$$

Assuming the deformation of surface as very small, a is approximately equal to r (Fig. 1.10), then

$$c = \frac{F \cdot r}{W} \text{ or } F = \frac{cW}{r} \qquad \ldots 1.9$$

The terms c and F may be defined as the coefficient of rolling resistance and rolling resistance respectively. It is evident from the equation that for rigid surfaces, small values of c is obtained resulting in smaller rolling resistance. The rolling resistance is directly proportional to the weight of the rolling object and to the coefficient of rolling resistance which is dependent on the rigidity of the supporting surface and indirectly proportional to the effective radius of the rolling object.

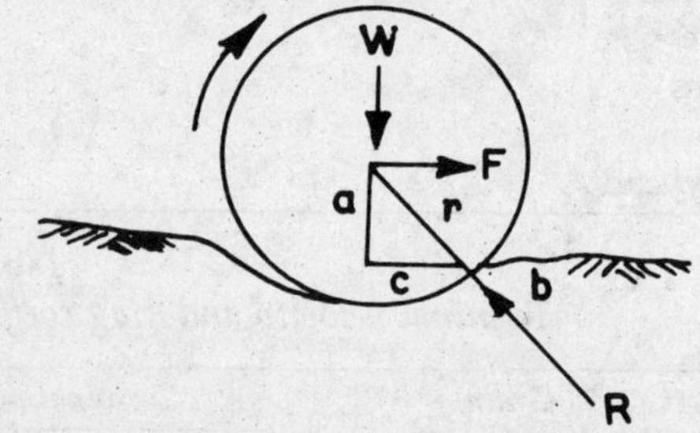

Fig. 1.10 : Force exerted on a wheel in motion

Various mechanisms have been designed by making use of difference in rolling resistance of the materials. One such example is the separation of potatoes and stones. The rolling resistance of stones differs from that of potatoes. The equal rolling resistance is observed only in the case of large potatoes and very small stones.

Angle of repose : The angle of repose is the angle between the base and the slope of the cone formed on a free vertical fall of the granular material to a horizontal plane. The size, shape, moisture content and orientation of the grains affect the angle of repose.

There are two angles of repose, (1) static angle of repose, and (2) dynamic angle of repose.

Static angle of repose : It is the angle of friction taken up by granular material to just slide upon itself.

Dynamic angle of repose : It comes in picture when bulk of the grain is in motion like discharge of grain from bins and hoppers. The dynamic angle of repose is more important than static angle.

Measurement of angle of repose

Method I : A wooden frame full of grain sample is mounted on a tilting top drafting table. The table top is tilted till the grain starts moving over the inclined surface Fig. 1.11. The angle of inclination is measured which is the angle of repose of the grain sample.

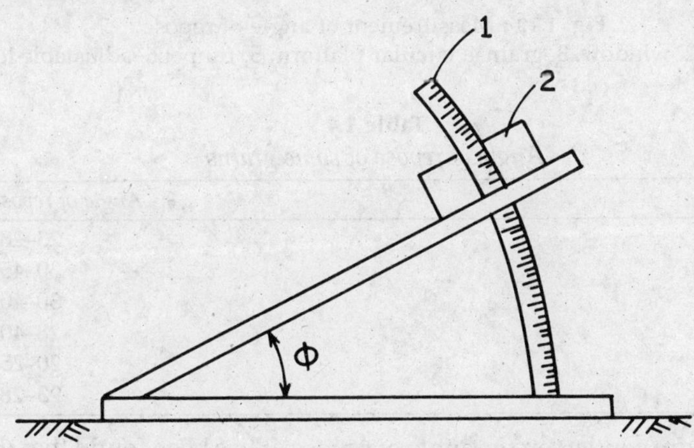

Fig. 1.11 : An apparatus for measurement of angle of repose
1. scale 2. grain container

Method II : This apparatus consists of a circular platform immersed in a box filled with grain and with a glass window in one side. The platform is supported by three adjustable screw legs and is surrounded by a metal funnel leading to a discharge hole. The grain is allowed to escape from the box, leaving a free-standing cone of grains on the platform. A travelling microscope is used to measure the heights as shown in Fig. 1.12. The angle of repose, ϕ, is obtained from the geometry of the cone as given below.

$$\phi = \tan^{-1} \frac{2\,(H_a - H_b)}{D_b} \qquad \ldots 1.9$$

Where H_a, H_b and D_b are the height of the cone, height of the platform and diameter of the platform respectively. The angle of repose for some of the grains is given in Table 1.4.

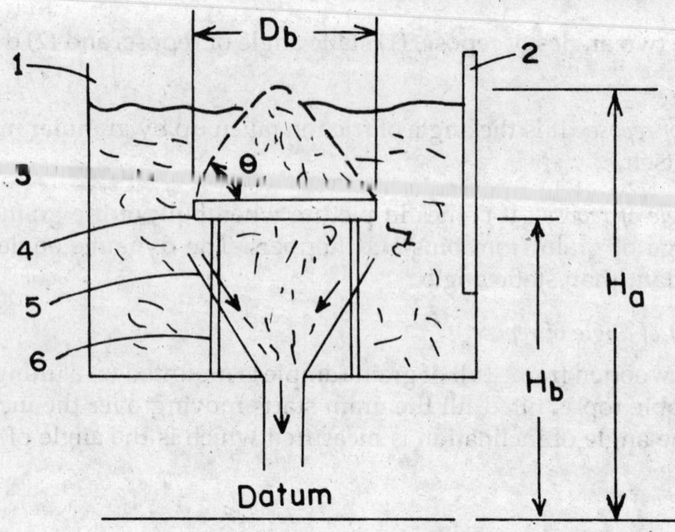

Fig. 1.12 : Measurement of angle of repose
1. box 2. window. 3. grain 4. circular platform 5. funnel 6. adjustable leg

Table 1.4
Angle of repose of some grains

Grain	*Angle of repose, degree*
Wheat	23–28
Paddy	30–45
Maize	30–40
Barley	28–40
Millets	20–25
Rye	23–28

Effect of moisture content on the angle of repose: It has been found that the angle of repose increases with the increase of moisture content of material. This variation of angle of the repose with moisture content occurs because surface layer of moisture surrounding the particle holds the aggregate of grain together by the surface tension.

An empirical equation has been developed to correlate the angle of repose and moisture content for rice. The equation was tested with the experimental values which gave the correlation coefficient of 97%. The equation is as follows.

$$\tan\phi = an^2 + b\left(\frac{M}{Dav}\right) + cs + d \qquad ...1.10$$

Where,

ϕ = angle of repose

n = shape factor = $\dfrac{\text{specific surface of solid}}{\text{specific surface of sphere}}$

M = percent moisture content

Dav = average screen particle diameter

s = specific gravity

a, b, c and d = constants

Angle of internal friction : The value of angle of internal friction is equal to the tangent of the coefficient of friction for the material.

The angle of internal friction is an important property which helps to estimate the lateral pressure in storage silos. Angle of internal friction values are also used in designing of storage bins and hopper for gravity discharge. The coefficient of friction between grains is required as a design parameter for design of shallow and deep bins.

The applications of angle of repose and angle of internal friction for designing storage structures have been described in chapter 4.

Difference between angle of repose and angle of internal friction

The engineers generally assume that both the angle of repose and angle of internal friction are same. Some investigators have indicated that for sorghum, the angle of repose and angle of internal friction are different. If one is used in place of other to design any system, it will lead to error.

Some investigators also attempted to develop a relationship between angle of repose and angle of internal friction so that by the simple test of angle of repose, the angle of internal friction could be estimated. But the results revealed that the two angles run almost parallel to each other for various moisture content levels, thus no simple relationship existed by which angle of internal friction could be estimated from angle of repose within a reasonable accuracy.

The angle of repose is generally higher than angle of the internal friction for the grains of approximately the same moisture content and density.

Mechanical properties

Mechanical properties may be defined as those which affect the behaviour of the agricultural material under the applied force. The mechanical properties such as hardness, compressive strength, impact and shear resistance and the rheological properties affect the various operations of agricultural processing. Data on these properties are useful for application in designing equipment for milling, handling, storage, transportation, food processing etc.

The mechanical damage to grain and seed in threshing and handling operations causes reduction in germination power and viability of seeds, increases the chances of insect and pest infestation and also affect the quality of final product.

The hardness of the grain effects the milling characteristics and it is also useful to live stock feeders and plant breeders.

The impact and shear resistance are important for size reduction of food grains. These information are useful in determination of the appropriate methods of crushing, breaking or grinding the grains. These properties also play important roles towards seed resistance to cracking under harvesting and threshing conditions.

Thermal properties

The thermal properties like specific heat, thermal conductivity, thermal diffusivity, enthalpy, surface heat transfer coefficient, emissivity etc. are important for the development of any thermal processing system. The thermal processing may include heating, cooling, freezing, drying etc. There are some other thermal properties like melting or freezing points, latent heats, heat of adsorption, heat of respiration, coefficient of thermal expansion etc., but these are not of much importance for the most heat transfer applications.

The heat treatment to cereals (wheat, maize etc.) and some of the pulses is given for stimulating germination. The heat treatment given to cereals like wheat, maize, sorghum and few millets for thermal killing of insect-pest in storage has been proved to be a promising technology in recent years.

Heat treatment is also given to the fruits for fruit fly quarantine disinfectation. In all the calculations of heat treatment the heat balance for heating is worked out with the knowledge of specific heat and thermal conductivity.

To design a dryer, the calculations of heat requirement is the most important step. For calculation of sensible and latent heat requirement, the thermal properties like specific heat and thermal conductivity should be known.

The enzymatic activities and attack of micro organisms on fruits and vegetables can be minimised for extending shelf life by lowering the storage temperature. There are many other thermal processing operations for which the values of thermal properties of agricultural materials should be known.

Some of the important thermal properties have been described here.

(i) Specific heat : The specific heat may be defined as the amount of heat in kilocalories that must be added to or removed from 1 kg of a substance to change its temperature by 1°C. The specific heat of wet agricultural material is the sum of specific heats of bone dry material and its moisture content. If C_d and C_w are the specific heats of bone dry material and water respectively, and m is the moisture content of the material in percent wet basis, then the specific heat can be expressed as given below.

$$C = \left(\frac{m}{100}\right) C_w + \left(\frac{100-m}{100}\right) C_d \ k\,cal/kg°C$$

Note : The above relationship exists above 8% moisture content of the grain only.

The specific heat of bone dry grain varies from 0.35 – 0.45 *k cal/kg°C*.

The specific heat is measured by calorimeter, generally a simple thermos vacuum bottle. The differential scanning–calorimeter (DSC) is suitable for measuring the specific heat.

(ii) Thermal conductivity : The thermal conductivity may be defined as the rate of heat flow through unit thickness of material per unit area normal to the direction of heat flow and per unit time for unit temperature difference. It is a measure of ability of the material to conduct heat. The thermal conductivity can be expressed by the following equation.

$$Q = K A \Delta T \qquad ...1.11$$

where Q = amount of heat flow, *k cal*

A = area, m^2

ΔT = temperature difference in the direction of heat flow, °C

K = thermal conductivity, *k cal/m.hr.°C*

If, m is the moisture content of the grain, then the thermal conductivity of wheat can be expressed as follows,

$$K = 0.060 + 0.002\, m, \ k\,cal/m.hr.°C$$

The above relationship exists for the wheat bulk with the moisture content range of 10–20% (db). The thermal conductivity of sorghum may be given as follows,

$$K = 0.564 + 0.0858\, w, \ W/m.°C$$

where, w = fraction of moisture present in material.

The thermal conductivity of fruits and vegetables may be given as,

$$K = 0.148 + 0.493\, w$$

The thermal conductivity of single grain ranges from 0.3 – 0.6 *k cal/m.hr.*°C and bulk grain varies from 0.10 – 0.15 *k cal/m.hr.*°C. The difference is due to the air spaces present in the bulk grain. The thermal conductivity of air is 0.02 *k cal/m.hr.*°C.

Determination of thermal conductivity of grains and food materials

The measurement of thermal conductivity of agricultural grains pose much problem because it depends on the structure and chemical composition of the grain. Thermal conductivity probe is used for measurement of thermal conductivity of most of the food materials. A typical line heat-source thermal conductivity probe is shown in Fig. 1.13. The probe consists of a hypodermic needle of 0.66 mm

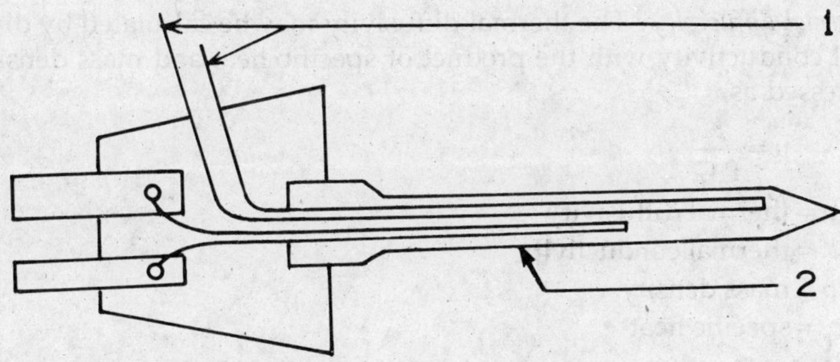

Fig. 1.13 : Diagram of a line heat source for measurement of thermal conductivity 1. heating wires 2. junction of thermocouple wires

diameter and connected to a handle. A 0.077 mm diameter constantan heater wire insulated with plastic spaghetti tubing is inside the needle. The wire is taken from the handle to the top and back. The purpose of using constantan wire is to avoid the change in electrical resistance with temperature. The insulated chromel constantan thermocouples wire 0.051 mm in diameter is wrapped around the upper half of the probe handle and the needle tip. The line-heat source probe is inserted into the sample having uniform initial temperature. The probe is heated at a constant rate, and the temperature adjacent to the line-heat source is monitored. After a brief transient period, the plot of natural logarithm of time *vs* temperature is plotted, which has a slope equal to $Q/4\pi K$. The thermal conductivity of material can be estimated by the following equation.

$$K = Q\,\frac{l_n\,[(t_2 - t_0)/(t_1 - t_0)]}{4\pi\,(T_2 - T_1)} \qquad \ldots 1.12$$

Where, K = thermal conductivity of the sample, $W/m°C$

Q = power generated by the probe heater, W/m

t_1 and t_2 = time since probe is energised, *s*

t_0 = time conductance factor, *s*

T_1 and T_2 = temperature of probe thermocouple at time t_1 and t_2 respectively, °C

(*iii*) *Enthalpy* : Enthalpy is the total heat content or energy level of a material. The enthalpy data are required for frozen foods that freeze over a range of temperatures below 0°C and not for those substances that freeze in a narrow temperature limit, as the case of pure substance like water. The enthalpy of the moist material can be estimated by using following expression.

$$h_2 - h_1 = m\, c_p\, (T_2 - T_1) + m\, X_w L \quad \text{...1.13}$$

Where, $h_2 - h_1$ = enthalpy difference

m = mass of the product

c_p = Specific heat of the product

X_w = water fraction

$T_2 - T_1$ = temperature difference

L = latent heat of fusion for water

(*iv*) *Thermal diffusivity* : The thermal diffusivity may be calculated by dividing the thermal conductivity with the product of specific heat and mass density. It may be expressed as,

$$\mu = \frac{K}{\rho\, c_p} \quad \text{...1.14}$$

where, μ = thermal diffusivity

K = thermal conductivity

ρ = mass density

c_p = specific heat

Thermal diffusivity is important in determination of heat transfer rates in solid food materials of any shape. Physically it shows the relationship between the ability of a material to conduct heat to its ability to store heat.

The common method of determination of thermal diffusivity is to calculate from experimentally measured values of thermal conductivity, specific heat and mass density. In most of the cases the specific heat can be estimated from the product composition, so only experimental measurements of thermal conductivity and mass density are required.

Rheological properties

The rheological properties may be defined as the science which deals with the deformation and flow of the material under action of the applied forces. Time is an important parameter during application of load to the body. Therefore, in rheology three important parameters such as force, deformation and time are used for expressing the mechanical behaviour of the material.

The knowledge of rheological properties is useful in designing mechanical handling systems of agricultural products where deformation and flow of the material occur. In liquid foods, the knowledge of rheological properties are essential for the proper design and operation of various machines and also for understanding the pertinent transport processes in the operations. Some of the foods like milk, syrups, filtered dilute juices and vegetable oils are Newtonian fluids. For such foods, the knowledge of viscosity function and its dependence upon temperature and concentration is sufficient for engineering design. Other examples of the rheological properties are textural attributes of foods like firmness, yielding quality, crispness and fibrousness.

The deformation can be divided into elastic deformation and flow. The flow may be further divided into plastic flow and viscous flow. Thus the three basic properties which describe the rheological behaviour of the material are elasticity, plasticity and viscosity. These properties are generally represented by three classical ideal bodies known as Hookean body, St. Venant body and Newtonian liquid. The behaviour of none of the real material is perfectly elastic or perfectly plastic, therefore, these three ideal bodies are used to serve as standards of comparison for analysing the behaviour of any real material.

Ideal elastic behaviour or Hookean body : The Hooke's law states that stress is directly proportional to strain within the elastic limit. This behaviour of the material is referred to as Hookean behaviour and it may hold true for small strains generally less than 0.1% in some of the solids. But in any real material, perfect elasticity may not be obtained. Also in elasticity, the complete recovery of strain takes place upon removal of stress. Thus, the Hookean or linear elasticity shown for steel can be differentiated with the non-linear elasticity shown for rubber in Fig. 1.14 (a) and (b).

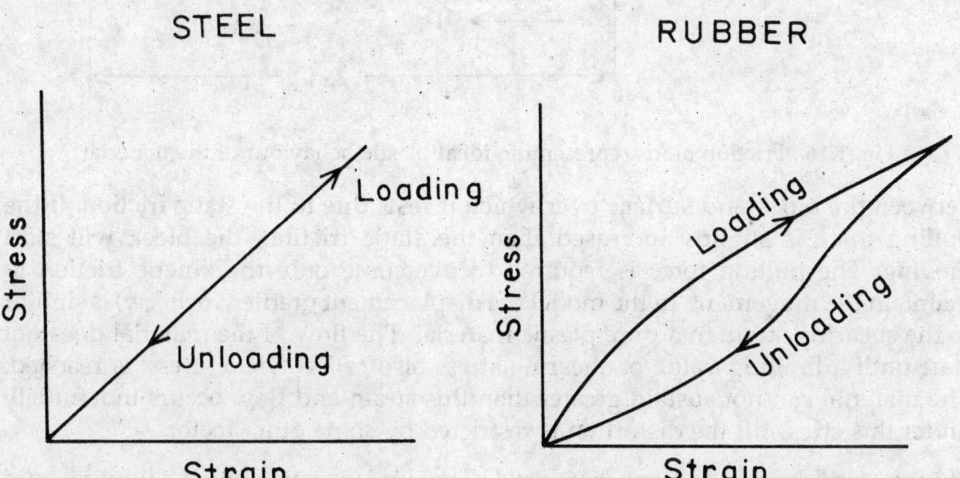

Fig. 1.14a : Diagram showing linear elastic behaviour for steel

Fig. 1.14b : Diagram showing non-linear elasticity in rubber

The compression tests conducted on some of the products like fruits and vegetables, cereal grains etc. show that Hookean elasticity does not exist in these biological materials even for very small strains. A typical example of first cycle of loading and unloading for the horny endosperm of dry maize kernel has been shown in Fig. 1.15. The curve shows the residual deformation upon unloading. Similar nature of curve is obtained for most of the food materials.

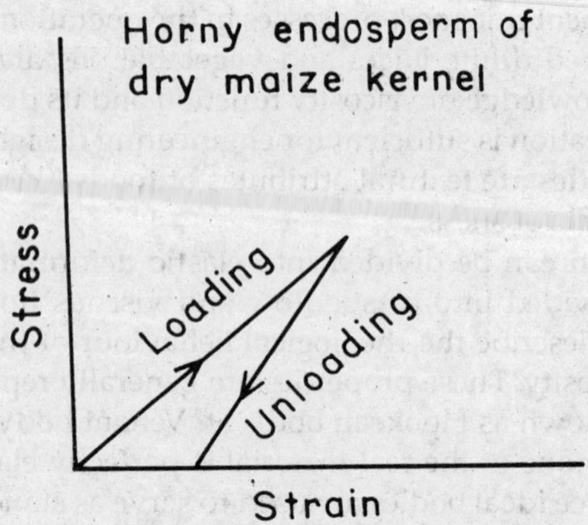

Fig. 1.15 : Diagram showing inelasticity in maize kernel

Ideal plastic behaviour or St. Venant body : The St. Venant body or pure plastic behaviour of the material can be represented by a mechanical model consists of a friction block as shown in Fig. 1.16. The model indicates that there is no movement

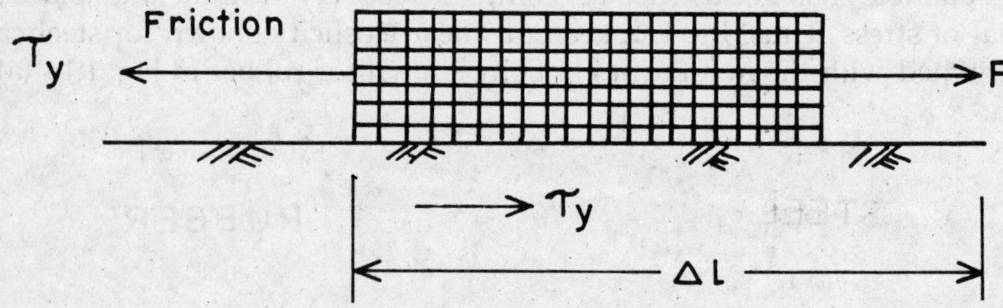

Fig. 1.16 : Friction block representing ideal plastic behaviour of the material

between the block and surface over which it rests due to the static friction. If the pulling force is slightly increased than the static friction, the block will start moving. The pulling force is required to overcome only the kinetic friction to maintain the movement. In the model the displacement gradient (*cm/cm*) is similar to the shearing strain in a pure plastic material. The flow of the material does not start until a limiting value of shearing stress also called 'yield stress' is reached. The material can not sustain greater than this strain and flow occurs indefinitely under this stress till the distortion is restricted by some other factor.

Pure viscous behaviour or Newtonian liquid : The pure viscous flow of a liquid is said to be the flow when slightest force can cause movement of liquid and when the rate of its flow is proportional to the force applied. The flow of the liquid continues infinitely until the force is removed, upon removal of force, like in the case of plastic flow, it will not regain its original state. Such a material is called a Newtonian liquid. This type of liquid can be represented by a dashpot, which may be assumed as a piston inside a cylinder as shown in Fig. 1.17a. When the force is applied to the piston, it starts moving in or out of the cylinder at a constant velocity, the rate generally depends upon the magnitude of the force applied. After removal of the force, the piston remains stationary and can not return to its original position. A material having such a nature has a rheological constant called the coefficient of viscosity (η). The coefficient of viscosity may be referred as the ratio of shearing stress applied to the resulting shear rate $\left(\eta = \frac{\tau}{r}\right)$.

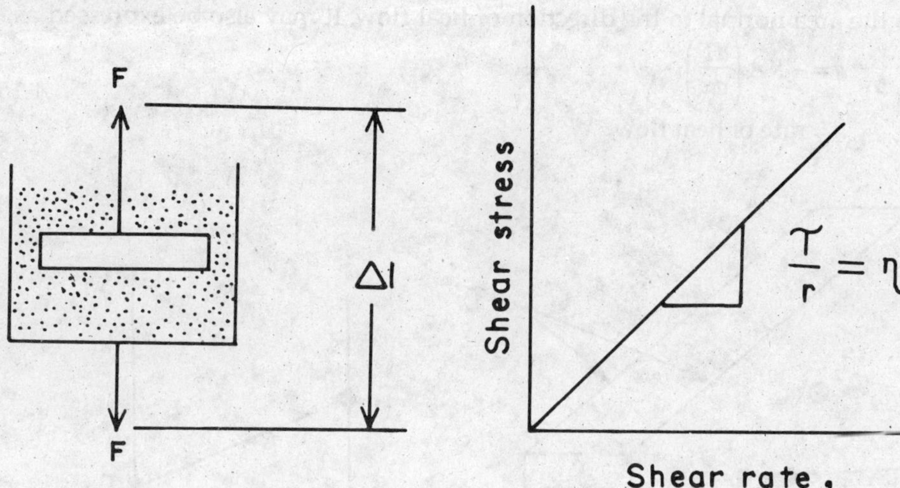

Fig 1.17a : Dashpot representing rheological model for Newtonian liquid

Fig. 1.17b : Relationship between shear rate and shear stress for Newtonian liquid

The Newton's law of viscosity and a simple dashpot representing rheological model for Newtonian liquid is shown in Fig. 1.17a. The relationship between the shear rate and shear stress is a straight line passing through the origin. The slope of the line is viscosity. It indicates that although the yield value of an ideal liquid is zero, the internal frictional resistance causes the flow which is called viscosity (Fig. 1.17b).

Heat transfer

Heat transfer is one of the important unit operations in the agricultural processing activities, for example; thermal treatment of food grains, cooling of perishables, pasteurisation and sterilisation of milk and fruit juices, freezing etc. It is also an essential operation to provide energy for evaporation of moisture during drying, solar heating or cooling, distillation process.

The various heat transfer processes can be categorised into three basic types or modes, conduction, convection and radiation. Sometimes in many systems, all the three modes of heat transfer occur simultaneously. These three modes of heat transfer have been described briefly in this chapter.

Conduction

Conduction is a mode of heat transfer in solids. Molecules are packed close together in solids. When one end of the solid is heated, the molecules of that spot absorb the heat energy and begin to vibrate rapidly. These molecules in turn, cause their adjacent molecules to vibrate. This process continues along the mass of solid. The heat energy is thus transferred from the hotter part to the colder part. Such a transfer of energy between different parts of a body, or from one body to another in contact with it is called conduction. The conduction of heat stops once the two objects have the same temperature.

According to Fourier's law of heat transfer the rate of heat transfer by conduction through a substance is directly proportional to the temperature gradient, $\frac{dT}{dx}$, and to the area normal to the direction of heat flow. It may also be expressed as,

$$q = -KA\left(\frac{dT}{dx}\right) \qquad ...1.15$$

where, q = rate of heat flow, W

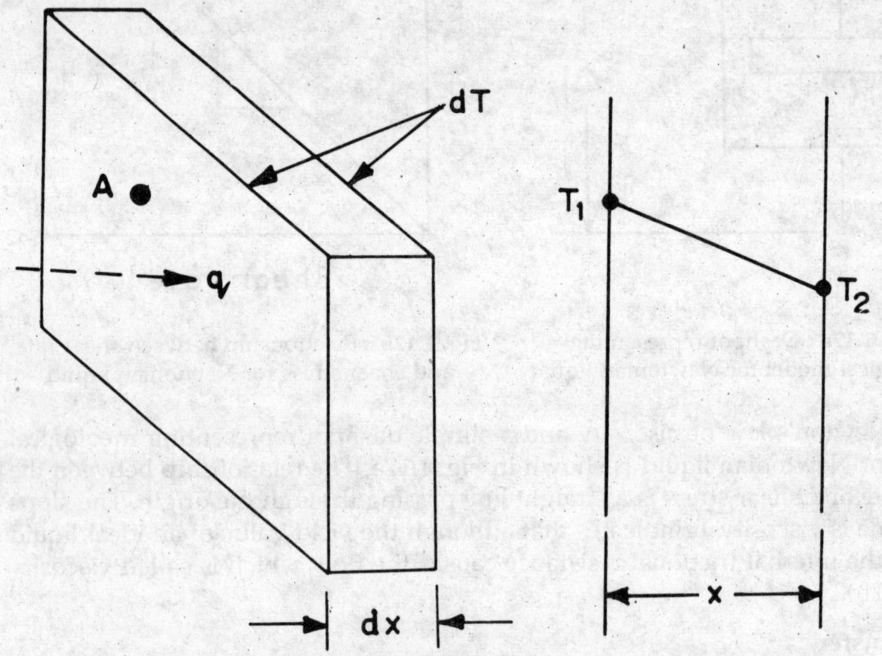

A = area normal to the direction of heat flow, m^2

$\frac{dT}{dx}$ = temperature gradient in the x direction, K/m

K = thermal conductivity, $W/m\,K$

The ratio q/A is referred to as the x-directional heat flux. Its unit is W/m^2. The heat flux may be expressed as.

$$\frac{q}{A} = -K\frac{dT}{dx} \qquad ...1.16$$

The negative (–) sign in equation 1.16 indicates that heat flow by conduction occurs in the direction of decreasing temperature gradient.

Conduction heat flow through slab : According to Eq. 1.15 the heat flow through a slab is expressed by:

$$q = -KA\frac{dT}{dx}$$

The above equation is integrated with the limits o to x,

$$q\int_0^x dx = -KA\int_{T_1}^{T_2} dT$$

or $$qx = -KA\,(T_2 - T_1)$$

$$\therefore \qquad q = \frac{KA\,(T_1 - T_2)}{x} \qquad \text{...1.17}$$

Heat transfer through a composite wall : Consider the heat flow through composite wall as shown in Fig. 1.18.

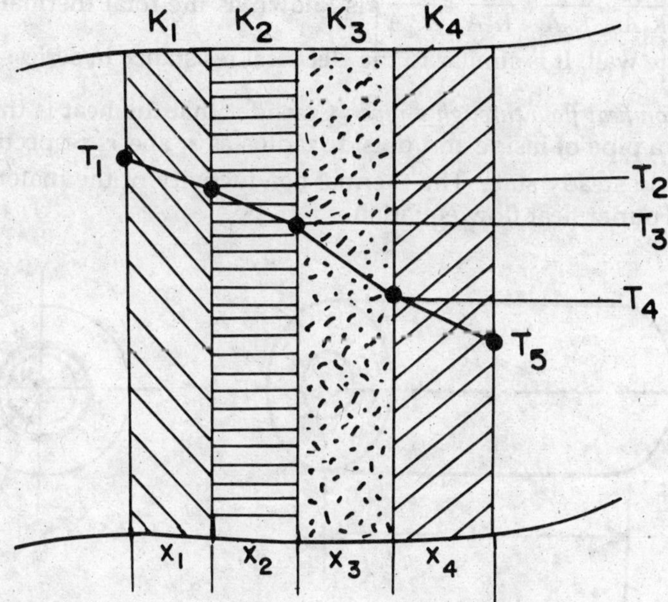

Fig. 1.18 : Diagram of heat flow through composite wall

The thickness of the walls are X_1, X_2, X_3 and X_4 and the thermal conductivities of the walls are K_1, K_2, K_3 and K_4 respectively. The temperatures at the contact surfaces of the walls are T_2, T_3 and T_4 then for steady state conduction of heat flow may be given as;

$$q = \frac{(T_1 - T_2)\,K_1A}{x_1} = \frac{(T_2 - T_3)\,K_2A}{x_2} = \frac{(T_3 - T_4)\,K_3A}{x_3} = \frac{(T_4 - T_5)\,K_4A}{x_4}$$

The above equation may also be written as,

$$T_1 - T_2 = \frac{x_1}{K_1A}\,q$$

$$T_2 - T_3 = \frac{x_2}{K_2A}\,q$$

$$T_3 - T_4 = \frac{x_3}{K_3A}\,q$$

$$T_4 - T_5 = \frac{x_4}{K_4A}\,q$$

All the above equations can be added to yield,

$$(T_1 - T_5) = q\left[\frac{x_1}{K_1A} + \frac{x_2}{K_2A} + \frac{x_3}{K_3A} + \frac{x_4}{K_4A}\right]$$

or,
$$q = \frac{T_1 - T_5}{\frac{x_1}{K_1A} + \frac{x_2}{K_2A} + \frac{x_3}{K_3A} + \frac{x_4}{K_4A}} \quad 1.18$$

where, $T_1 - T_5$ = thermal potential responsible for heat flow

The $\left[\frac{x_1}{K_1A} + \frac{x_2}{K_2A} + \frac{x_3}{K_3A} + \frac{x_4}{K_4A}\right]$ is known as the total thermal resistance of the composite wall. It is similar to the electrical resistance in series.

Conduction heat flow through a pipe : Consider that the heat is transferred radially through a pipe of inside and outside radius as r_1 and r_2 respectively as shown in Fig. 1.19 in steady state. The thermal conductivity of the material is K, hence as per the Fourier heat flow equation.

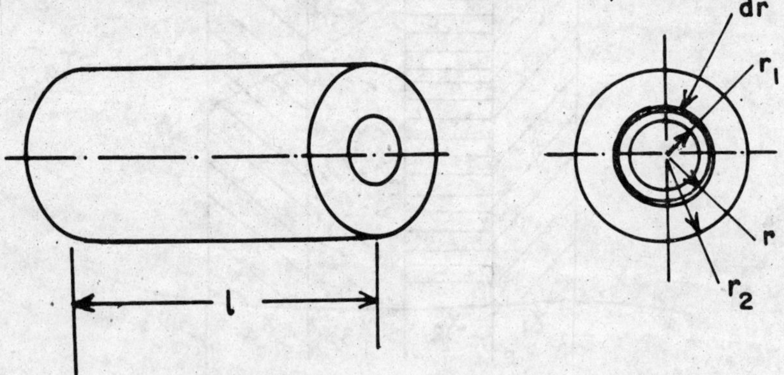

Fig. 1.19 : Conduction heat flow through a pipe

$$q = -KA\frac{dT}{dr}$$

area $A = 2\pi r l$ [at distance r and thickness dr]

$$q = -K(2\pi r l)\frac{dT}{dr}$$

where l is the length of the pipe. Integrating the above equation within limits of r_1 and r_2.

$$q\int_{r_1}^{r_2}\frac{dr}{r} = -2\pi r l\int_{T_1}^{T_2} dT$$

or,
$$q\,ln\frac{r_2}{r_1} = -2\pi r l\,(T_2 - T_1)$$

$$= 2\pi r l\,(T_1 - T_2)$$

or,
$$q = \frac{2\pi r l\,(T_1 - T_2)}{ln\frac{r_2}{r_1}} \quad ...1.19$$

or, $$q = \frac{T_1 - T_2}{\frac{1}{2\pi r l} ln \frac{r_2}{r_1}} \qquad ...1.20$$

The $\left[\frac{1}{2\pi rl} ln \frac{r_2}{r_1}\right]$ is known as thermal resistance of the pipe.

Conduction heat flow through a hollow sphere : The Fourier equation of heat flow through a hollow sphere can be given as under,

$$q = -KA \frac{dT}{dr}$$

Where area $A = 4\pi r^2$ (at distance r and thickness dr)

or, $$q = -K(4\pi r^2) \frac{dT}{dr}$$

Integrating the above equation within the limits r_1 to r_2 and T_1 to T_2 yields,

$$q\int_{r_1}^{r_2} \frac{dr}{r^2} = -4\pi K \int_{T_1}^{T_2} dT$$

or, $$q\left[-\frac{1}{r}\right]_{r_2}^{r_1} = -4\pi K (T_2 - T_1)$$

or, $$q\left[\frac{1}{r_1} - \frac{1}{r_2}\right] = 4\pi K (T_1 - T_2)$$

or , $$q = \frac{4\pi K}{\left(\frac{1}{r_1} - \frac{1}{r_2}\right)} (T_1 - T_2) \qquad ...1.21$$

or, $$q = \frac{(T_1 - T_2)}{\frac{1}{4\pi K}\left(\frac{1}{r_1} - \frac{1}{r_2}\right)} \qquad ...1.22$$

$\frac{1}{4\pi K}\left(\frac{1}{r_1} - \frac{1}{r_2}\right)$ is known as thermal resistance of the hollow sphere.

Equation 1.22 can also be written as

$$q = 4\pi K\left[\frac{r_1 r_2}{r_2 - r_1}\right] [T_1 - T_2]$$

If the thickness of the material is very small, then $r_1 r_2 = r^2$ and $r_2 - r_1 = x$ (thickness then,

$$q = 4\pi r^2 K \frac{T_1 - T_2}{x}$$

or, $$q = \frac{AK(T_1 - T_2)}{x} \qquad ...1.23$$

Problem : A 5 *m* high and 12 *m* long composite wall of a cold storage is made up of 100 *mm* thick brick wall as the outside wall. The inner wall surface is of fibre glass of 60 *mm* thick. In between the two walls an insulating board 20 *mm* thick is placed. The coefficients of thermal conductivity for the three layers are given below:

Brick wall = 1.15 $W/m-K$
Fibre glass = 0.04 $W/m-K$
Insulating board = 0.06 $W/m-K$

It the outside atmospheric temperature is 27°C and cold room temperature is 8°C. Calculate the heat loss per hour through the wall. Also determine the interface temperatures.

Solution : The heat flow through the composite wall can be given by:

$$q = \frac{T_1 - T_4}{\frac{x_1}{K_1A} + \frac{x_2}{K_2A} + \frac{x_3}{K_3A}}$$

$$= \frac{27-8}{\frac{0.1}{1.15 \times 60} + \frac{0.02}{0.06 \times 60} + \frac{0.06}{0.04 \times 60}}$$

$$= \frac{19}{1.45 \times 10^{-3} + 5.55 \times 10^{-3} + 25 \times 10^{-3}}$$

$$= \frac{19}{32 \times 10^{-3}}$$

$$= 593 \text{ W}$$

The interface temperatures can be calculated as under

$$T_1 - T_2 = \frac{x_1}{K_1A} q$$

$$= 1.45 \times 10^{-3} \times 593$$

$$= 0.86$$

or, $$T_2 = 26.14°C$$

Similarly, $$T_2 - T_3 = \frac{x_2}{K_2A} q$$

$$= 5.55 \times 10^{-3} \times 593$$

$$= 3.30$$

or, $$T_3 = 22.84°C$$

Convection

Convection is a process of heat transfer in which the whole mass of heated fluid particles flows towards a region of lower temperature and mixes with it, thus the heat energy is transferred to cold fluid mass. For example when water is heated from bottom in a flask, the warm liquid lighter than the surrounding cool water rises up and the colder water moves down, gets heated and in turn rises again.

The heat is transferred from the hot water at the bottom of the flask to the cooler top by the actual movement of the water itself. This process of the hotter fluid moving and transferring heat to the colder surroundings is called convection.

In the convection process, the heat is transferred from the hot fluid to the relatively cold fluid through a solid layer as shown in Fig. 1.20. The heat is first transferred from the hot fluid at temperature T_i to the solid layer at temperature T_1 by convection. The rate of heat transfer is directly proportional to the temperature difference between the mass of the fluid and the solid layer and the surface

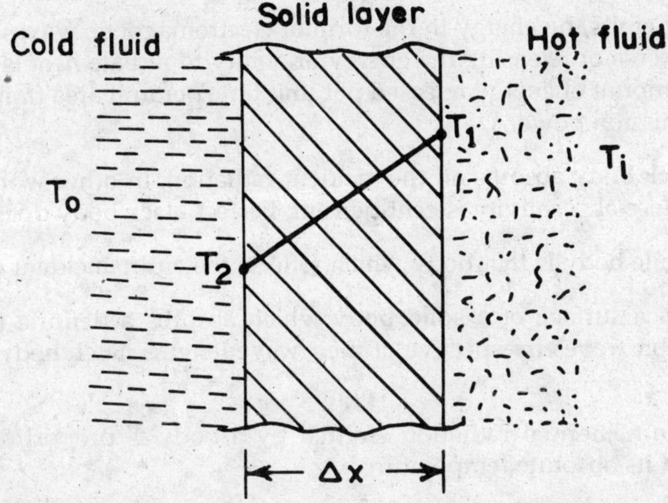

Fig. 1.20 : Diagram showing heat transfer by convection

area. The following relationship was indicated by Newton, which takes into account the heat flow from a solid surface to a fluid.

$$q = h_i A (T_i - T_1) \qquad \text{...1.24}$$

The heat is then transferred through the solid layer of thickness '*x*' with temperature difference of $T_1 - T_2$ by conduction. Finally the heat is transferred from the other surface of the solid layer to the cold fluid by the convection process.

The heat flow for steady state condition may be represented by the following equation

$$q = h_i A (T_i - T_1) = \frac{KA (T_1 - T_2)}{\Delta x}$$

$$= h_o A (T_2 - T_o) \qquad \text{...1.25}$$

where, T_i and T_o = temperatures of hot and cold fluid respectively

T_1 and T_2 = temperatures of hot and cold surfaces of the solid layer, respectively

h_i and h_o = inside and outside heat transfer coefficient respectively.

The Equation 1.25 can also be given as;

$$q = U A \Delta T \qquad \text{...1.26}$$

where, U = overall heat transfer coefficient, $k\,cal/m^2 \cdot hr - °C$

ΔT = temperature difference of two fluids $(T_i - T_0)$

Radiation

Radiation is a process of heat transfer in which heat flows from a hotter body to cold body in the form of electromagnetic waves. Radiation does not need any medium between the two bodies. It lets out heat rays to the surrounding just as a bulb or a candle lets out light rays. The hot object radiates heat rays to the colder surroundings. All hot objects transfer heat by radiation, whether they are solids, liquids or gases. Radiation needs neither a conducting medium nor convecting fluids for transfer. It occurs even in vacuum.

Emission : Body emits the energy in the form of electromagnetic waves when it is heated. The process of releasing the energy or ability to radiate heat is known as emission and amount of energy released per unit time per unit area from the body is known as emission power.

Black body : Black body absorbs all the incident radiation. In other words a black body is one whose absorptivity is cent percent. Perfect black body does not exist.

White body : While body is that body which reflects the entire incident energy.

Grey body : It is a surface of a solid body which absorbs a definite fraction of incident radiation waves irrespective of their wave lengths. Such body is known as 'Grey body'.

Thermal radiation : Thermal radiation emitted by a body is proportional to the fourth power of its absolute temperature, viz,

$$q = \frac{Q}{A} \propto T^4$$

For an ideal or a black body, the radiation emitted is given by;

$$Q = \sigma A T^4 \quad \text{...1.27}$$

Where, Q = thermal radiation emission rate, *k cal/hr*

A = surface area, m^2

σ = Stefan–Boltzman constant, $k\,cal/m^2\text{–}hr\text{–}K^4$

T = absolute temperature, K

If two black bodies are at temperature T_1 and T_2, having same area A, exchange heat with each other only, then the net radiant heat exchange between them will be;

$$Q = \sigma A (T_1^4 - T_2^4) \quad \text{...1.28}$$

Refrigeration

Refrigeration may be defined as the process of removing heat from a space at a temperature lower than the surrounding temperature. In other terms, refrigeration is the process of transfering heat energy from a lower to higher temperature region. The basic terms and processes related with a refrigeration system are described below.

Work : Work is said to be done when on applying force a body moves through a distance or the force causes displacement in a system. The amount of work is equal to the product of force and component of displacement parallel to the force.

Heat : It is a form of energy possessed by a substance in the form of kinetic energy or atomic or molecular vibration or rotation. When heat enters a body, it raises the kinetic energy and the molecular motion. It can be changed into other forms of energy.

Specific heat : The specific heat may be defined as the amount of heat in kilocalories that must be added to, or removed from 1 kg of a substance to change its temperature by 1°C. In refrigeration, the specific heat is used to determine the amount of heat that must be removed to cool various food products.

Sensible heat : The heat which causes an increase or decrease in the temperature of a body without changing its state is known as sensible heat. If the temperature of water is raised from 30 to 70°C, the heat absorbed to make this change is the sensible heat.

The change in sensible heat of a substance is given by the following equation

$$Q = m\,c_p\,(T_2 - T_1);\ \text{where } T_2 > T_1 \qquad \text{...1.29}$$

Where, Q = sensible heat, *k cal*

m = mass of substance, *kg*

C_p = specific heat of substance, *k cal/kg*°C

T_2 and T_1 = temperatures, °C

Latent heat : The latent heat does not cause a temperature rise within a body, but it affects a change in its state of being. The latent heat is defined as the amount of heat required to change the state of a substance. There are two types of latent heats; (*i*) latent heat of fusion and (*ii*) latent heat of vaporisation.

If 1 *kg* ice is converted into water at a constant temperature of 0°C, 80 *k cal* heat is absorbed from the surroundings. Similarly 80 *k cal* heat is to be removed from 1 *kg* of water at 0°C to freeze it into ice at the same temperature. The heat involved in above processes is known as latent heat of fusion.

When the temperature is raised to boiling point, upon further addition of heat the temperature of boiling water does not rises. Water at boiling temperature is converted into steam by heat. The amount of heat required to change 1 *kg* of water into steam at atmospheric pressure is 540 *k cal*. This heat is known as latent heat of vaporisation.

Total heat : Total heat is the sum of the sensible heat and latent heat.

First law of thermodynamics

According to fundamental principle, the matter can neither be created nor destroyed, however, it may be converted into different forms. Likewise, the energy can neither be created nor destroyed but its form can be changed. The sum of all forms of energy in a system remains always the same.

The first law of thermodynamics can be represented in following form,

$$E_1 + Q_a - Q_t = E_2 \qquad \text{...1.30}$$

Where, E_1 and E_2 = Initial and final energy of the system

Q_a and Q_t = energy added to and energy taken away from the system

If $Q = Q_a - Q_t$

then, $E_1 + Q = E_2$

or $Q = E_2 - E_1$ ...1.31

For the system which develop work the equation 1.31 can be written as

$$Q = (E_2 - E_1) + W \qquad \text{...1.32}$$

Where, W = work developed by the system

Enthalpy : Enthalpy is the sum of internal energy, U and the product of pressure, p and volume v.

$$H = u\,.\,c + p \cdot v \qquad \text{...1.33}$$

or per unit mass $h = u + p \cdot v_s$

where, v_s = specific volume

Open system and steady state condition : An open system through which fluid passes under steady flow condition is called a steady flow system. Considering the unit mass of fluid, it can be written as,

$$E_1 + p_1 v_{s1} + Q - W = E_2 + p_2 v_{s2} \qquad ...1.34$$

where, 1 and 2 = conditions at inlet and outlet.

The equation 1.34 can also be written as,

$$E_1 + p_1 v_{s1} \pm Q \pm W = E_2 + p_2 v_{s2}$$

$$U_1 + \frac{Z_1}{J} + \frac{V_1^2}{2gJ} + \frac{p_1 v_{s1}}{J} \pm Q \pm W = U_2 + \frac{Z_2}{J} + \frac{V_2^2}{2gJ} + \frac{p_2 v_{s2}}{J} \qquad ...1.35$$

Where,

J = mechanical equivalent of heat or Joule's constant

V_1 and V_2 = volumes at stage 1 and 2

g = acceleration due to gravity

The equation 1.35 is applied for the following equipment as under

(i) Centrifugal compressor :

$$h_1 + \frac{V_1^2}{2gJ} - Q + W = h_2 + \frac{V_2^2}{2gJ}$$

In compressors $V_1 << V_2$ and $\Delta Z = 0$

$$\therefore \quad h_1 - Q + W = h_2 + \frac{V_2^2}{2gJ} \qquad ...1.36$$

(ii) Blowers : Assuming pressure increase in a blower is an isentropic process.

$\Delta Z = 0$, $Q = 0$, $p_1 \approx p_2$ and $V_1 << V_2$

$$\therefore \quad U_1 + W = U_2 + \frac{V_2^2}{2gJ} \qquad ...1.37$$

Second law of thermodynamics

The second law of thermodynamics states that heat flows from a higher temperature to lower temperature irrespective of material and mass of the bodies through which heat transfer takes places. Flow of heat from lower temperature body to a higher temperature body is not possible without application of an external force.

Refrigeration is the removal of heat from lower temperature to higher temperature region. Since heat has a natural tendency to flow from higher to lower temperatures, therefore, work input is essential according to second law of thermodynamics in refrigeration systems. The refrigeration system works on second law of thermodynamics.

Reversible and irreversible processes

A process is said to be reversible, if both the system and surroundings by any means whatsoever be returned to their initial states after the completion of process.

In irreversible process the system and surroundings do not come back to their initial states. In other words, a process is irreversible if it involves heat transfer across a finite temperature difference between the system and surroundings.

Entropy : A property used to measure the quality of energy or irreversibility of the process is called entropy.

According to Clausius, the German physicist, entropy is the measure of unavailable thermal energy within a substance for conversion into mechanical work. It is not a physical property of substance.

The sum of available energy and unavailable energy is the total energy of a body. Entropy is a numerical factor that indicates relationship of two kinds of energy during work processes when considered in conjunction with changes in thermodynamic properties of the fluid.

An increase in entropy indicates an increase in the degradation of energy or a decrease in the availability of energy. Mathematically, the change in entropy between two sates a and b is defined as;

$$d\phi = \int_a^b \frac{dQ}{T}$$

where, Q = heat, T = absolute temperature, ϕ = entropy

The other processes that take place in the cycle of a closed system are briefly described below.

1. *Constant volume process* : In this process the volume of the system remains constant during the process. The work done by the system is given below.

$$W = p.dv = 0 \text{ Since } dv = 0 \qquad \text{...1.38}$$

The change in entropy is given by;

$$d\phi = c_v \log_e \left(\frac{T_2}{T_1}\right) \qquad \text{...1.39}$$

Where, c_v = Specific heat at constant volume

T_2 and T_1 = temperature

2. *Constant pressure process* : In this process the pressure of the system remains constant during the entire process. The work done by the system may be given as under

$$W = pdv = \frac{P(V_2 - V_1)}{J} \qquad \text{...1.40}$$

The change in entropy can be given by,

$$d\phi = c_p \log_e \left(\frac{T_2}{T_1}\right) \qquad \text{...1.41}$$

Where, c_p = specific heat at constant pressure.

3. *Constant temperature or isothermal process* : In this the temperature of the system remains constant during the process.

$$W = \frac{P_1 V_1}{T} \log_e \left(\frac{V_2}{V_1}\right) = \frac{P_1 V_1}{T} \log_e \left(\frac{P_1}{P_2}\right) \qquad \text{...1.42}$$

When $dT = 0$; $Q = W$

Therefore, the total heat supplied during the process is converted into work.

The change in entropy is given by,

$$d\phi = \frac{R}{J} \log_e \left(\frac{V_2}{V_1}\right) \qquad \text{...1.43}$$

4. *Adiabatic process* : In this process, heat exchange between the system and surrounding does not take place. The process may be given as under

$$P.V^{\nu} = \text{constant}$$

A reversible adiabatic process is known as Isentropic process.

The work during the process may be given by the following equation

$$W = \frac{P_1V_1 - P_2V_2}{(\nu - 1)J} = c_p\,(T_1 - T_2) \qquad ...1.44$$

It indicates that the work done during the process is equal to change in internal energy (when heat due to friction is zero) for isentropic process.

Since the heat transfer $dQ = 0$ for irreversible adiabatic process, the change in entropy is also equal to zero, or $d\phi = 0$.

5. *Polytropic process* : Various processes which occur in practice may be represented by an equation

$$P\,V^n = \text{constant}$$

Where, n = constant known as polytropic index.

The work done during the process may be given by the following equation

$$W = \frac{P_1V_1 - P_2V_2}{(n - 1)J} \qquad ...1.45$$

The transfer of heat between the system and surrounding is represented by,

$$Q = W\left(\frac{\nu - n}{\nu - 1}\right) \qquad ...1.46$$

The change in entropy is given by

$$d\phi = c_v \log_e\left(\frac{T_2}{T_1}\right) + \frac{R}{J}\log_e\left(\frac{V_2}{V_1}\right)$$

$$= \frac{\nu - n}{\nu - 1}\,\frac{R}{J}\log_e\left(\frac{V_2}{V_1}\right) \qquad ...1.47$$

Carnot refrigeration system

In the refrigeration system, the Carnot cycle is considered as reversed Carnot cycle being used for heat engines. It is a theoretical concept but serves as an ideal cycle. It is not achieved in practice. The physical properties of the working medium do not affect performance of the Carnot cycle, due to this it is considered valuable as an indication of the highest coefficient of performance achievable for given conditions.

The P–V and T–ϕ diagrams of Carnot cycle for air as refrigerating medium is shown in Fig. 1.21.

T_a is the temperature to be maintained by refrigeration and T_b is the atmospheric temperature where the heat is rejected.

The assumption of the Carnot cycle refrigeration system is that the heat absorption and rejection occur at constant temperatures and the compression and expansion are isentropic.

The Carnot cycle consists of all reversible process, therefore, it is known as a "reversible process".

In the cycle process

$$\oint Q = \oint W$$

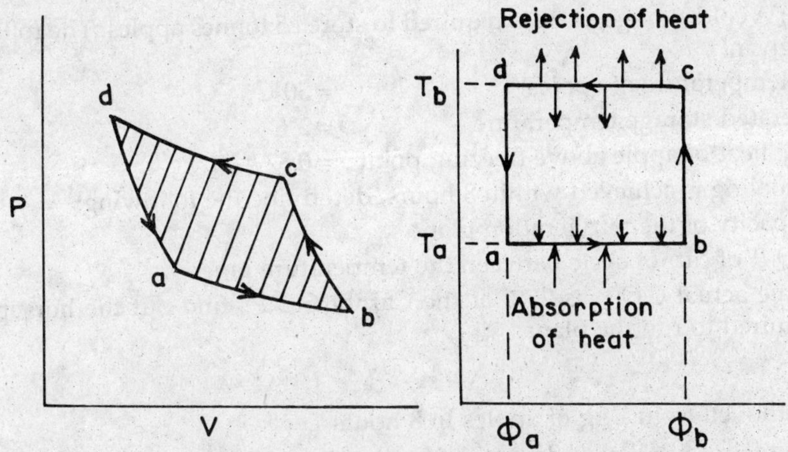

Fig. 1.21 : Carnot air refrigeration cycle
a-b: isothermal expansion c-d: isothermal compression
d-a: isentropic expansion b-c: isentropic compression

$\therefore$ Net Work done, $W = \oint w = \oint Q$ = Heat rejected – Heat absorbed

$$= Q_r - Q_a$$

$$T_b(\phi_b - \phi_a) - T_a(\phi_b - \phi_a) = (T_b - T_a)(\phi_b - \phi_a)$$

$$\therefore C.O.P. = \frac{\text{refrigeration/min}}{\text{work done/min}} = \frac{Q_a}{W}$$

$$= \frac{T_a(\phi_b - \phi_a)}{(T_b - T_a)(\phi_b - \phi_a)} = \frac{T_a}{T_b - T_a}$$

where $T_b > T_a$.

Carnot cycle for vapour refrigeration system (two phase system) is shown in Fig. 1.22

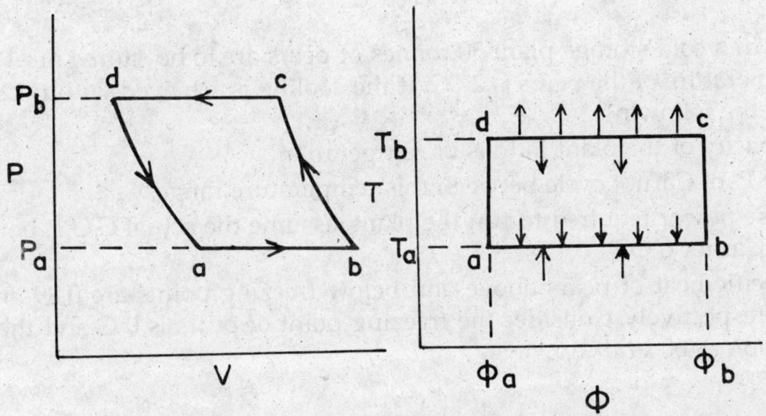

Fig. 1.22 : Carnot vapour-refrigeration cycle
a-b: isothermal expansion c-d: isothermal compression
d-a: adiabatic expansion b-c: adiabatic compression

Problem : A cold storage plant is required to store 25 tonnes apples. The following data are given.

Initial temperature of apples = 30°C
Refrigerated storage temperature = 2°C
Specific heat of apple above freezing point = 0.87 k cal/kg°C

If the cooling is achieved within 8 hours, determine the following:

(*i*) Capacity of the refrigeration plant
(*ii*) C.O.P. of carnot cycle between the temperature range
(*iii*) If the actual C.O.P. is 25% of the Carnot C.O.P., find out the horsepower required to run the plant.

Solution :

Heat removed from 1 kg of apples in 8 hours

$$= 0.87\,(30 - 2) = 24.36\ k\ cal/kg$$

Heat removed per minute by the plant

$$= \frac{24.36 \times 25 \times 1000}{8 \times 60} = 1268.75\ k\ cal$$

Capacity of the plant

$$= \frac{1268.75}{50} = 25.375\ tons$$

$$\text{C.O.P. of Carnot cycle} = \frac{T_a}{T_b - T_a} = \frac{(273+2)}{(273+30)-(273+2)}$$

$$= 9.82$$

Actual $C.O.P. = 9.82 \times 0.25 = 2.455$

$$\therefore \quad 2.455 = \frac{\text{refrigeration/min}}{\text{work done/min}} = \frac{1268.75}{W}$$

or $\quad W = 516.80\ k\ cal/min$

$$\therefore \text{ Horse power required} = \frac{516.80}{10.54} = 49.06\ hp.$$

Problem : In a cold storage plant 50 tonnes of pears are to be stored at –1°C. The initial temperature of the pears is 27°C. If the cooling is achieved within 12 hours, determine the following:

(*i*) Capacity of the plant in tons of refrigeration.
(*ii*) C.O.P. of Carnot cycle between this temperature range.
(*iii*) horse power required to run the plant, assume the actual C.O.P. be 30% of the Carnot C.O.P.

The specific heat of pears above and below freezing points are 0.89 and 0.43 *k cal/kg*°C respectively. Consider the freezing point of pears is 0°C and the latent heat of fusion as 60 *k cal/kg*.

Solution :

Heat removed from 1 kg of pears in 12 hours

$$= 0.89\,[30 - (-0)] + 60 + 0.43\,[0 - (-1)]$$

$$= 87.13\ k\ cal$$

Heat removed by the plant per minute

$$= \frac{87.13 \times 50 \times 1000}{12 \times 60} = 6050.69\ k\ cal$$

(*a*) Capacity of the plant $= \frac{6050.69}{50} = 121.01$ *ton*

(*b*) *COP* of Carnot cycle $= \frac{T_a}{T_b - T_a} = \frac{272}{300 - 272}$

$$= 9.714$$

(*c*) Actual $COP = 9.714 \times 0.3 = 2.914$

or, $$2.914 = \frac{6050.69}{W}$$

$\therefore$ $$W = \frac{6059.69}{2.914} = 2076.42\ k\ cal/\text{min.}$$

$\therefore$ horsepower required $= \frac{2076.42}{10.54} = 197$ *hp*.

There are two types of refrigeration methods, (1) natural refrigeration and (2) mechanical refrigeration. The natural refrigeration is produced by using natural ice. The mechanical refrigeration is achieved by refrigerating engines which work on the principles of thermodynamics.

Natural refrigeration : The temperature in the range of 5–10°C can be achieved by using ice as refrigerating medium. The most common ice refrigerator is made up of an insulated cabinet equipped with trays at the top in which blocks of ice pieces are kept. The shelves on which food products are kept are placed below the ice trays. The air moves downward passing through the ice compartments and cools the food products placed below the ice trays. The air picks up heat from the products and the warmer air rises and comes in contact with ice compartment at above and the cool air move downwards to refrigerate the food products.

Mechanical refrigeration : The application of mechanical devices and electrical or other form of energy for the refrigeration purpose are called mechanical refrigeration. The mechanical refrigeration can broadly be classified into, (*a*) vapour compression systems and (*b*) absorption refrigeration system.

Vapour compression system : The schematic diagram of vapour compression refrigeration cycle is shown in Fig. 1.23. All the major parts and the path of the refrigerant flow are also shown in the diagram. The expansion valve (high side float valve) maintains the pressure at different levels in two parts of the system. The main purpose of the expansion valve is to allow the liquid refrigerant under high pressure to pass at a controlled rate into the low pressure of the system. Some part of the liquid refrigerant evaporates while passing through the expansion valve, but a greater portion is vaporised in the evaporator at low pressure (and low temperature). The liquid refrigerant absorbs its latent heat of vaporisation from the air, water or other material being cooled. The compressor increases the pressure and temperature of the refrigerant above atmospheric. The refrigerant at high pressure and temperature dissipates its latent heat in the condenser. While passing through the condenser, the refrigerant discharges the heat absorbed in the evaporator and the heat equivalent of the work done upon it by the compressor. This

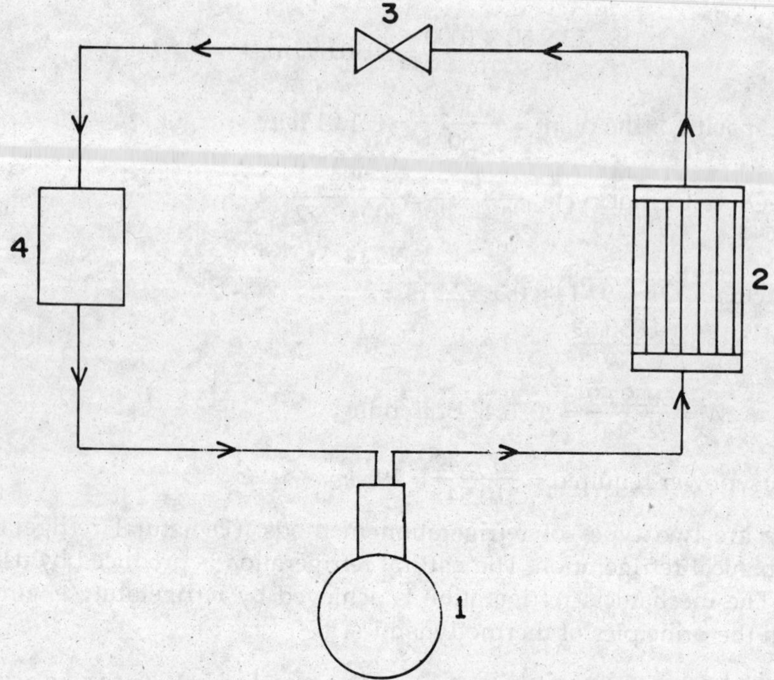

Fig. 1.23: Simple vapour compression refrigeration cycle
1. compressor 2. condenser 3. expansion valve 4. evaporator

heat is transferred to the air or water which is used as cooling medium in the condenser.

The vapour compression cycle described in the Fig. 1.24 can be represented by temperature-entropy (T–ϕ) and pressure-enthalpy (P-h) diagram as shown in

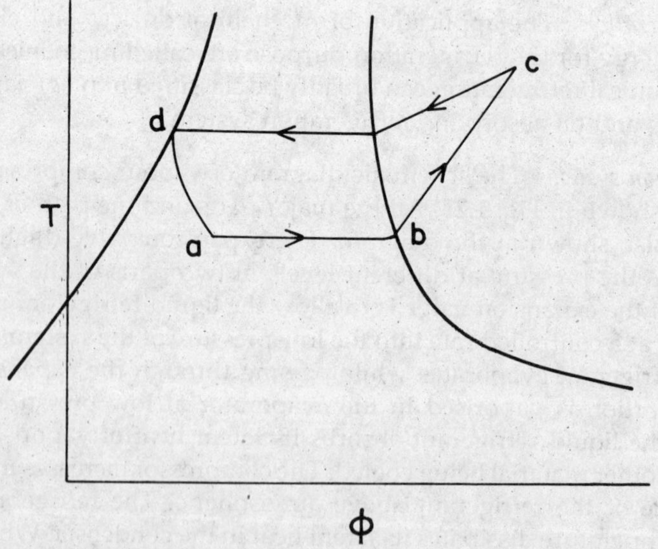

Fig. 1.24 : Temperature-entropy diagram for vapour-compression refrigeration cycle

Fig. 1.25. With the condition that the liquid coming out of condenser is saturated liquid and vapour coming out of evaporator is saturated vapour.

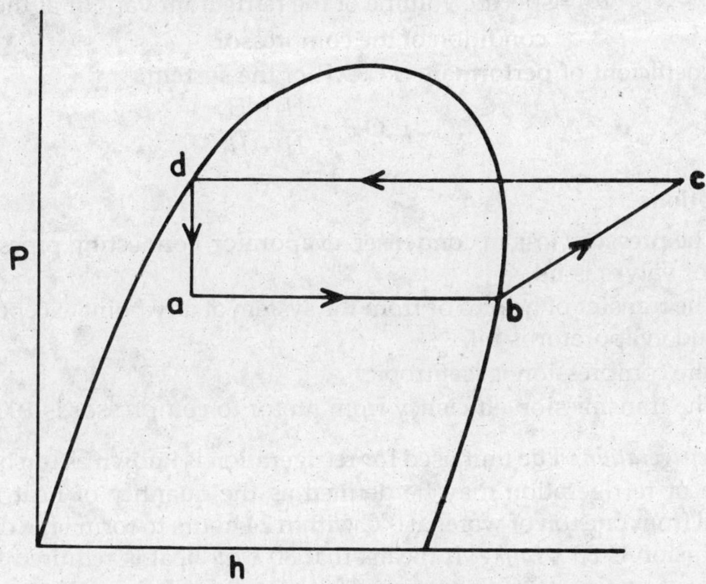

Fig. 1.25 : Pressure-enthalpy diagram for vapour-compression refrigeration cycle

Following factors should be known for the complete analysis of refrigeration system;

(*i*) Load on refrigeration plant in Tons of refrigeration.

(*ii*) Temperature at which the refrigeration occurs.

(*iii*) Temperature of the atmosphere (sink).

The refrigeration temperature and atmospheric temperature decide the temperature of the refrigerant entering into the evaporator and also in condenser and it also decides the required pressure ratio.

For designing the refrigeration system, following factors are calculated.

Weight of refrigerant required/min, $W_r = \dfrac{50\,T}{H_b - H_a}$

where, $T =$ tons of refrigeration load

Horsepower required to run the compressor,

$$H.P. = \frac{W_r\,(H_c - H_b)}{10.54} \qquad \ldots 1.48$$

If m_c and c_p are the weight of coolent and specific heat respectively and ΔT is the temperature rise then, the weight of cooling medium required in condenser is,

$$m_c\, c_p\, (\Delta T) = W_r\,(H_c - H_d)$$

For the design of the compressor,

$$\left(\frac{\pi}{4} D^2 . L\right) \eta_v\, N = W_r\, v_{sb} \qquad \ldots 1.49$$

Where, L and D = stroke and diameter of the cylinder

N = speed of compressor, rpm

η_v = volumetric efficiency of the compressor

v_{sb} = specific volume of the refrigerant vapour at the suction condition of the compressor.

The coefficient of performance, *C.O.P.* of the system,

$$C.O.P. = \frac{H_b - H_a}{H_c - H_b} \qquad \text{...1.50}$$

Assumptions

(*i*) The pressure loss in condenser, evaporator, connecting pipes and compressor valves is nil.

(*ii*) The transfer of heat to or from the system at any point except the condenser and evaporator is nil.

(*iii*) The compression is isentropic.

(*iv*) The transmission efficiency from motor to compressor is 100%.

Unit of refrigeration : The unit used for refrigeration is known as ton of refrigeration.

A ton of refrigeration may be defined as the quantity of heat required to be removed from one ton of water at 0°C within 24 hours to form one ton of ice. Latent heat of fusion is 80 *k cal/kg*. It means that 80 *k cal* heat is required to be removed from 1 *kg* of water at 0°C to convert it into ice at 0°C.

Thus ton of refrigeration in metric unit

$$\frac{1000 \times 80}{24 \times 60} = 55.4 \, k\, cal/min$$

The commonly used unit of refrigeration is equivalent to 50 *k cal/min.*

Heat Pump

The heat pump is basically a refrigeration unit which is capable of extracting heat from low grade heat sources like atmosphere, sea, ground and upgrading the same to a useful temperature. This absorbed heat is upgraded with the help of a compressor and the fluid carrying the heat is discharged to the required space through the condenser. The heating system is same as refrigeration unit, see Fig. 1.26.

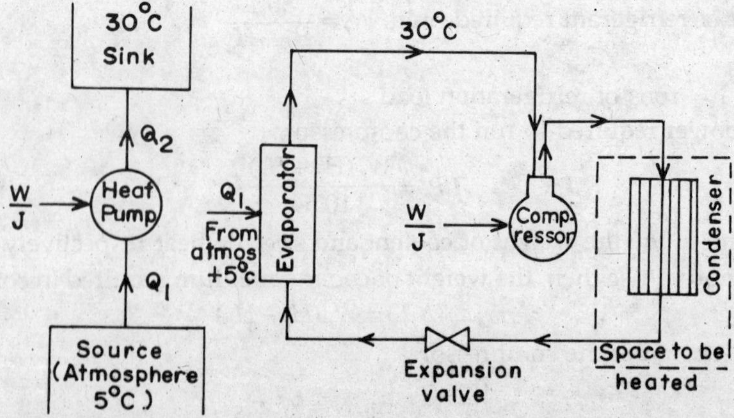

Fig. 1.26 : Diagram of a heat pump and space heating system

The refrigerant which passes out of the expansion valve extracts heat Q_1 from atmosphere as its temperature (– 5°C) is lower than the atmospheric temperature (+ 5°C). This absorbed heat by refrigerant is upgraded by addition the work through the compressor and the total heat Q_2 $\left(Q_2 = Q + \frac{W}{J}\right)$ is discharged in the room through condenser. Thus the device that lifts or "pumps" heat energy from a low temperature source for use at higher temperature is known as "Heat-pump" The refrigeration cycle is used for heating purposes when the atmospheric temperature is below the required temperature for comfort is called heat pump.

The heat pump performs a number of operations, some of them are listed below.

(*i*) Heating and cooling of buildings for comfort conditions.
(*ii*) Evaporating and concentrating the juices.
(*iii*) Drying.
(*iv*) Cooling of water and heating of space.
(*v*) Heating of water and cooling of space etc.

The coefficient of performance of a heat pump refers as the ratio of heat delivered to the work supplied. This ratio is known as "Effective Performance Ratio" (EPR) or "Heat Pump Performance Ratio" and is given by as under

$$EPR = \frac{Q_2}{\frac{W_1}{J}} = \frac{Q_1 + \frac{W_1}{J}}{\frac{W_1}{J}} = \frac{Q_1}{\frac{W_1}{J}} + 1$$

$$= C.O.P. + 1$$

Desirable properties of a refrigerant

The desirable properties of a refrigerant may be classified into three main groups:

(*i*) Thermodynamic properties
(*ii*) Safe working properties
(*iii*) Physical properties

Thermodynamic properties

(a) Freezing point : The freezing point of the refrigerant should be low enough so that it must not freeze under required evaporator's temperature. Therefore, it is necessary to have a refrigerant with freezing point well below the operating evaporator temperature. However, the freezing point of most of the refrigerants are – 30°C which meet the requirement of normal refrigeration system.

(b) Boiling point : The boiling point of the refrigerant should be low at atmospheric pressure. If the boiling point of the refrigerant is high at atmospheric pressure, it becomes necessary for a compressor to be operated at high vacuums. The refrigerants having high boiling point at atmospheric pressure reduces capacity of the refrigeration system and also lowers the operating cost.

(c) Pressure of evaporator and condenser : The positive pressures in evaporator and condenser are always desirable for the required temperatures, but it should not be higher than atmosphere. The positive pressures prevent the leakage of air and moisture into refrigeration system. Thus the operating pressure range is one of the

main criteria in selecting a refrigerant for the economical operation of the refrigeration system.

(d) Critical temperature : The critical temperature is the temperature of vapour above which it can not be condensed irrespective of high pressures. The critical temperature of the refrigerant being used should be higher than the temperature within the condenser which assists easy condensation of the refrigerant vapour.

(e) Latent heat of refrigerant : The high latent heat of refrigerant at evaporator's temperature gives high refrigeration effect per kg of refrigerant. Thus the requirement of the refrigerant for the system per ton of refrigeration also reduces which in turn reduces the initial cost of the refrigerant used in the system.

Safe working properties

The safe working properties of the refrigerant form basis and important criteria for the selection of refrigerant. Some refrigerants may be highly desirable from thermodynamic point of view but these may not be suitable due to unsafe working properties like; ammonia, which is used only for few applications. The safe working properties are briefly described below.

(a) Toxicity : The toxic refrigerant may cause injury to the human body. If the concentration is high it may become fatal. The use of toxic refrigerants are also limited in cold storage systems because it influences the qualities of many food materials.

(b) Flammability : The refrigerant should be non-flammable and non-explosive in the presence of air or lubricating oil. Most of the commonly used refrigerants like; Freon, CO_2 etc. are non-flammable.

(c) Corrosive properties : Although the chemical reaction of the refrigerant with the materials of the refrigeration system is not primarily considered for the selection of the refrigerant, but the material to be used for the construction of the system is decided on the basis of the selected refrigerant.

However, the refrigerant should have non-corrosive property towards common materials used in the refrigeration system.

(d) Effect on stored products : The quality like colour, test etc. of the food materials stored in cold storage should not be affected by the refrigerant when it comes in contact with the stored product.

Ammonia becomes alkaline in nature when dissolved in water and reacts with most vegetables and fruits which are slightly acidic in nature and spoils the test.

Freons do not affect colour, test or texture of the food materials exposed to them.

Sulphur dioxide becomes acidic in nature when dissolved in water and it does not affect the quality of the food materials as most of the foods are acidic in nature. But it has bleaching action on flowers, plants etc.

Physical properties

The following physical properties are desirable for an ideal refrigerant:

(*i*) Low specific volume of the refrigerant is desired as it reduces the size of the compressor.

(*ii*) Low specific heat of liquid and high specific heat of vapour refrigerant are desired as both increase the refrigeration effect per kg of refrigerant.

(*iii*) High thermal conductivity of refrigerant is desired.

(*iv*) Low viscosity of refrigerant is found to be helpful in better heat transfer and low pumping power.

The other properties considered for selection of refrigerant are odour, leak tendency, its relationship with oil, *COP* and horsepower requirement, cost and availability.

Refrigeration Equipment

The main components of vapour compression refrigeration system are described below:

1. Compressor
2. Condenser
3. Evaporator
4. Expansion valve
5. Piping and Pumps
6. Electric motors

Compressor : The compressor pumps the refrigerant through the system is like our heart which pumps the blood throughout the body, therefore, the compressor is considered as heart of the refrigeration system.

Compressors commonly used in a refrigeration plant is of three types; (1) reciprocating compressor, (2) rotary compressor and (3) centrifugal compressor.

(a) Reciprocating compressor : The reciprocating compressor sucks the refrigerant at low pressure and low temperature during its suction stroke and delivers it at high pressure and high temperature. The reciprocating compressors are available in wide range of sizes from fraction of horsepower to several hundred horse power. These are used for refrigeration plant ranging in sizes from 0.25 ton to 1000 tons. The reciprocating compressors are further classified as, (1) single acting vertical compressors and (2) double acting horizontal compressors.

(b) Rotary compressor : The rotary compressor are generally preferred with fractional tonnage refrigeration applications. The rotary compressors are further classified as, (1) rotary compressors with one stationary sealing blade and eccentric motor and (2) rotary compressors with sealing blade which rotate with eccentric shaft.

(c) Centrifugal compressor : The centrifugal compressor draws the refrigerant vapour and discharges it with a considerable high velocity at the outside edge of impeller. This high velocity head is further converted into pressure head by passing through the diffuser. The rise in pressure per stage of centrifugal compressor is considerably smaller as compared to single stage reciprocating compressor. Therefore, when high compression ratio is required, a multistage centrifugal compressor system is used.

Condenser : The condenser removes the heat from refrigerant carried from evaporator and added by compressor to convert the vaporised refrigerant into liquid form.

It is a heat exchanger in which heat transfer takes place from high temperature vapour to low temperature air or water used as cooling medium. There are two important factors which are considered in design for effective performance of the condenser. They are, (1) effective temperature differential and (2) high heat transfer coefficient.

The condensers are classified on the basis of he cooling medium used, (1) air cooled condensers, (2) water cooled condensers and (3) evaporative condensers.

Evaporators : The refrigerant from the expansion valve comes into the evaporator below the temperature required to be maintained in the evaporator and carries the heat from the evaporator. It is also known as cooler or freezer. The evaporators are manufactured in various shapes, sizes and types as per requirement.

The evaporators are generally grouped into two categories; (1) flooded evaporators and (2) dry expansion evaporators.

Expansion devices : It is one of the basic components of the refrigeration system. There are two functions of the expansion devices; (1) to reduce the pressure and temperature of the refrigerant received from the condenser as per the requirement of refrigeration system, and (2) to regulate the flow of the refrigerant as per the load on the evaporator.

Piping and pump : The main purpose of the piping is to carry the different fluids from one point to another in a refrigeration system.

Pumps are required for the circulation of water and brine in the secondary systems and water circulation through the condensers. The pumps are further classified as; (1) reciprocating pumps, (2) gear pumps or rotary pumps and (3) centrifugal pumps.

Electric motors : Electric motors are used to run the compressor, fans and pumps. The electric motors are classified as; (1) fractional horsepower motor and (2) motors of 1 hp and above.

Fans and Blowers

The air flow is a fundamental requirement for grain winnowing, drying, aeration and pneumatic conveying systems. Fans and blowers are used to deliver the specified volume of air to perform these operations. For selection and comparison among various types of fans, knowledge of fan characteristics is required. The characteristics of a fan are the relationship among its air flow rate, pressure, speed, power requirement and efficiency. The performance charts of fans are calculated on the assumption that air entering the fan is of standard density at 16°C, 1000 mb and 55% relative humidity.

Fan power : In S.I. base unit, power is represented by 'watt'. However mechanical power output of motors are expressed in horsepower (hp) whereas electrical power input is expressed in watt.

The power of a fan may be represented by following equation

$$P = \frac{Q \times h}{102} \qquad \text{...1.51}$$

Where, P = power of fan, kW
Q = air flow rate, kg/s
h = air pressure, m

Efficiency of fan : The efficiency of fan may be referred as the ratio of energy output to input. The energy output of a fan is expressed in air flow rate and pressure at which it is delivered.

The heat generated in the process is not taken into consideration.

$$\text{Fan total efficiency} = \frac{\text{Air flow rate, } \frac{m^3}{s} \times \text{Fan total pressure, } mm \text{ swg}}{\text{Power input, } kW \times 1.02}$$

$$\text{Fan static efficiency} = \frac{\text{Air flow rate, } \frac{m^3}{s} \times \text{Fan static pressure, } mm \text{ swg}}{\text{Power input, } kW \times 1.02}$$

Fan total efficiency is more important criterion because it relates to the total amount of work done by the fan.

Problem : A fan delivers air at a flow rate of 15 m^3/s against static pressure of 150 mm water gauge. Determine the static efficiency of the fan. The power required to operate the fan is 30 kW.

Solution :

$$\text{Fan static efficiency} = \frac{\text{Air flow rate} \times \text{Static pressure}}{\text{Power input} \times 1.02}$$

$$= \frac{15 \times 150}{30 \times 1.02}$$

$$= 73.53\%$$

Problem : The particulars of a fan used in a drying system are given below:

Air flow rate = 12 m^3/s
Total pressure head = 600 pascal
Pressure loss/m = 11.2 pascal
Specific volume of air = 0.88 m^3/s
Efficiency of motor = 65%

Determine the power requirement of the fan.

Solution :

$$\text{Mass flow rate of air} = Q = \frac{12}{0.88} = 13.63 \; kg/s$$

$$\text{Pressure head of air, } h = \frac{600}{11.2} = 53.57 \; m$$

$$\text{Power required} = \frac{Q \times h}{102}$$

$$= \frac{13.63 \times 53.57}{102}$$

$$= 7.16 \; kW$$

Efficiency of the motor $= 65\%$

$\therefore$ Total power required $= \dfrac{7.16}{0.65} = 11.01\ kW$

Fan Laws

The performance of geometrically similar fans is governed by a number of laws. The laws of fan are described below.

Law No. 1 : The volume of air delivered by a fan is directly proportional to its speed of rotation. This may be represented as,

$$Q_2 = Q_1 \frac{N_2}{N_1} \qquad \text{...1.52}$$

Where, Q_2 and Q_1 = air flow rates

N_2 and N_1 = speed of rotation of fan

Law No. 2 : The pressure developed by a fan is directly proportional to the square of the fan speed. It can be given as;

$$P_2 = P_1 \left(\frac{N_2}{N_1}\right)^2 \qquad \text{...1.53}$$

Where, P_2 and P_1 = static pressure developed by fan at speeds N_2 and N_1

Law No. 3 : The power absorbed by a fan varies in direct proportion to the cube of the fan speed.

or, $$hp_2 = hp_1 \left(\frac{N_2}{N_1}\right)^3 \qquad \text{...1.54}$$

Where hp_2 and hp_1 = power absorbed by a fan.

The above laws hold good for a given size and type of fan at constant density.

When fluid density changes and for a given size and type of fan, the following laws hold good at constant speed.

1. $Q_2 = Q_1$
2. The pressure developed by a fan is directly proportional to the density.

$$P_2 = P_1 \frac{\rho_2}{\rho_1}$$

where, ρ_2 and ρ_1 = densities of air

3. The power absorbed by a fan is directly proportional to the density.

$$hp_2 = hp_1 \frac{\rho_2}{\rho_1}$$

When size of impeller changes and the fan is working at constant speed with density of air remaining unchanged, the following laws hold good.

1. The volume of air delivered by a fan varies directly proportional to the cube of the diameter of the impeller.

$$Q_2 = Q_1 \left(\frac{d_2}{d_1}\right)^3$$

where, d_2 and d_1 = diameters of impeller

2. The pressure developed by a fan is directly proportional to the square of the impeller diameter.

$$P_2 = P_1 \left(\frac{d_2}{d_1}\right)^2$$

3. The power developed by a fan varies directly as the fifth power of the impeller diameter.

$$hp_2 = hp_1 \left(\frac{d_2}{d_1}\right)^5$$

An approximate relation for centrifugal fans with respect to width is given by,

$$Q_2 = Q_1 \frac{w_2}{w_1}$$

Where, w_2 and w_1 = width of fans.

Types of fans

Three types of fans, namely propeller, axial flow and centrifugal are commonly used in grain processing units. Each type has different characteristics due to which it is suitable for a specific function.

Propeller fan : Propeller fans move a large air flow against very low resistance (maximum is 15 *mm* water gauge pressure). It consumes low power. Propeller fans

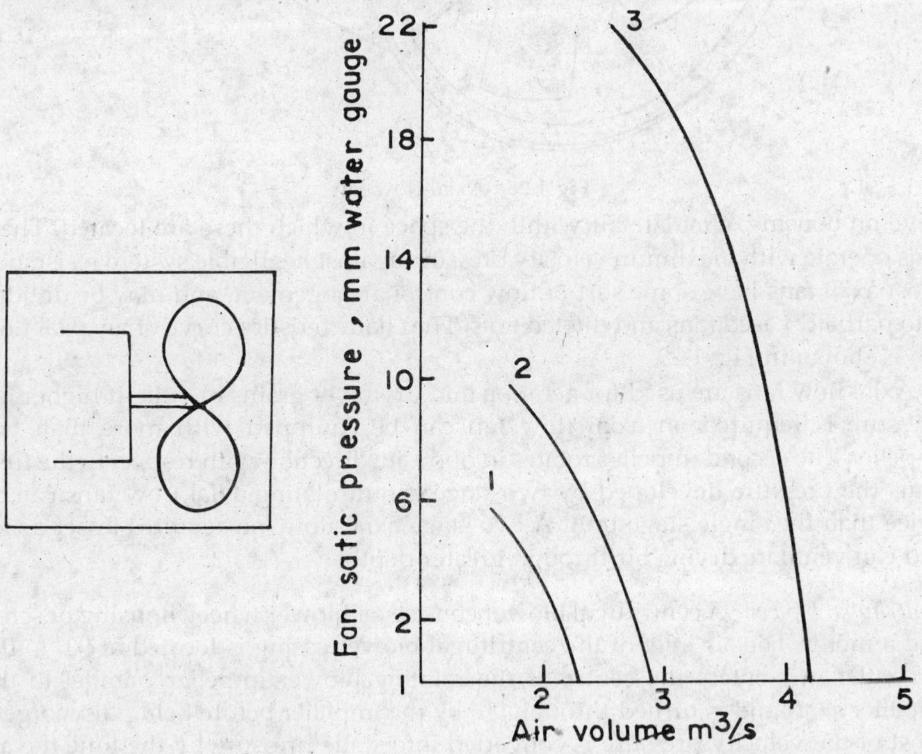

Fig. 1.27 : Diagram of propeller fan and its characteristics

are rarely used with ducts as ducts create resistance to air flow. Therefore, propeller fans are not used to dry or aerate grains. Propeller fans are only used for ventilation of air in stores. The performance characteristics of propeller fan is shown in Fig. 1.27.

Axial flow fan : An axial flow fan propels the air in an axial direction from inlet to discharge end. The air follows a course parallel to the axis of the fan shaft and does not turn through right angle between the inlet and discharge, thus a high fan efficiency is achieved. The fan consists of a number of blades mounted on a spider or hub at an angle with the plane of rotation. (Fig. 1.28). An axial flow fan may or may not be furnished with motors and flow control housings. Free air fans, which have no housing, circulate air within the space in which these are located. These fans operate with maximum velocity pressure against negligible system resistance. Most axial fans have some sort of flow control arrangement, and may be divided into partially cased fans and ducted fans. The characteristics curve of an axial flow fan is shown in Fig. 1.29.

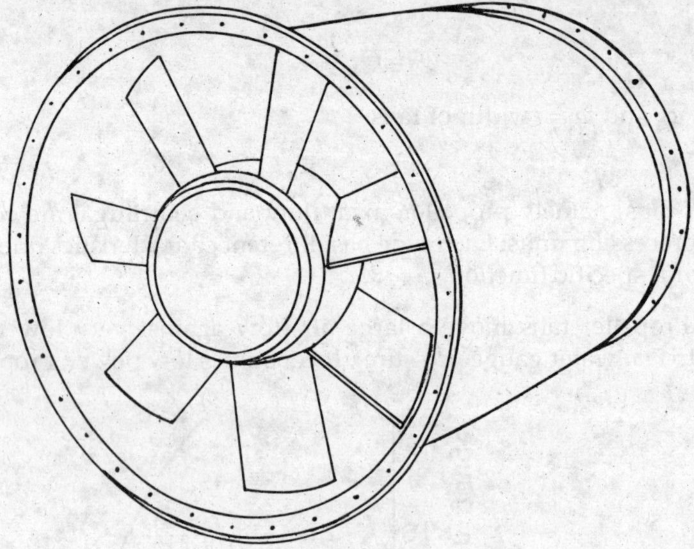

Fig 1.28 : Axial-flow fan

Axial flow fans are used for aeration and drying of grains in bulk. If higher fan pressure is required an axial flow fan can be equipped with more than one impeller. The second impeller rotates in opposite direction with respect to the first. Thus the pressure developed by two-stage contrarotating axial flow fans can be twice than the single stage unit. A two stage axial flow fan requires lesser space and can ventilate drying air through greater depths.

Centrifugal blower : A centrifugal blower consists of blower wheel, housing or scroll and a motor. The air inlet of the centrifugal blower casing is located at 90° to the air outlet. Air enters the centre of the rotating blower impeller parallel to the impeller shaft and is turned through 90° by the impeller before being discharged. Most of the velocity pressure is converted into static pressure by the time the air is discharged from the outlet opening. Energy is required to change the direction

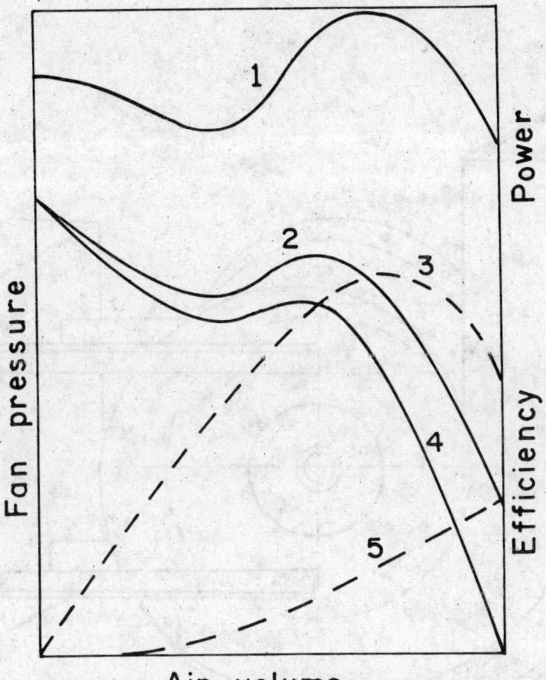

Fig. 1.29 : Characteristic curves for axial-flow fan
1. fan power 2. total pressure of fan 3. fan efficiency 4. fan static pressure
5. fan dynamic pressure

of air flow, therefore, slightly lower efficiency is achieved as compared to axial flow fans. Air is thrown from the blades due to centrifugal forces generated by the rotor (Fig 1.30 a).

Performance of a centrifugal blower depends on the configuration of the blower housing, the size of the inlet opening relative to the impeller, the location and shape of the cutoff, the shape and dimension of the scroll and the shape of the outlet into the air system.

The centrifugal blower can be divided into three broad types on the basis of shape of impeller. These are forward curved, backward curved and radial or straight bladed (Fig. 1.30 b).

Forward curved blower : A centrifugal blower with forward blade is basically an impulse device. The impeller carries a large number of shallow blades with curved surfaces, the concave side facing the direction of rotation. It accelerates the air to a high velocity, while rotating at a speed that is usually low in comparison to a backward curved blower. A forward curved blower requires high power for generating large volume of air. Hence care must be taken to ensure that sufficient power is available to drive the blower while it is delivering larger than the normal volume of air.

Backward curved blower : The blades of the impeller of backward curved blower are curved backwards. The convex side of the blade face the direction of rotation. The backward curved centrifugal blowers are the most efficient blowers which make

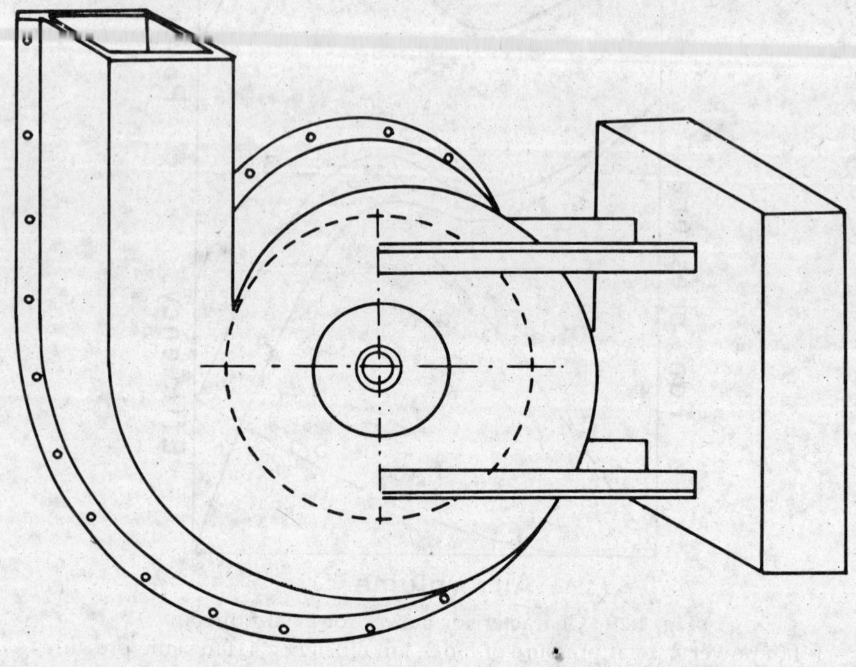

Fig. 1.30a : Centrifugal blower

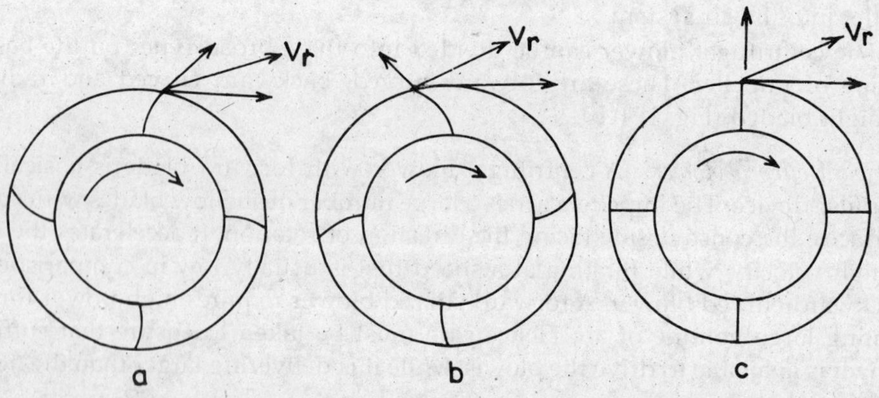

Fig. 1.30b : Types of centrifugal blower impellers
a. forward curved b. backward curved c. straight bladed

them suitable for drying grains in bulk. The characteristic curves of a backward curved blower is shown in Fig. 1.31.

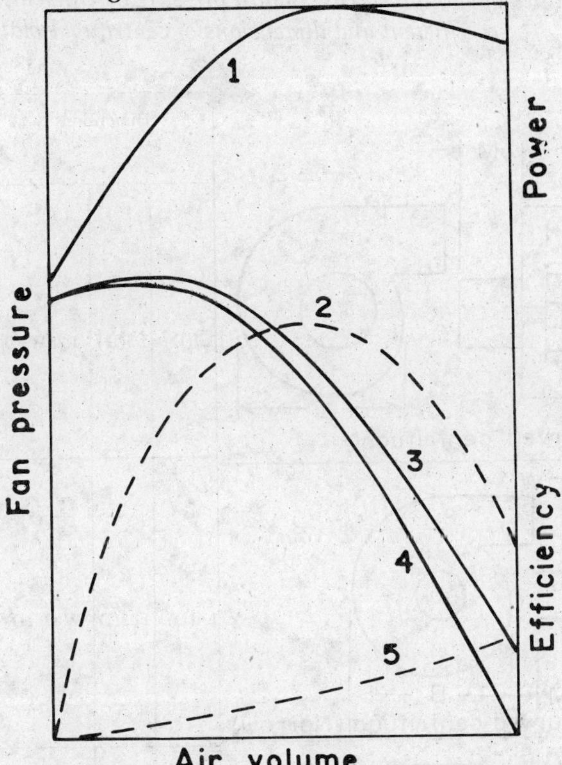

Fig. 1.31 : Characteristics of a back-ward curved centrifugal blower
1. fan power 2. fan efficiency 3. fan pressure
4. static pressure of fan 5. dynamic pressure of fan

Selection/Design of a centrifugal blower

For a given set of airflow rate and static pressure, the impeller diameter and width of blower may be calculated as below,

1. Calculate the specific speed (N_s) for the specific static pressure, air flow rate and designed motor speed.

$$N_s = \frac{N\sqrt{Q}}{p_s^{0.75}},\ rpm \qquad \ldots 1.55$$

Where, N = speed of motor, *rpm*
Q = air flow rate, *cfm*
p_s = static pressure, inches water gauge.

2. From Fig. 1.32 determine the type of air moving unit which would operate at high efficiency, at or near peak efficiency at the calculated specific speed. If more than one type of air moving unit has good efficiency, make the final selection on other factors involved in the application such as relative cost, size and shape of space available and the characteristics of the air flow path.
3. From Table 1.5 find the typical value of pressure coefficient ψ by interpolation, for the type of fan or blower selected and the value of N_s.

Table 1.5

Values of specific speeds, typical pressure co-efficient, typical flow co-efficient and dimensions of centrifugal blowers

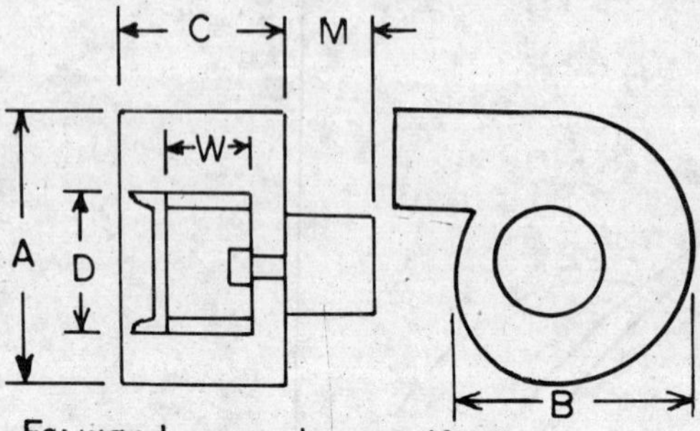

	Dimensions			Specific speed	Typical pressure coeff.	Typical flow coeff.
	A	B	C	Ns		
Forward curved centrifugal	1.7(D)	1.5(D)	1.25W + 0.1D	13,000	1.0	0.15
				20,000	2.0	0.5
				40,000	1.0	0.75
Backward curved centrifugal (Narrow)	1.4(D)	1.35(D)	W + 0.1D	4,000	1.4	0.002
				8,000	1.0	0.01
				20,000	0.8	0.10
Backward curved centrifugal (Wide)	2.0(D)	1.6(D)	W + 0.16D	15,000	1.0	0.08
				30,000	0.75	0.3
				45,000	0.5	0.5

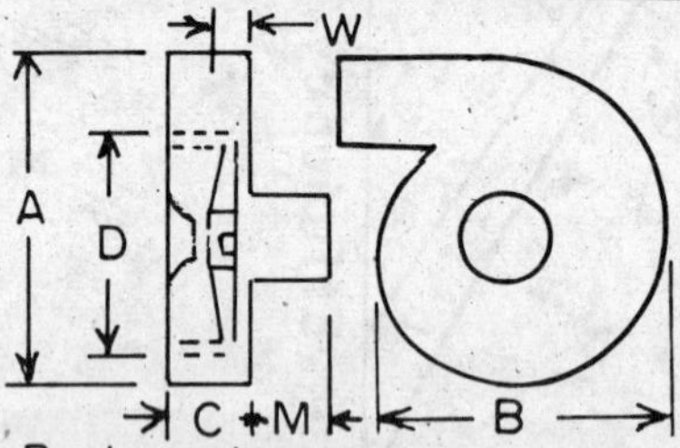

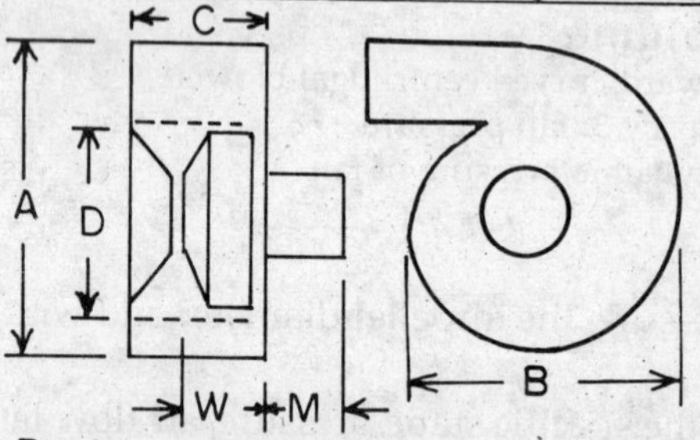

4. Calculate the diameter of the impeller.

$$\psi = \frac{2.35 \times 10^8 \, p_s}{N^2 d^2} \qquad \ldots 1.56$$

Where, ψ = pressure coefficient

d = diameter of the impeller, inch

5. Find out the typical value of flow coefficient ϕ from Table 1.5 and then calculate the width

$$W = \frac{175\, Q}{\phi\, N\, d^2} \qquad \ldots 1.57$$

Where, ϕ = flow coefficient

W = width of the impeller, inch

Problem : A blower is required to deliver 100 m^3/min of air against a static pressure of 100 *mm* water gauge. The rated Speed of motor is 1440 *rpm*, determine the diameter and width of the impeller.

Solution :

1. The specific speed is calculated by the equation:

$$N_s = \frac{N\sqrt{Q}}{p_s^{0.75}}$$

$Q = 100\ m^3/\text{min} = 3546\ cfm$

$p_s = 100\ mm = 4$ inch

$$\therefore \quad N_s = \frac{1440\sqrt{3546}}{4^{0.75}}$$

$= 30313$

2. Fig. 1.32 indicates that two types of air moving units are quit efficient at $N_s = 30313$; a forward curved centrifugal blower and a backward curved (wide) centrifugal blower. While the forward curved centrifugal blower seems to have a substantially higher static efficiency. Table 1.5 shows that both the pressure and flow coefficient of forward curved centrifugal blower are high. Therefore, a forward curved unit is selected

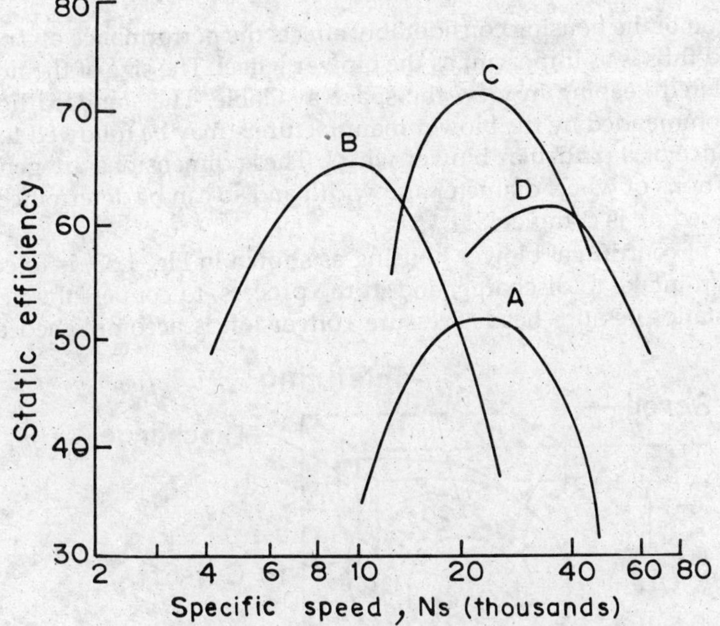

Fig. 1.32 : Specific speed vs. static efficiency of various impellers

3. From Table 1.5, $\psi = 1.5$
4. Since, $\psi = \dfrac{2.35 \times 10^8 \, p_s}{N^2 d^2}$

or,
$$d = \frac{1.53 \times 10^4}{N} \sqrt{\frac{p_s}{\psi}}$$
$$= \frac{1.53 \times 10^4}{1440} \sqrt{\frac{4}{1.5}}$$
$$= 17.35'' \text{ or say } 18''$$
$$= 450 \; mm$$

5. From Table 1.5, $\phi = 0.62$

$$\therefore \quad W = 175 \frac{Q}{\phi N d^2}$$
$$= \frac{175 \times 3546}{0.62 \times 1440 \times (17.35)^2}$$
$$= 2.3'' \text{ or } 58 \; mm$$

A review of manufacturer's literature is done to see that whether blower wheel of calculated dimensions is available. If it is not available, then slightly smaller or larger wheel is selected and the performance requirement is revised.

Blower housing

The configuration of the housing considerably affects the performance of a centrifugal blower and thus is as important as the blower wheel. The size of the housing must be considered keeping in view the space available. The standard housing dimensions recommended by the blower manufacturers may be followed to maximise performance of a particular blower wheel. These dimensions are generally given as proportions of wheel diameter and width and so can be determined after the selected procedure is completed.

The purpose of centrifugal blower housing as shown in Fig. 1.33, is to control the air flow from intake to discharge, and in the process, to convert the velocity head into the static pressure head. Pressure conversion is accomplished as the

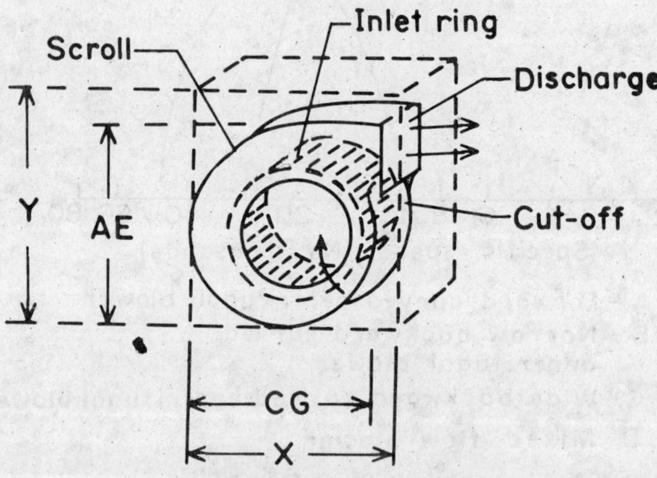

Fig. 1.33 : Housing of centrifugal blower

cross-section of the air stream increases in the increasing annular space on the periphery of the blower wheel from cutoff to discharge. Since the amount of pressure conversion is determined by the scroll configuration, the shape of the housing considerably affects air performance. The cutoff eliminates almost all free circulation of air within the housing.

Diffuser angle

The increase in annular cross-section in the scroll around a blower wheel is proportional to the developed length of the wheel periphery (Fig 1.34). The angle between the developed scroll surface and blower wheel periphery is called the diffuser angle. Wheel diameter and diffuser angle determines the shape and dimensions of the scroll.

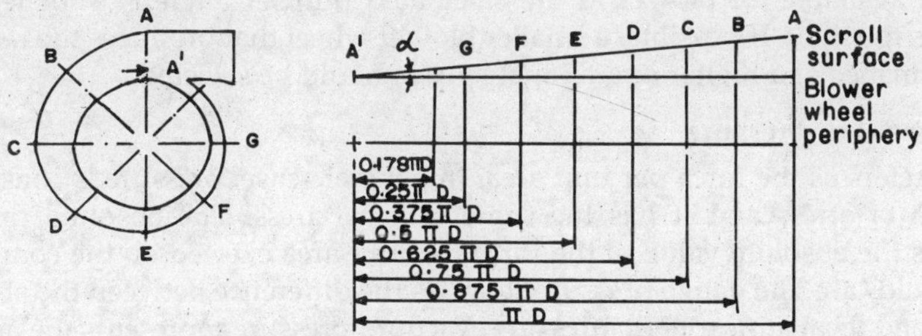

Fig. 1. 34 : Scroll development of a centrifugal blower

The diffuser angle can be determined graphically and expressed in terms of impeller diameter and either the maximum height or maximum width of the housing. Diffuser angle may be given as,

$$\alpha_h = 12\left(\frac{H_M}{d} - 1\right) \quad ...1.58$$

and

$$\alpha_w = 12\left(\frac{W_M}{d} - 1\right) \quad ...1.59$$

As shown by above equations, the diffuser angle decreases if either dimensions *AE* or *CG* decreases. However, it is less sensitive to change in *AE*.

As the diffuser angle increases, the flow rate increases significantly at any particular static pressure. Diffuser angle also affects performance of the blower in a particular system.

The diffuser angle generally used as the basis for blower performance data is 10°. Although large diffuser angle improve performance, the relative amount of improvement gradually diminishes and the size of the housing with respect to the diameter of the blower wheel becomes too large.

The housing width may be determined by the following equation

$$M = 1.25\,W + 0.1\,d \quad ...1.60$$

where, M = housing width

W = impeller width

d = impeller diameter

The optimum diameter is based on a blower wheel mounted close to the inlet ring and minimal clearance between the wheel back plate and side of the housing.

If the width of the housing recommended for the standard blower wheel is too large, a narrower housing should be selected. If either dimension *AE* or *CG* of the recommended housing is too large for the space available, a housing with a smaller diffuser angle should be selected. The resulting reduction in air flow rate should then be determined and compared to the original requirement.

Calculating diffuser angle

The diffuser angle, α_h, equation (1.58) may be calculated with dimensions *AE* equal to the maximum dimension *Y* of the space available (Fig. 1.33).

The diffuser angle, α_w, equation (1.59) may be calculated with dimension *CG* equal to the maximum dimension X of the space available.

The smaller of two diffuser angles corresponds to the housing that will fit into the space available for blower. If the calculated diffuser angle is 4° or less, the housing is too tight. If possible, a smaller blower wheel that produces the required air performance at a higher operational speed should be selected.

Measurement of pressure

Pressure refers as the force per unit area. In general sense, pressure is considered as a form of stress and it has the same unit as stress. The absolute pressure represents the absolute value of the force per unit area exerted on the containing wall by fluid/air. The gauge pressure refers as the difference between the absolute pressure and local atmospheric pressure. Vacuum pressure represents the amount by which the atmospheric pressure exceeds the absolute pressure.

In this book, the pressure measurement is restricted to fluid and gas systems. The knowledge of accurate measurement of pressure is necessary for agricultural process engineer to design, select, install or operate a system where air/fluid flow occurs.

A number of devices are used for the measurement of pressure. Static or steady state pressure may be measured with considerably fair accuracy. But the dynamic measurement of pressure is more complicated because these are affected by many factors such as characteristics of the fluid used in the measurement devices etc. In many cases a pressure measuring device which gives fairly accurate readings for static pressures, may give unsatisfactory results for dynamic pressures. Some of the commonly used pressure measuring devices are described in this chapter.

Manometers : The manometer is the simplest, accurate and widely used device for pressure measurement under steady-state conditions. Some of the commonly used manometers in agricultural processes are described here.

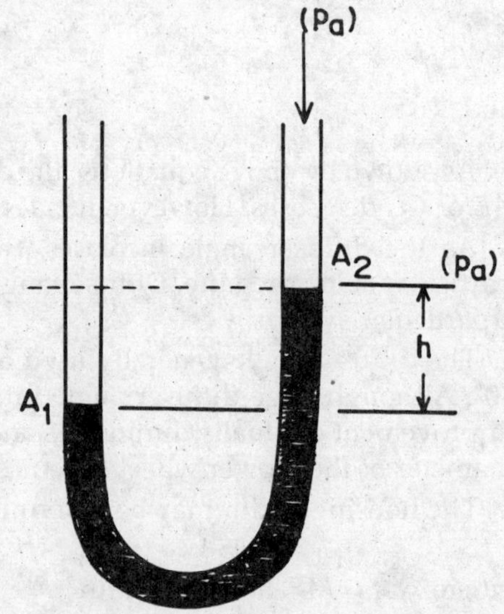

Fig. 1.35 : U-tube manometer

In Fig 1.35, a U-tube manometer is shown. The pressure difference between the unknown pressure 'p' and the atmosphere 'p_a' is determined as a function of difference in height 'h' of the tube columns in mm of fluid filled in the manometer. If the density of the air transmitting the pressure 'p' is 'ρ', and the density of the fluid filled in manometer is denoted as 'ρ'_m. The pressure balance of the two columns may be represented as

$$p_a + \frac{g}{g_c} h \rho_m = p + \frac{g}{g_c} h \rho \qquad \text{...1.61}$$

or

$$p - p_a = \frac{g}{g_c} h \; (\rho_m - \rho) \qquad \text{...1.62}$$

Equation 1.62 gives the basic principle of the U-tube manometer. It is important to note that the height h is measured parallel to the gravitational force and the pressure difference $p - p_a$ is measured at the location shown by the dashed line. If this is not followed, there could be significant error in pressure measurement.

A well type manometer as shown in Fig. 1.36 is similar to the U-tube manometer in operation but it is different in construction. The well is large in comparison to

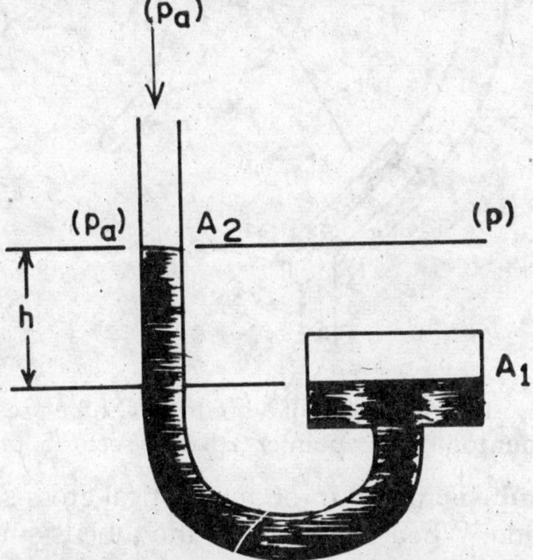

Fig. 1.36 : Well type manometer

the tube so that the change in fluid level in the well is not appreciable for a fluid elevation in the column. Thus the fluid displacement 'h' can be measured directly from a scale without any significant correction. In this case the pressure balance of Equation 1.62 still holds

$$p - p_a = \frac{g}{g_c} h \, (\rho_m - \rho)$$

Since height h is not the fluid displacement which is generally measured, the above equation is not generally used. The well type manometer is filled to a certain level at a zero-pressure differential conditions, measurement is then taken of the displacement of the small column from this zero level. If this displacement is denoted by h' then we have

$$h' A_2 = (h \quad h') A_1 \qquad \text{...1.63}$$

Since the volume displacement are the same on both sides of the manometer. Combining the Equations 1.62 and 1.63 give:

$$p - p_a = \frac{g}{g_c} h' \left(\frac{A_2}{A_1} + 1 \right) (\rho_m - \rho) \qquad \text{...1.64}$$

Bourdon-tube : The Bourdon type pressure gauge (Fig. 1.37) is used for a wide range of applications where consistent, inexpensive measurements of static pressure are desired. The accuracy of Bourdon gauge is approximately 2% for moderately high pressures.

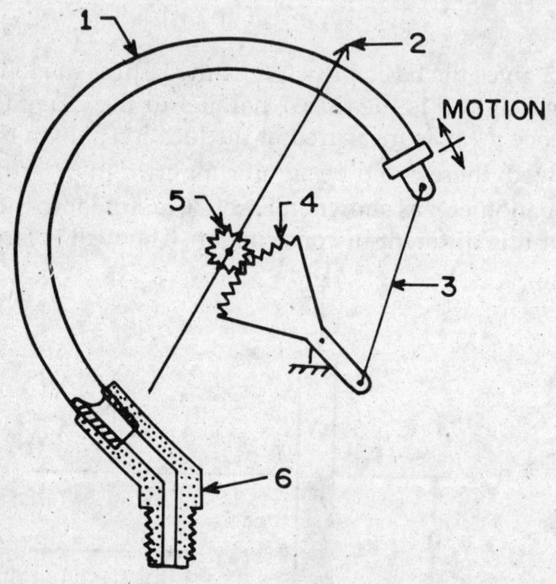

Fig. 1.37 : Diagram of a bourdon C-tube pressure gauge
1. bourdon tube 2. pointer 3. link 4. sector 5. pinion 6. socket

The bourdon tube is made up of an elliptical cross-sectional tube having a 'C' shape configuration. When the inside of the tube is subjected to the pressure, an elastic deformation proportional to the pressure takes place. The degree of linearity depends on the quality of gauge. The end portion of the gauge is connected to a spring loaded linkage, which amplifies the displacement and transforms it to an angular rotation of the pointer.

Diaphragm : The diaphragm type of gauges are useful for low pressure operation to approximately 125 mm of water. These are as accurate as inclined water manometer. Apart from this factor, these are simple in operation which makes it superior than inclined water manometer for many applications.

Considering the flat diaphragm subjected to the pressure difference $p_a - p_b$ as shown in Fig. 1.38 a. The deflection in diaphragm will occur according to the pressure differential and the deflection will be sensed by an appropriate displacement transducer. The electrical strain gauges may be installed on the diaphragm as shown in Fig. 1.38 b. The output of these gauges is a function of local strain, which in turn, may be related to the diaphragm deflection and pressure differential. The diaphragm gauge is very stiff, involves small displacement, and is suitable for high frequency pressure measurements.

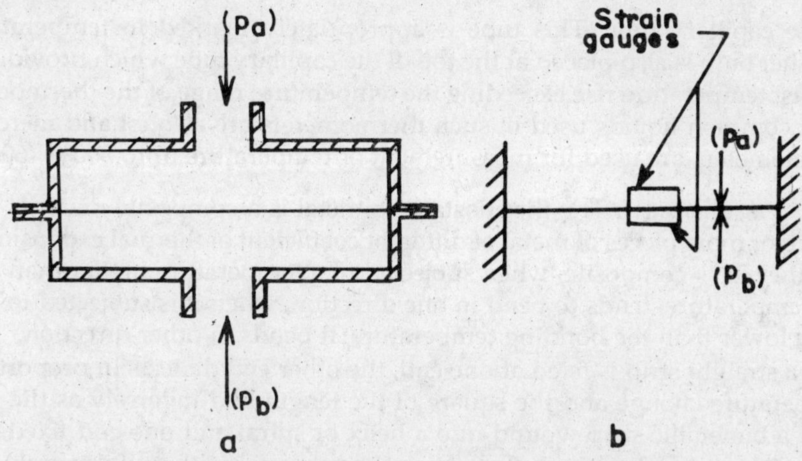

Fig. 1.38 : Diagram of a diaphragm gauge
a. diaphragm gauge b. strain gauges mounted on a diaphragm gauge

Bellows : The bellow type of pressure gauge is shown in Fig. 1.39. The differential pressure force causes a displacement of the bellows, which may be converted to an electric signal or undergo a mechanical amplification to allow display of the output on an indicator dial. The bellow is not suitable for measurement of pressure in transient conditions due to the larger relative motion and mass involved.

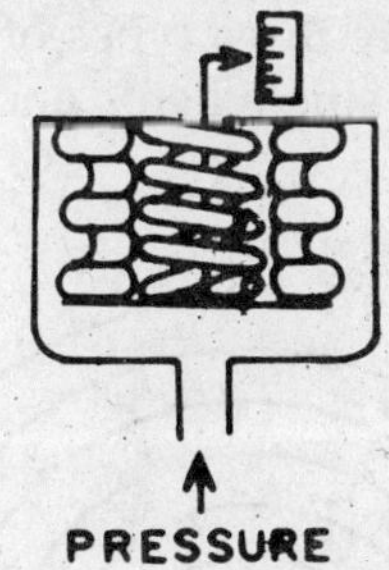

Fig. 1.39 : Bellow type pressure gauge

Temperature measurement

In any heat transfer operation, an accurate measurement of temperature is necessary. The heat transfer operations may be heating or cooling of fruits and vegetables, drying or dehydration of agricultural products, heat and mass transfer in storage bins, pasteurization and sterilisation, canning, extrusion etc. Some of the temperature measuring devices are described below.

Liquid-in-Glass thermometer : The liquid-in-glass thermometer is one of the most common devices for the measurement of temperature (Fig. 1.40). It is not used in

Fig. 1.40 : Liquid in glass thermometer

automatic pressure control systems, but generally used as measuring device for manual control. It has a relatively large bulb at the lower end which contains the major portion of the liquid. The liquid expands when subjected to heat and it rises in the capillary tube. This tube is appropriately marked in temperature scale. Another bulb is also placed at the top of the capillary tube which provides a safety against temperature rise exceeding the temperature range of the thermometer. The most common liquids used in such thermometers are alcohol and mercury. Such thermometers are used for measurement of temperature upto 300°C (600°F).

Bimetal thermometer : The thermostatic bimetal is a composite material, made up of two or more pieces of metal of different coefficient of thermal expansion bonded together. This composite, when subjected to a temperature higher than the bonding temperature, tends to bend in one direction; when it is subjected to a temperature lower than the bonding temperature, it bends in other direction.

If a straight strip is fixed at one end, the other end deflects in proportion to the temperature change and the square of the length and inversely as the thickness. With a bimetallic strip wound into a helix or spiral and one end fixed, the other end rotates when heat is applied. For a thermometer with uniform scale divisions, a bimetal must be designed to have linear deflection over the desired temperature range.

There are three types of elements which are commonly used in thermometers; flat spiral, single helix, and the multiple helix (Fig. 1.41).

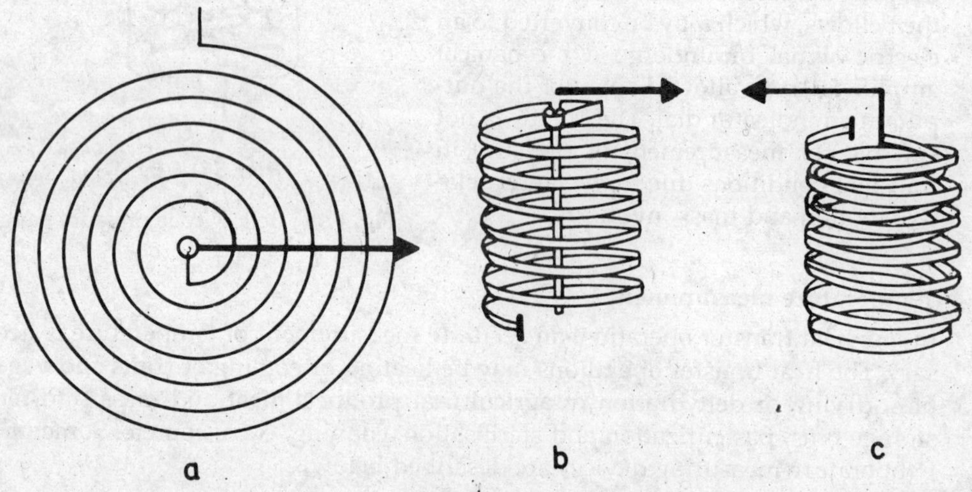

Fig. 1.41 : Elements of bimetal thermometers
a. flat spiral b. single helix c. multiple helix

The bimetal thermometers are used at temperatures ranging from 538°C (1000°F) down to – 185°C (– 300°F) and lower.

Electrical-resistance thermometers : The electrical resistance thermometer is a quite accurate device for temperature measurement. The principle of resistance thermometer is based on the characteristics of metals to change in electrical resistance when they undergo a change in temperature. It consists of some type of resistive element, which is exposed to the environment whose temperature is to be measured. The temperature is indicated through a measurement of the change in resistance of the element. The industrial resistance thermometers are generally made of platinum, copper or nickel but the semi conducting material like thermistors is finding wide applications.

Basically a resistance thermometer is an instrument for measurement of electrical resistance which is calibrated in units of temperature rather than units of resistance. The resistance measurement may be performed with some type of bridge circuit like wheatstone bridge. For steady-state measurement a null condition will suffice, while transient measurements will generally require the use of deflection bridge.

Thermocouples : The thermocouple is a quite common electrical method of temperature measurement. It is based on the theory that an electrical current flows in a continuous circuit of two different metallic wires if the two junctions of the wire are at different temperatures. As shown in Fig. 1.42, *A* and *B* are two metals, and T_1 and T_2 are the temperatures of the junctions. If T_1 is the colder junction and the thermoelectric current *'i'* flows in the direction indicated in Fig. 1.42 a, metal *A* is referred as thermoelectrically positive to metal *B*.

In electric circuits, the current is dependent on the e.m.f. (electro motive force) developed and the resistance of the circuit. For accurate temperature measurements, the measuring instrument is constructed so that a no-current e.m.f. is measured to eliminate the effects of circuit resistance.

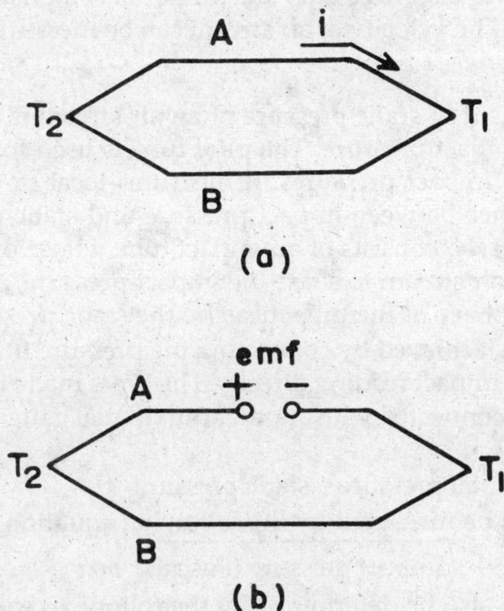

Fig. 1.42 : Simple thermocouple circuit

The thermal e.m.f. as indicated in Fig. 1.42b, is a measure of the difference in temperature between T_1 and T_2. In control systems the reference junction (cold junction) is usually located at the e.m.f. measuring device. The reference junction may be held at constant temperature such as in an ice bath or a thermostatically controlled oven, or it may be at ambient temperature but electrically compensated so that it appears to be held at a constant temperature.

Series combinations of several identical thermocouples can be used to provide a large temperature measurement signal or to achieve the average value of the temperatures at several locations. The cold junctions as shown in Fig. 1.43 must be at the same temperature.

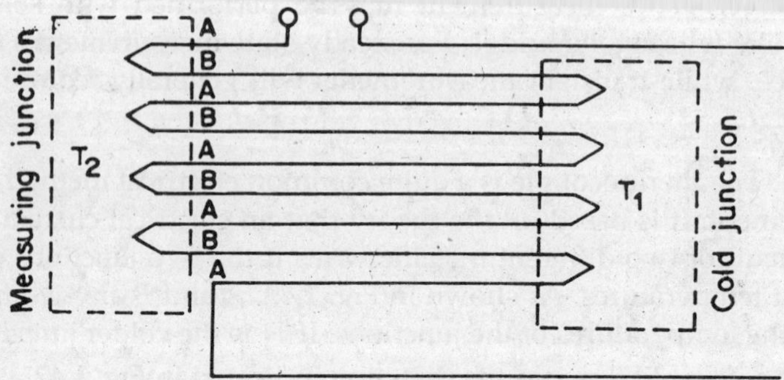

Fig. 1.43 : Diagram of a series combination of identical thermocouples

Measurement of air velocity

Measurement of air stream velocity in any processing job or operation is necessary as it affects the performance of operation. In cleaning of grains by air screen cleaners terminal velocity is considered for separation of different constituents. Optimum air velocity is also necessary for proper accomplishment of drying of agricultural products. The velocity of air stream can be measured by the following devices.

Pitot tube : Measurement of static pressure of an air stream is more difficult than the measurement of impact pressure. The pitot tube is used for the measurement of both the static and impact pressures. It measures local or point velocities by measuring the difference between impact pressure and static pressure. The pitot tube as shown in Fig. 1.44, consists of an impact tube whose opening in the front faces directly into the air stream to sense the impact pressure, and the small holes around the outer periphery of the tube measure the static pressure.

The measurement is achieved by connecting the pressure fittings to differential gauge which gives the impact reading directly. The holes made in the outer periphery of the pitot tube is connected with another differential gauge which gives static pressure.

Impact pressure = Total pressure – static pressure

The air velocity can be measured by the following equation

$$\text{Air velocity} = 1.291\sqrt{\text{Impact pressure (Pascal)}},\ m/s$$

The air velocity can also be determined by the following equation:

$$\text{Air velocity} = \sqrt{2gh}$$

where, g = acceleration due to gravity, m/s^2

h = total pressure, m

Hot wire anemometer : The hot wire anemometer is used for the measurement of low air velocities and varying flow conditions. It consists of an electrically heated fine wire (normally platinum) exposed to the air stream whose velocity is to be measured. If other things remain constant, change in air velocity changes the rate

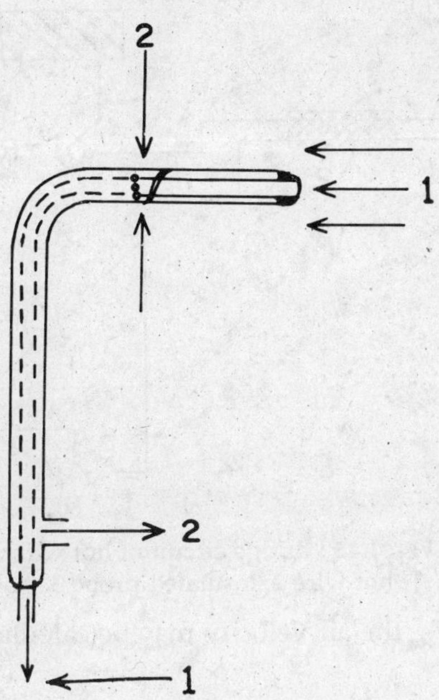

Fig. 1.44 : Pitot static tube
1. total pressure 2. static pressure

of heat transfer from the wire to the flowing air, thereby affects the rate of cooling of wire and in turn changes wire's electrical resistance. The rate of heat transfer from the wire can be given as,

$$q = (a + b\sqrt{\rho u})\ (T_w - T_\infty),\ k\ cal/hr.m^2 \qquad ...1.65$$

Where, q = rate of heat transfer
T_w = wire temperature
T_∞ = air temperature
u = air velocity
a and b = constants, obtainable by calibration of device.

The heat transfer rate may also be given as,

$$q = i^2 R_w = i^2 R_0 [1 + \alpha (T_w - T_0)] \qquad ...1.66$$

where, i = electrical current
R_0 = resistance of the wire at the reference temperature, T_0
α = temperature coefficient of resistance

In a constant current anemometer, air velocity is determined by measuring the final wire resistance, while in the constant resistance type, air velocity is determined from the current required to maintain the wire temperature at constant level.

For measurement of the air velocity, the hot wire is connected to a bridge circuit as shown in Fig. 1.45. The current is determined by measuring the voltage drop across the standard resistance R_s, and the wire resistance is determined from the bridge circuit. For steady state measurement thé null condition is used, whereas for the transient measurement an oscillography output may be used. With deter-

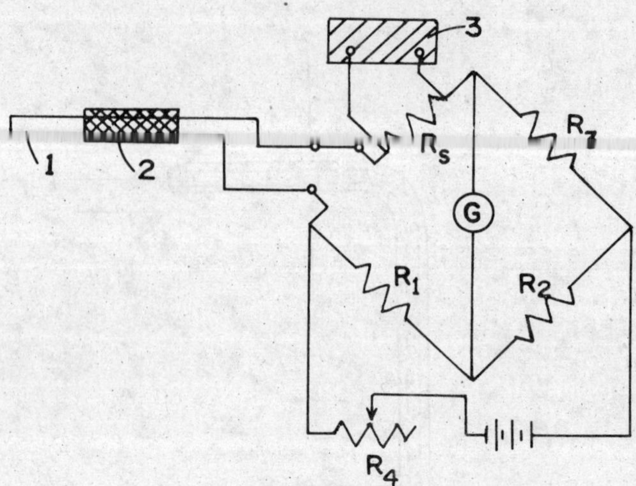

Fig. 1.45 : Bridge circuit of hot wire anemometer
1. hot wire 2. insulated probe 3. potentiometer

mination of i and R_w the air velocity may be calculated with the help of Equation 1.65 and 1.66.

Measurement of air flow

Flow refers to quantity of air in m^3 or kg flowing per unit time. The volumetric air flow can be given by the following equation,

$$Q = av \qquad \text{...1.67}$$

Where, Q = air flow, m^3/min

a = cross–sectional area, m^2

v = air velocity, m/min

Rotameter : The rotameter is a common device for measurement of air flow (Fig. 1.46). It consists of a bob or float which is free to move up or down in a tapered vertical tube having its smaller end downwards. The air enters from bottom through smaller end and move upwards, forcing the float to move up. The drag forces are balanced by the float weight and buoyancy forces. The position of the float in the tapered tubes of glass is taken as an indication of the air flow rate. The force balance on the float may be given as,

$$F_d + \rho_f v_b \frac{g}{g_c} = \rho_b v_b \frac{g}{g_c} \qquad \text{...1.68}$$

where, ρ_f = density of the fluid

ρ_b = density of the float/bob

v_b = total volume of the bob

g = acceleration due to gravity

F_d = drag force

The drag coefficient may be given as

$$F_d = c_d A_b \frac{\rho_f u_{m^2}}{2g_c} \qquad \text{...1.69}$$

where, c_d = drag coefficient

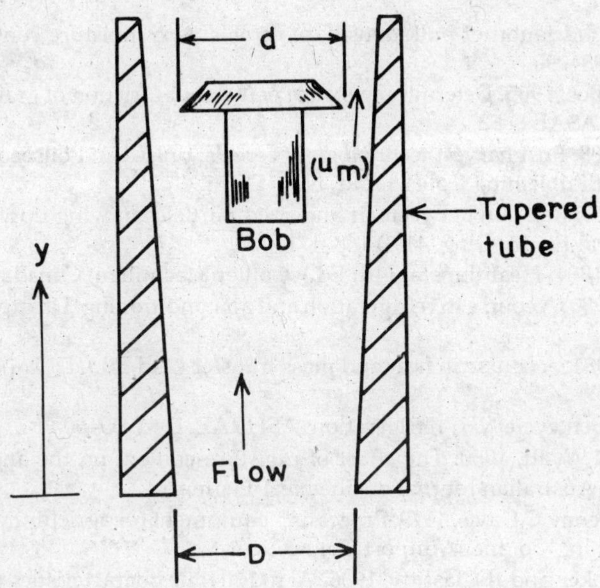

Fig. 1.46 : Schematic diagram of a rotameter

A_b = frontal area of the bob

u_m = mean flow velocity in the annular space between the bob and the tube

Combining Equation 1.68 and 1.69 yields,

$$u_m = \left[\frac{1}{c_d}\ \frac{2g\ v_b}{A_b}\left(\frac{\rho_b}{\rho_f} - 1\right)\right]^{1/2} \quad \text{...1.70}$$

or,

$$Q = A\ u_m = A\left[\frac{1}{c_d}\ \frac{2g\ v_b}{A_b}\left(\frac{\rho_b}{\rho_f} - 1\right)\right]^{1/2} \quad \text{...1.71}$$

where A = annular area of tube

$$= \frac{\pi}{4}[(D + ay)^2 - d^2]$$

where, D = inlet diameter of tube

d = maximum bob diameter

y = vertical distance of bob from inlet

a = constant, depends upon tube taperness

Bibliography

Anon. 1980. Engineering properties of food materials. Technical bulletin No. CIAE/80/15, CIAE, Bhopal.

Arnberg, BT, 1962. Review of critical flow meters for gas flow measurements. Trans. AS ME, 84D:447.

Arnold, PC and AW Roberts, 1969. Fundamental aspects of load-deformation behaviour of wheat grain. Trans. ASAE 12(1):104–108.

Bilanski, WK, SH Collins and P Chu, 1962. Aerodynamic properties of seed grains. Agric. Eng. 43 (4):216–219.

Browne, DA, 1962. Variation of bulk density of cereals with moisture content. J. of Agric. Eng. Res. 7(4):288–290.

Brubaker, JE and PPos, 1965. Determining static coefficient of friction of grains on structural surfaces. Trans. ASAE 8: 53.

Chakraverty, A, 1988. Post harvest technology of cereals, pulses and oilseeds (Revised Ed.). Oxford and IBH Publishing Co. Pvt. Ltd. New Delhi.

Chancellor, WJ, 1960. Relation between air and solid particles moving upward in a vertical pipe. Agricultural Engineering, 41(3):168–171.

Chapman, Alan J, 1984. Heat transfer (4th Ed.), Collier Macmillan, Canada, Inc.

Domkundwar, S, 1980. A course in refrigeration and air conditioning. Dhanpat Rai and Sons, Delhi.

Domkundwar S, 1981. A course in heat and mass transfer (2nd Ed.), Dhanpat Rai and Sons, Delhi.

Elliot, DE, 1971. Open cycle NH_3 refrigeration. ASHRAE, Dec.: 55–60.

Fowler, RT and FA Wyatt, 1960. The effect of moisture content on the angle of repose of granular solids. Australian Journal of Chemical Engineers.

Hall, Carl W and Denny C Davis, 1979. Processing equipment for agricultural products (2nd Ed.) AVI Publishing Co. Inc. Westport Connecticut.

Hawk, AL,DB Brooker and JJ Classidy, 1966. Aerodynamic characteristics of selected farm grains. Trans ASAE 9(1):48–51.

Henderson, SM and RL Perry, 1966. Agricultural process engineering. Edwards Brothers Inc. Ann Arbor. Michigan.

Henderson, SM and RL Perry, 1976. Agricultural process engineering (3rd Ed.) The AVI Publishing Co. Inc. Westport, Connecticut.

Holman, JP, 1971. Experimental methods for engineers (2nd Ed.) Mc Graw Hill, Kogakusha Ltd.

Holman, JP, 1981. Heat transfer (5th Ed.), Mc Graw-Hill, Kogkusha Ltd.

Houston, RK, 1957b. New criterion of size for agricultural products. Agricultural Engineering, 39(12):856–858.

Kashyap, MM and BS Pathak, 1974. Air velocity requirement for winnowing of wheat grains. Paper presented at XIII meeting of ISAE.

Keck, H and JR Goss, 1965. Determing aerodynamic drag and terminal velocities of agronomic seeds in free fall. Trans ASAE 8(4):1181–1184.

Mc Lean, KA. 1980. Drying and storing combinable crops. Farming Press Ltd. Suffolk.

Mohsenin, Nuri N. Physical properties of plant and animal materials. Gordon and Breach Science Publishers, New York.

Nag, PK, 1985. Engineering thermodynamics, Tata Mc Graw-Hill Publishing Co. Ltd., New Delhi.

Ozisik, M. Necati, 1985. Heat transfer—a basic approach. Mc Graw-Hill, Singapore.

Perry, RH and CH Chilton, 1973. Chemical engineering hand book. Mc Graw-Hill, Kogakusha Ltd.

Siwaswami, S, 1971. Terminal velocity of solids. M. Tech. Thesis, GBPUA & T, Pantnagar.

Snyder, LH, WL Roller and GE Hall 1967. Coefficients of kinetic friction of wheat on various metal surfaces. Trans ASAE 10(3):411–413, 419.

Stephens, LE and GH Foster, 1978. Bulk properties of wheat and grain sorghum as affected by a mechanical grain spreader. Trans ASAE 21:1217–1218, 1221.

Stewart, BR, 1968. Effect of moisture content and specific weight on internal friction properties of sorghum grain. Trans ASAE 11(2):260–266.

Welty, James R, 1978. Engineering heat transfer, John Wiley and Sons, Inc.

White, RK and NN Mohsenin, 1967. Apparatus for determination of bulk modulus and compressibility of materials. Trans ASAE 10(5):670–671.

Uhl, JB and BJ Lamp, 1966. Pneumatic separation of grain and straw mixtures. Trans. ASAE 9(2):244–246.

Zoerb, GC and CW Hall, 1960. Some mechanical and rheological properties of grains. J. of Agric. Eng. Res. 5:83.

Zoerb, GC, 1967. Instrumentation and measurement techniques for determination of physical properties of agricultural products. Trans. ASAE 10(1):100–109.

Chapter 2

Cleaning and Grading

Cleaning and grading are the first and most important post harvest operations undertaken to remove foreign and undesirable materials from the threshed crops/grains and to separate the grains/products into various fractions. The comparative commercial value of agricultural products is dependent on their grade factors. These grade factors further depend upon. (1) physical characteristics like size, shape, moisture content, colour etc., (2) chemical characteristics like odour, freefatty acid content and (3) biological factors like germination, insect damage.

A mixture of seeds can be separated on the basis of difference in length, width/thickness, specific gravity, surface texture, drag in moving air, colour, shape, electrical conductivity and magnetic properties.

Cleaning in agricultural processing generally means the removal of foreign and undesirable matters from the desired grains/products. This may be accomplished by washing, screening, hand picking etc.

Grading refers to the classification of cleaned products into various quality fractions depending upon the various commercial values and other usage.

Sorting refers to the separation of cleaned product into various quality fractions that may be defined on the basis of size, shape, density, texture and colour.

Scalping refers to the removal of few large particles in an initial process.

Screening

Screening is a method of separating grain/seed into two or more fractions according to size alone. For cleaning and separation of seeds, the most widely used device is screen. When solid particles are dropped over a screen, the particles smaller than the size of screen openings pass through it, whereas larger particles are retained over the screen or sieve. A single screen can thus make separation into two fractions. When the feed is passed through a set of different sizes of sieves, it is separated into different fractions according to the size of openings of sieves. Screens alongwith an air blast (air screen) can satisfactorily clean and sort most of the granular materials. The screens are generally suspended by hangers, and when this unit is oscillated by an eccentric unit they have a horizontal oscillating motion and at the same time a smaller vertical motion. These two motions cause grains to travel downward to the screen and at the same time the grains are thoroughly stirred during the passage.

Screen motions

The purposes of screen motion are,

1. to spread the material over the surface of screen.
2. to cause the fine particles to settle at the sub- surface.
3. to discharge the oversize particles.

Screen fractions

The material which is retained over a screen surface is called the oversize or plus material, while the material which passes through the screen is the undersize or minus material. The material which passes through a particular screen and retained over a subsequent screen is the intermediate material.

Mesh and Space cloth

The wiremesh is usually specified by 'mesh' which is the number of openings per linear inch, counted from the centre of any wire to a point exactly 1 inch distant, or by an opening specified in inches or millimeters, which is clear opening or space between the wires. When the openings are limited to 2 or more in an inch it is called a mesh, whereas in space cloth the openings are ½ inch or more than ½ inch.

Aperture

It is the minimum clear space between the edges of the opening in the screening surface and is generally specified in inches or millimeters.

The open area of square mesh wire cloth may be determined by the following equation

$$P = \frac{O^2}{(O+D)^2} \times 100 = (OM)^2 \times 100 \qquad ...2.1$$

where, P = open area, %
M = mesh
O = size of opening
D = wire diameter

Types of screens

In most screens the grain/seed drops through the screen opening by gravity. Coarse grains drop quickly and easily through large opening in a stationary surface. With finer particles, the screening surface must be agitated in some way. The common ways are, (1) revolving a cylindrical screen about a horizontal axis and (2) shaking, gyrating or vibrating the flat screens.

Grizzly

The grizzly is a simple device consisting of a grid made up of metal bars, usually built on a slope, across which the material is passed. The path of material flow is parallel to the length of bars. The bars are usually so shaped that the top is wider than the bottom. The grizzly is often constructed in the form of a short endless belt so that the oversize is dumped over the end while the sized material passes through. In this case bar length is transverse to the path of materials. The grizzly is used for coarsest and rough separations.

Revolving screen/cylinder sorter

Trommel or revolving screen is a cylinder that rotates about its longitudinal axis. The wall of the cylinder is made of perforated steel plate or sometime the cloth wire on a frame, through which the material falls as the screen rotates. The axis of cylinder is inclined along with the feed end to the discharge end. Sizing is achieved by having smallest opening screen at the feed end with progressively larger opening screens towards the discharge end. This type of sorter is simple and compact with no vibration problem. But the capacity of cylinder sorter is lesser than the vibrating screen of same size. Although it is an accurate sizer, it does not perform well with friable material or in cases where particle degradation is undesirable because tumbling produces some autogeneous grinding. The speed of rotation of the trommel be kept within the limit at which the material is carried from bottom to a distance equal to the radius of cylinder before it starts tumbling. The inclination of cylinder sorter for dry granular materials is kept upto 125 *mm/m*. The capacity, bed depth and efficiency of these screens can be changed by changing the speed of operation and the inclination of cylinder. Effective screening area (not the total surface of cylinder) is calculated by multiplying the length of cylinder by ⅓ of the diameter.

Shaking screen

Like the vibrating screen, shaker is a rectangular surface over which material moves down on an inclined plane. Motion of the screen is back and forth in a straight line. Although in some cases vibration is also given to the screen. Unlike the vibrating screen, the shaker does not tumble or turn material enroute except that some shaking screens have a step-off between surfaces having different size openings, so that there may be two or three tumbles over the full length of the screen. The shaker is widely used as combined screen and conveyor for many types of bulk material.

Rotary screen

Rotary and gyratory screens are either circular or rectangular decked. Their motion is almost circular and affects sifting action. These are capable of accurate and complete separation of very fine sizes but their capacity is limited. These screens are further classified into two categories.

1. *Gyratory screens* : This is generally a single decked machine. It has horizontal plane motion, which is circular at feed end and reciprocating at the discharge end. The drive mechanism is at the feed end and is either a V-belt or direct coupling.

The shaft that imparts motion to the screen is a counter balanced eccentric. The shaft moves about a vertical axis. At the discharge end most rotary screens have linkage to the base frame, usually a self aligning bearing. Gyratory screens operate with screening surface nearly horizontal.

2. *Circular screens* : These are also rotary screens but their motion in horizontal plane is circular over the entire surface. Similar to the gyratory screens, the screening surface of circular screens are also little bit tilted for allowing the material to move over them.

Vibratory Screen

For cleaning, grading and separation of agricultural granular materials vibratory screen cleaners are very popular. For separation of undesirable foreign materials from feed mass and sorting the materials into various size groups according to desired quality standard or grades, the vibratory screens are extensively used. Accuracy in separation based on size is very important as small difference in size may be able to separate undesirable or low quality matter.

The vibratory screens are agitated by an eccentric unit. When materials to be separated are put on a vibratory screen, because of its vibration, materials are also agitated and separated during their transit over the screen. The eccentricity is usually of two types, (1) a shaft to which off centre weights are attached, and (2) a shaft that itself is eccentric or off centred. In the latter case the eccentricity is balanced by a fly wheel for providing uniform vibration. Most vibrating screens are inclined downward from the feed end. Vibration is provided to the screen assembly only, and the body and other surrounding structures are isolated from vibration. Generally, up to three decks are used in vibrating screens. The capacity of vibrating screen is higher than any other similar sized screen and is very popular for cleaning and grading of granular agricultural products.

The vibration of screen is responsible for performance of the following functions, (1) it helps in providing passage to particles through the openings of screen, (2) it restricts clogging of the screen by particles that become trapped in the opening, (3) because of vibration the particles are stratified over the screen surface and each particle has a chance to meet the screen opening and (4) a continuous flow of particles along the screen is possible.

Horizontal screen

Horizontal screens are special case of vibrating screen. These are designed for operation with low head room. These operate absolutely flat without the aid of gravity. All sorting, stratification and materials transportation take place on the strength of a sharp forward thurst which imparts motion to particles with a missile like trajectory, while the return stroke pulls the deck out from underneath the bed. Effectiveness of these screens is higher because material is kept on the screen for a longer period in comparison to inclined screens.

Various other types of screen used for cleaning and separation are listed below: 1. Rotex screens 2. Hummer screens 3. Circular vibrators 4. Symon's rod deck screens 5. Resonant vibrant screens 6. Centrifugal screens.

Particle motions

There are four different regime motions that can take place for the rigid particles placed on a moving trough depending upon the frequency of oscillation: 1. Particles stationary with respect to trough. 2. Particles slip during part of cycle and remain adherent to the trough during the rest of cycle. 3. As regime (2) with slip and gliding motion. 4. Particles purely in stick and slip motion.

The particle paths in cleaner and separators are described below:

1. *Reciprocatory :* This can be either purely in horizontal plane or in an inclined plane. Depending upon the plane, the motion of particles is different as given below:

(a) Horizontal reciprocating motion : This motion is obtained with an eccentric and a connecting rod and is usually in a path parallel to the horizontal projection of the path of material as it moves from inlet to outlet as shown in Fig 2.1. If the pitch (amplitude) is steep enough the motion has a substantial vertical component to the screen sufrace which initiates sliding motion and a large quantity of material can be moved.

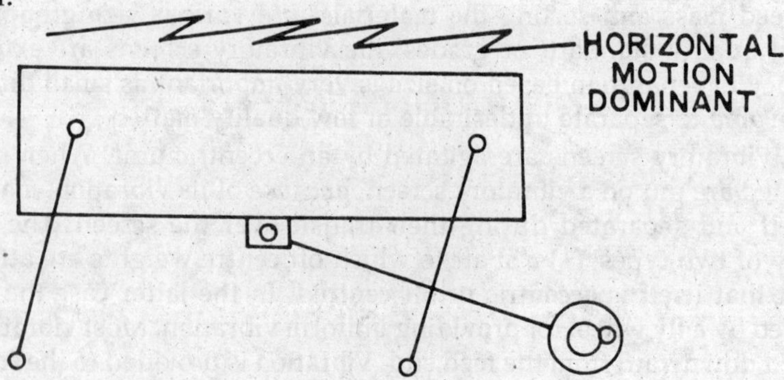

Fig. 2.1 : Horizontal reciprocating motion

(b) Inclined reciprocating motion : The screen can be moved in an inclined plane reciprocatingly by eccentric and connecting rod unit (Fig 2.2). Such motion would

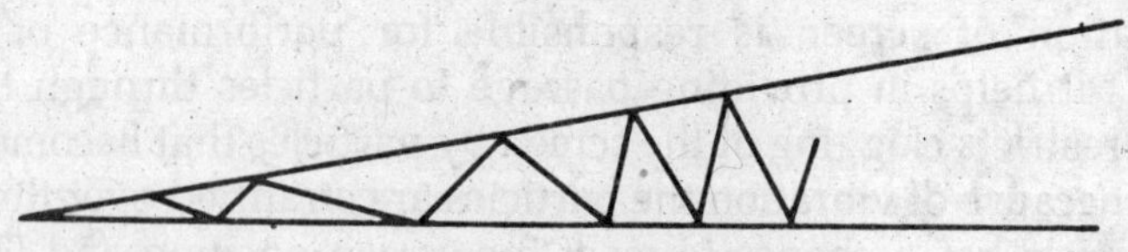

Fig. 2.2 : Inclined reciprocatory motion

have a vertical and horizontal components parallel to the horizontal projection of path of moving material. Combined horizontal and vertical motion can also be achieved by a rotary drive attached directly to the screen and operating in a vertical plane parallel to the path of the material flow. The horizontal motion is shown in Fig 2.3b and the vertical motion is shown in Fig 2.3a. The vertical component lifts the material from the screen surface for a fraction of time. The vertical motion can loosen the mass of material thus dislodging finer particles to settle at the bottom. The combined vertical and horizontal motion is also effective in moving large volume of material rapidly over the surface. This technique is useful in coarse sifting or where the screen openings are substantially larger than the particle size.

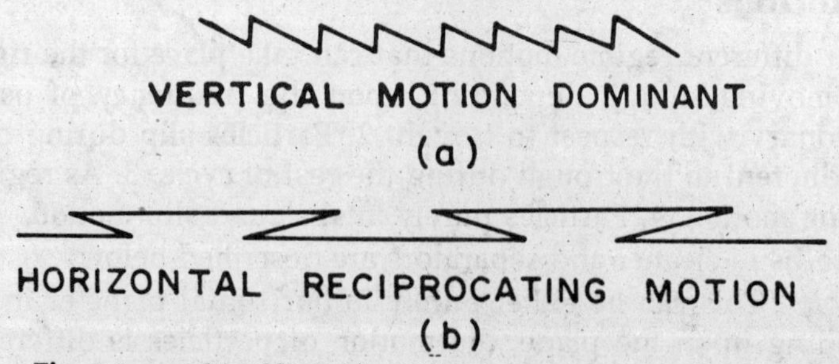

Fig. 2.3 a and b : Vertical and horizontal reciprocating motion

2. *Combined horizontal and reciprocating motion* : In this, an eccentric drive is used to change the horizontal reciprocating motion into a rotary motion in a horizontal plane at inlet end, whereas the other end has reciprocating motion. In such arrangement tnere is a component of reciprocating motion both in parallel and perpendicular to the direction of material flow at inlet and this gradually changes into a elliptical motion at the central section of the screen and finally becomes a true reciprocating motion at the discharge end as shown in Fig. 2.4. This section is effective in spreading the material to the sides of screen at the inlet end. The looping path of the material also presents more screen openings since particle moves not only back and forth but also from side to side across the screen surface.

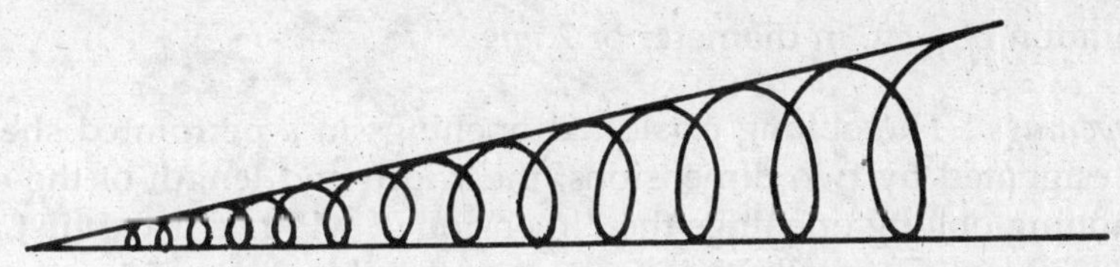

Fig. 2.4 : Combined horizontal and reciprocatory motion

3. *Horizontal rotary motion* : If the screen surface is rotated in a horizontal plane the material will travel in the overlapping loop path during the passage from inlet to discharge end as shown in Fig. 2.5. The multidirectional motion overcomes inter

Fig. 2.5 : Horizontal rotary motion

particle friction and maximum number of openings are available, thus screening can be accomplished on a relatively shorter distance. Rotary motion is provided by an off-centred weight attached to the screen through the frame and it is rotated at proper speed in horizontal plane. This imparts desired frequency and motion. The amplitude of movement is controlled by the magnitude and eccentricity of weight as well as the position which affects its leverage on the screen. Sifters may also be driven by means of one or more directly connected eccentrics. The speed of eccentric and the amount of eccentricity control the frequency and amplitude of motion.

Screen openings

Screens are generally constructed by perforated sheet metal or woven wiremesh. The openings in perforated metal sheets may be round, oblong or triangular as shown in Fig. 2.6. The openings in wire mesh are square or rectangular. The size and shape and their combination of the screens available in market are identified by some trade numbers.

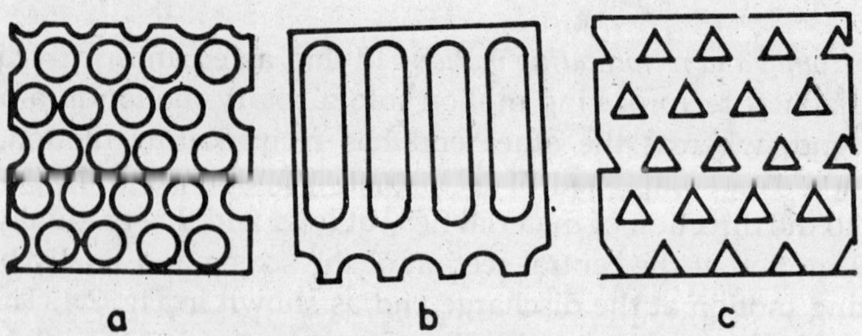

Fig. 2.6 : Perforated metal screens
a. round holes, b. oblong holes c. triangular holes

Perforated metal screens

(i) Round openings : The round openings in a perforated sheet metal screen are measured by the diameter (*mm* or in.) of the openings. For example, $\frac{1}{18}$ screen has round perforation of $\frac{1}{18}$ in. in diameter or 2 *mm*.

(ii) Oblong openings : The oblong or slotted openings in a perforated sheet metal screen are designated by two dimensions, the width and length of the opening. While mentioning oblong openings the dimension of width is listed first then the length as 1.8×20 *mm*. Generally, the direction of the oblong opening is kept in the direction of the grain flow over the screen.

(iii) Triangular openings : There are two different systems used to measure triangular perforations. The most commonly used system is to mention the length of each side of the triangle in *mm*, it means, 9 *mm* triangle has 3 equal sides each 9 *mm* long.

The second system is to mention openings according to the diameter in *mm* that can be inscribed inside the triangle. This system is identified by the letter *V* as *9V*, *10V* etc.

Wiremesh screens

(i) Square mesh : The square openings in wire mesh are measured by the number of openings per inch in each direction. *A* 9×9 screen has 9 openings per inch (Fig. 2.7).

(ii) Rectangular mesh : the rectangular openings in wiremesh screens are measured in the same way as square wiremesh screen. A 3×6 rectangular wiremesh screen will have 3 openings per inch in one direction and 6 openings per inch in the other direction. The rectangles formed by the wiremesh are kept parallel to the direction of grain flow (Fig. 2.8).

Fig. 2.7 : Wire mesh screen
(square openings)

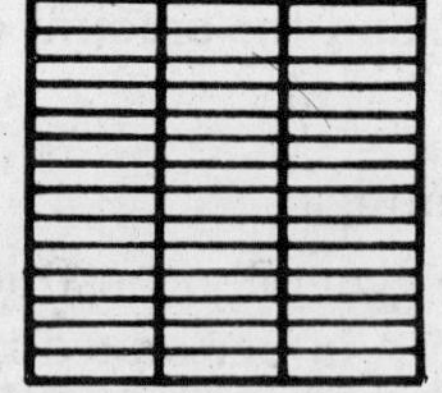

Fig. 2.8 : Wire mesh screen
(rectangular openings)

Ideal and Actual screens

The basic purpose of any screen is to separate the feed consisting of a mixture of particles of different sizes into two distinct fractions. These fractions are, (1) the underflow, the particles which pass through the screen, and (2) the overflow or oversize, the materials that are retained over the screen. A screen can be termed as ideal screen which separates the feed mixture in such a way that the largest particle of underflow is just smaller than screen opening, while the smallest particle of overflow is just larger than the screen opening. But in practice no screen gives perfect separation as stated above, and is called actual screen. The underflow may contain material coarser than screen size, whereas the overflow may contain smaller particles than screen size as shown in Fig. 2.9.

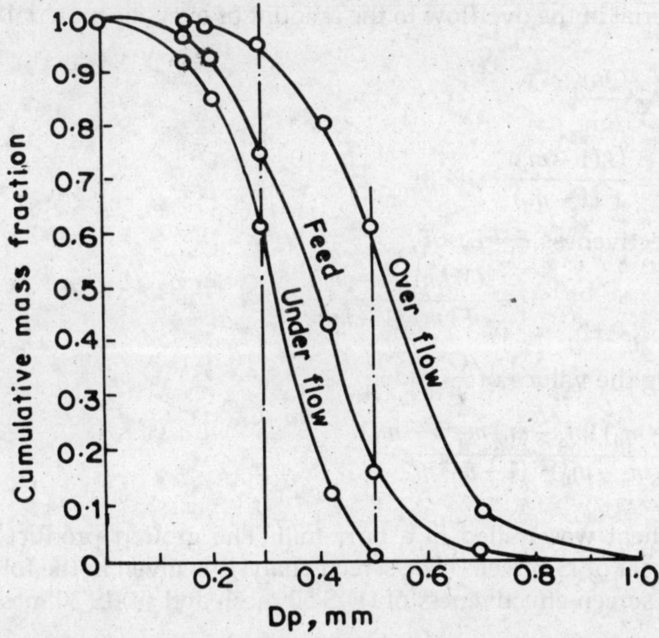

Fig. 2.9 : Graphical representation of various flows of a screen

Effectiveness of screen

The screen effectiveness may be defined as the ability of a screen in closely separating the feed into overflow and underflow according to its size. If the screen functions properly, all material 'O' would be in the overflow, while all the material 'U' would be in the underflow. The material balance in a screening operation can be derived as follows:

F = mass flow rate of feed, *kg/hr*.

O = mass flow rate of oversize, *kg/hr*.

U = mass flow rate of undersize, *kg/hr*.

m_f = mass fraction of material in feed

m_0 = mass fraction of material in overflow

m_u = mass fraction of material in underflow

The total quantity of feed is the sum of overflow and underflow

$$F = O + U$$

$$Fm_f = Om_0 + Um_u \quad \text{...2.2}$$

Substituting $O = F - U$

and $U = F - O$

$$\frac{O}{F} = \frac{m_f - m_u}{m_o - m_u} \quad \text{...2.3}$$

and

$$\frac{U}{F} = \frac{m_o - m_f}{m_o - m_u} \quad \text{...2.4}$$

A common measure of screen effectiveness is the ratio of actual amount of oversize material in the overflow to the amount of oversize material entering with the feed.

Thus, $$E_o = \frac{Om_o}{Fm_f} \quad \text{...2.5}$$

and $$E_u = \frac{U(1 - m_u)}{F(1 - m_f)} \quad \text{...2.6}$$

Overall effectiveness $E = E_o \times E_u$

$$= \frac{OUm_o\ (1 - m_u)}{F^2 m_f\ (1 - m_f)} \quad \text{...2.7}$$

Substituting the values of $\frac{O}{F}$ and $\frac{U}{F}$

$$E = \frac{(m_f - m_u)(m_o - m_f)\, m_o\, (1 - m_u)}{(m_o - m_u)^2\ (1 - m_f)\, m_f} \quad \text{...2.8}$$

Problem : Wheat was milled in a burr mill. The ground product was later on analysed in a set of IS screens. The screen analysis is given in the following Table. Calculate the screen effectiveness of (1)IS 50 mesh and (ii) IS 30 mesh.

Screen analysis for milled wheat

IS Screen	*Width of opening, mm*	*% material retained over each screen*		
		Feed	*Overflow*	*Underflow*
100	1.00	—	—	—
70	0.708	2	9.8	—
50	0.500	14	51.7	3.0
40	0.420	27	19.0	9.0
30	0.296	32	15.0	49.0
20	0.211	18	3.5	23.2
15	0.1157	4	1.0	7.8
Pan	—	3	—	8.0

Solution : Solving for the cumulative mass fraction of material smaller than screen opening retained over each screen. (Fig. 2.9)

This is presented in the following Table.

Cumulative mass fraction of material retained over each screen

IS Screen	*Width of opening, mm*	*Cumulative mass fraction smaller than screen opening*		
		Feed	*Overflow*	*Underflow*
100	1.000	—	—	—
70	0.708	0.02	0.098	—
50	0.500	0.16	0.615	0.030
40	0.420	0.43	0.805	0.120
30	0.296	0.75	0.955	0.610
20	0.211	0.93	0.990	0.842
15	0.157	0.97	1.000	0.920
Pan	0.000	1.00	—	1.000

(*i*) For *IS* 50 mesh the cumulative mass fractions of feed, overflow and underflow are 0.16, 0.615 and 0.030 respectively.

or $m_f = 0.16$; $m_o = 0.615$ and $m_u = 0.03$

The screen effectiveness $E = \dfrac{(m_f - m_u)\,(m_o - m_f)\,m_o\,(1 - m_u)}{(m_o - m_u)^2\,(1 - m_f)\,m_f}$

$$= \frac{(0.16 - 0.03)\,(0.615 - 0.16)\,0.615\,(1 - 0.03)}{(0.615 - 0.03)^2\,(1 - 0.16)\,0.16}$$

$= 0.767$

(*ii*) For IS 30 mesh;

$m_f = 0.75$; $m_o = 0.955$ and $m_u = 0.610$

The screen effectiveness $= \dfrac{(0.75 - 0.61)\,(0.955 - 0.75)\,(1 - 0.61)\,0.955}{(0.955 - 0.61)^2\,(1 - 0.75)\,0.75}$

$= 0.478$

The effectiveness of screening or cleaning efficiency for an air screen grain cleaner as suggested by the Bureau of Indian Standards (BIS) is given below (IS 5817:1980)

$$\text{Cleaning efficiency} = \frac{E\,(F - G)\,(E - F)\,(1 - G)}{F\,(E - G)^2\,(1 - F)} \qquad ...2.9$$

where, E = fraction of clean seed at clean seed outlet

F = fraction of clean seed in feed

and G = fraction of clean seed at foreign matter outlets

Problem : During the evaluation of an air screen grain cleaner with two screens the followings were observed.

(*i*) The impurities present in feed were 6.5%

(*ii*) The impurities present in clean grain were 0.5%

(*iii*) The outflow of blower contained 0.2% clean seed

(*iv*) The overflow of the 1st screen contained 1% clean seed

(*v*) The underflow contained 0.5% clean seed

Compute the cleaning efficiency of the cleaner.

Solution:

(*i*) fraction of clean seed in feed = 100 – 6.5

= 93.5

or 0.935

(*ii*) fraction of clean seed in clean grain outlet

$$= 100 - 0.5$$
$$= 0.995$$

(*iii*) fraction of clean seed in foreign matter outlets

$$= \frac{0.2}{100} + \frac{1}{100} + \frac{0.5}{100}$$
$$= 0.002 + 0.01 + 0.005 = 0.017$$

Then $E = 0.995,\ F = 0.935$ and $G = 0.017$

$$\text{Cleaning efficiency} = \frac{0.995\,(0.935 - 0.017)\,(0.995 - 0.935)\,(1 - 0.017)}{0.935\,(0.995 - 0.017)^2\,(1 - 0.935)}$$
$$= 91.18\%$$

Problem : During evaluation of an air screen grain cleaner with 2 screens 250 *g* samples were collected for analysis of clean seed fraction from different outlets. The data are presented in the following Table. Calculate the cleaning efficiency of the cleaner.

Sample fraction	*Feed, g*	*Cleangrain outlet, g*	*Blower outlet, g*	*Oversize outlet, g*	*Undersize outlet, g*
Cleaned seed, g	231.25	246.5	1.25	4.5	2.0
Impurities, g	18.75	3.5	248.75	245.5	248.0

Solution :

(1) Fraction of clean seed at clean seed out let,

$$= E = \frac{246.5}{250.0} = 0.986$$

(2) Fraction of clean seed in feed $= F = \dfrac{231.25}{250.0}$

$$= 0.925$$

(3) Fraction of clean seed in foreign matter outlets

$$= G = \frac{1.25}{250.0} + \frac{4.5}{250.0} + \frac{2.0}{250.0}$$
$$= 0.031$$

$$\therefore \text{Cleaning efficiency} = \frac{0.986\,(0.925 - 0.031)\,(0.986 - 0.925)\,(1 - 0.031)}{0.925\,(0.986 - 0.031)^2\,(1 - 0.925)}$$
$$= 82.34\%$$

Equipment for cleaning, grading and separation

It is very difficult to clearly differenciate among the processes of cleaning, grading and separation because all of these are carried out simultaneously with the common procedures.

The operation of cleaning, grading and separation of the products are performed by exploiting the difference in Engineering properties of the materials. These products may be used either for food or seed purposes. Various types of cleaning, grading and separation equipment have been designed and developed on the basis of properties of product to be handled. Thus, these equipment can be classified based upon following characteristics of the material.

1. Size
2. Shape
3. Specific gravity or weight
4. Surface roughness
5. Aerodynamic properties
6. Ferro-magnetic properties
7. Colour
8. Electrical properties

1. Separation based upon size

Screen cleaners/graders

It performs the separation according to size alone. The mixture of grain and foreign matter is dropped on a screening surface which is vibrated either manually or mechanically. A single screen can make the separation into two fractions. The screening unit may be composed of two or more screens as per the cleaning requirement.

A hand operated screen cleaner is shown in Fig. 2.10. This equipment is made of mild steel. The separation takes place due to difference in size of grain and foreign matter. The cleaner is operated by hanging on an elevated point with the

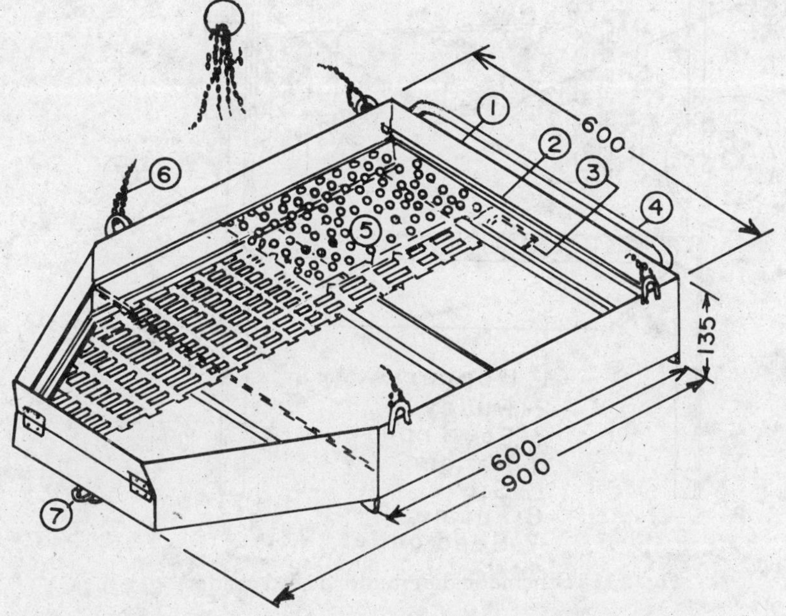

Fig. 2.10 : Hand operated double screen grain cleaner
1. frame 2. draper rod 3. screen handle 4. handle 5. scalper and grader screens 6. rope attachment 7. shutter opening attachment

help of four ropes. Grain is fed on the screening surface in batches. The screens can be changed as per the grain to be handled. The cleaner is swung to and fro till all the grain is screened. The cleaned grain is retained by the bottom sieve which can be discharged by pulling a spring loaded shutter. Impurities of larger size, stubbles, chaff etc. are retained on the top sieve and can be removed easily. Down stream from the bottom sieve consists of dust, dirt, brokens, shrivelled grain etc. drop down during the operation.

A typical seed grader is shown in Fig. 2.11. It consist of a hopper, seed, roller for controlling the feed rate, set of three sieves, pulley, eccentric system, outlets, frame and electric motor. The sieves are detachable and can be replaced by suitable sieves if other round grains are to be graded. The seed is put into the hopper and it is dropped onto the sieve through feed rollers. Sieves are vibrated through an eccentric system. Graded seeds are collected through three different spouts. The machine is suitable for grading of food grains.

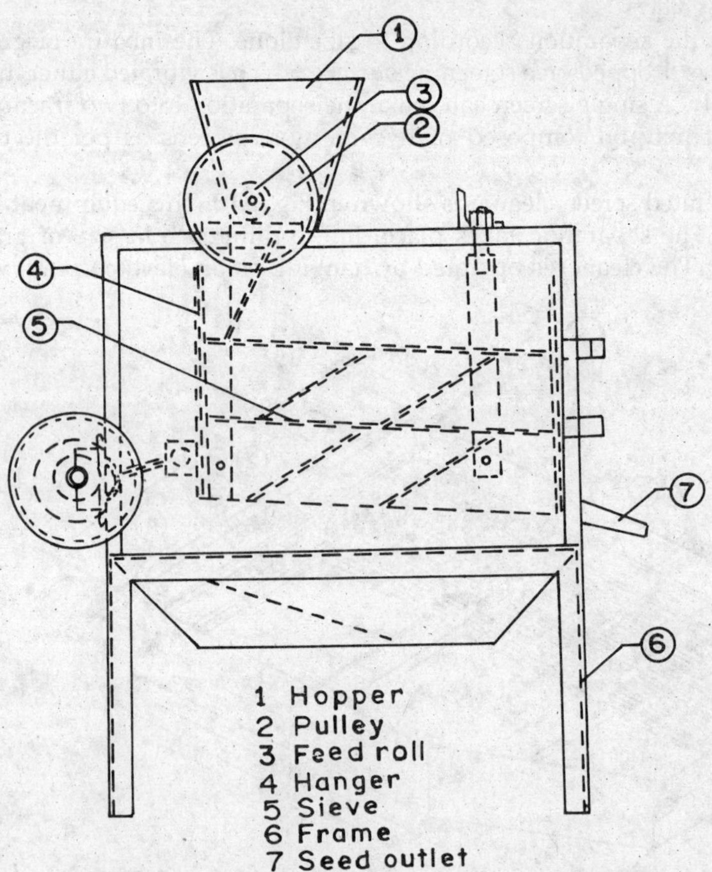

Fig. 2.11 : Schematic diagram of a seed grader

The schematic diagram of a groundnut grader is shown in Fig. 2.12. It consists of a feeding hopper, two oscillating sieves, brushes below the sieves to avoid clogging, frame and an electric motor. There are two oscillating sieves which are oscillated by the eccentric mechanism. At the head end of the sorting sieve, a spreader is provided to create uniform layer for efficiency in grading. The machine grades groundnut pods/kernels into three distinct grades according to size e.g. Grade-I, Grade-II and rejects. Oscillating sieves are replaceable by different grades depending upon the groundnut varieties to be graded.

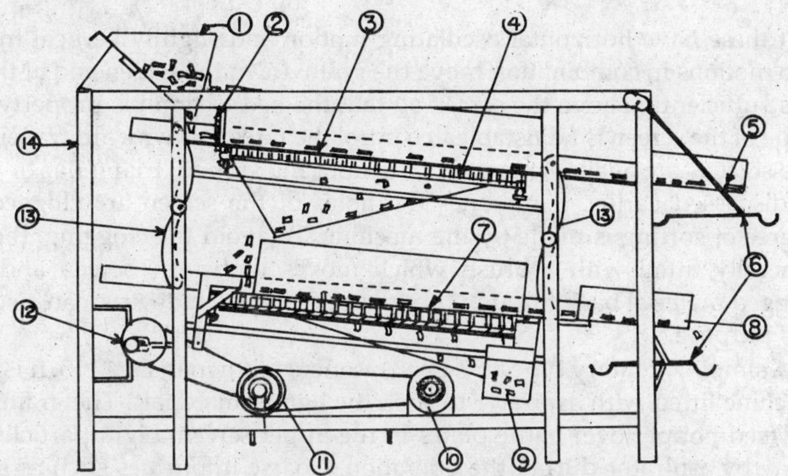

Fig. 2.12 : Schematic diagram of a groundnut grader
(1) Hopper (2) Damper (3) Top Sieve (4) Sliding Brush (5) Grade I
(6) Gunny Bag Holder (7) Bottom Sieve (8) Grade II (9) Rejects (10) Electric Motor
(11) Stepped Cone Pulley (12) Eccentric (13) Bell Crank Lever (14) Frame

Air-screen cleaners

The screens used in combination with air blast performs satisfactory cleaning and separation operations for most of the granular materials. The air-screen cleaner uses three cleaning systems; blowing or aspiration, scalping screens and grading lower screens. The air-screen grain cleaner can be classified in two distinct types: (*i*) vibratory screen, (*ii*) rotary screen, based on movement of the screening surface.

(i) Vibratory air-screen cleaner

The screening unit is composed of double or multiple (upto 8 number) screens. These screens are tightened together and suspended by hangers in such a manner

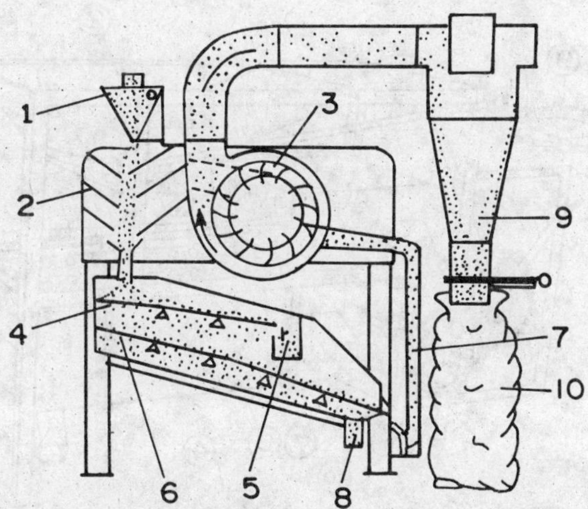

Fig. 2.13 : Schematic diagram of a vibratory air screen cleaner
1. Feed Hopper 2. Baffle plate 3. Blower 4. Upper screen 5. Discharge channel
6. Sand sifter 7. Ascending separator 8. Discharge funnel 9. Centriclone 10. Dust Bag

that these have horizontal oscillating motion and slightly vertical motion. These two motions in combination move the grain down the screen and at the same time toss sufficiently above the screen so that the bed of grain is properly stirred. The slope of the screen is adjustable to control the rate of downward travel of the grain. The screens are available in various shapes like, round, triangular or slotted holes as discussed earlier. Sometimes the holes of the screen are clogged when fine degree of sorting is made by the machine. To avoid the clogging, the screens are generally fitted with a brush which moves under the screen and pushes the clogged material back through the screen. Other such devices can also be used for this purpose.

A simple vibratory type air-screen cleaner is shown in Fig. 2.13. It is a two screen machine fitted with aspirator to suck the lighter materials. The grain passes from the feed-hopper over baffle plates to the upper screen. Light particles are sucked away by aspirator during the operation. Coarse impurities such as stones, straw particles etc are screened off by upper screen and discharged out through an outlet. The grain falls through on to the lower screen, where the sand and the dust particles are screened off. The grains leave the machine through discharge funnel. While passing through the funnel the grains are again cleaned by upward draft of air in the ascending separator. During this process the remaining light impurities and shrivelled grains are sucked away and the light impurities are removed by a cyclone separator to which dust bags are attached for the collection of the impurities.

Such cleaners can also be operated manually either using hand or pedal system. But the capacity of such cleaners is lower than the power operated machines. A pedal-cum-power operated cleaner is shown in Fig. 2.14. The machine is made up of mild steel and consists of a grain hopper, with feeding mechanism, sieve box on hanging shoes, blower unit, driving and eccentric unit. The machine can be operated either manually (pedal) or by electric motor.

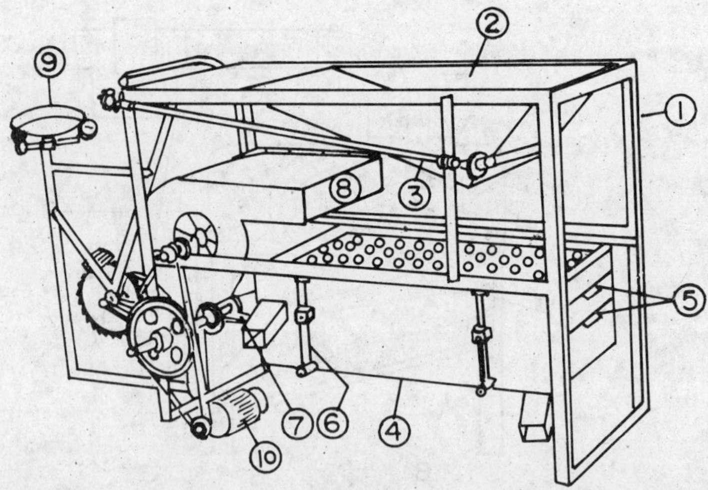

Fig. 2.14 : Pedal cum power operated air screen grain cleaner
1. main frame 2. hopper 3. feeding mechanism 4. sieve box 5. scalping and grading sieves 6. shoe for sieve box 7. eccentric unit 8. blower 9. cycle seat 10. electric motor

(ii) Rotary screen cleaner

The rotary screen cleaner has normally circular decks. Their motion is circular in horizontal plane. These have either single or double drum. A single drum rotary screen cleaner is shown in Fig. 2.15a.

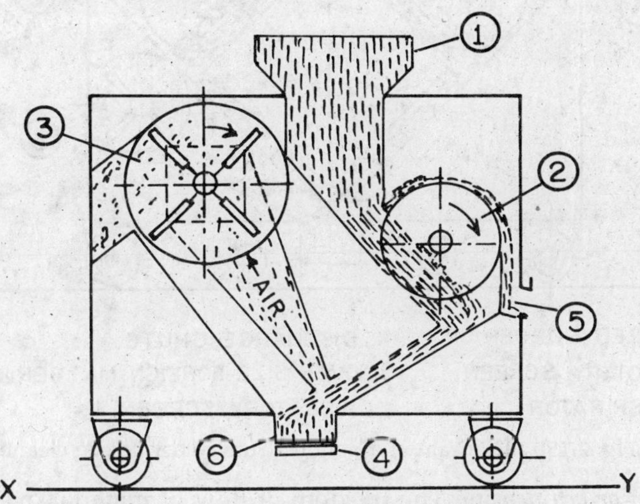

Fig. 2.15a : Diagram of a single drum rotary screen cleaner

The machine consists of a rotary screen, aspirator and hopper and equipped with an electric motor which gives drive to the rotary screen and the aspirator. The mixture is fed into the hopper. The sound grains pass through the screen perforation into the centre of the screen drum, where as oversized material is retained above and pass out though an outlet. The sound grains come out at the centre side of the screen drum rotating at low speed and fall onto the vibratory screen which remove the dirt particles. The light particles like straw and dust are sucked away by the aspirator and discharged through the aspirator outlet. The cleaned grains are delivered through the discharge chute.

A double drum rotary screen cleaner is shown in Fig. 2.15b. It has two rotary screens and other components are same as in the case of single drum rotary screen cleaner. The two screens rotate in opposite direction to each other.

Design consideration of an air-screen grain cleaner

The factors which influence the design of an air screen grain cleaner can be classified into two categories:

(*i*) Material factors

(*ii*) Machine and operational factors

(i) Material factors :

The various material factors which are taken into account while designing a cleaner are described below:

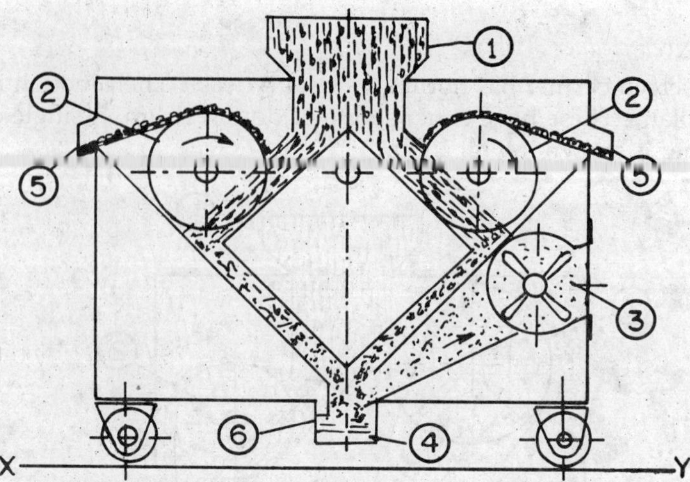

1. FEED HOPPER
2. ROTARY SCREEN
3. ASPIRATOR
4. DISCHARGE CHUTE
5. OVER SIZE FOREIGN MATTER OUTLET
6. VIBRATORY SCREEN

Fig. 2.15b : Diagram of a double drum rotary screen cleaner

(*a*) *Compaction or bed density:* The freedom of flow of material over the screen is greatly affected by compaction or bed density. If the bed of material over the screen is more compact, its normal flow over screen is restricted. The normal flow does not occur without constant changes in the relative position of a particle with respect to its adjacent particles. At the same time materials having low density also pose problem in screening due to fluffy nature.

(*b*) *Particle size and shape:* There are two ways in which particle shape is related to its size that passed or retained on a particular sieve. First, if all the openings in the sieve are square and having exactly the same size, the flattened grains would pass diagonally through the opening. Secondly, some variation in size and shape of opening is unavoidable in the manufacturing. Consequently, the size and shape of grain passed or retained would depend upon the number, size and shape of the off-size openings.

The shape and size of a particle are the deciding factors towards identification of proper sieve configuration e.g. round, oblong, triangular etc.

The size of spherical particles can be designated by diameter but in practice it is not true in case of dry matter. There is too much variation in the shape. The particles vary from spheres to long fibers which pose difficulty in exact sizing. The shape of particle affect the passage of undersize through the screen and also conveyance of oversize particle. It affects the interparticle friction and therefore, the case with which the finer ones settle down to the screen surface. When the long particles are involved, it is some times desirable to retain the longer particles whereas the particles whose width, length and thickness are approximately equal, those are allowed to pass through. Irregularly shaped particles are some times clogged in the screen apperture and thus cause blinding. Various types of brushes and cleaners can be placed beneath the screen to dislodge particles.

(*c*) *Distribution of various fraction sizes:* The distribution of various fractions of (straw, node, chaff etc) of materials in the mixture greatly influence the separation.

(*d*) *Stickiness and abrasiveness:* If the particles tend to stick together these must be separated. Material abrasiveness affects the selection of screen. Electrostatic charges built up by mechanical action of sifting will cause the particles to coat or blind the screen. Such particles can not be dislodged by mechanical action. Besides the above, the moisture content and crop species also affect the design of the cleaner.

(*ii*) *Machine factors :*

Capacity and efficiency in screening operations are closely related to each other. In cleaning operation, when the rated capacity is increased, its capacity is decreased.

Some of the machine factors which are taken into account while designing a grain cleaner, are described below:

(*a*) *Feeding mechanism:* The feeding of material over the screen should be done properly in order to obtain the maximum capacity and efficiency. The material should be spread uniformly over the full width of screen surface. It should approach the screen surface in a direction parallel to the longitudinal axis of the screen and at as low practical velocity as possible.

(*b*) *Size of screen surface:* The size of screen surface is dictated by the capacity requirement of the machine. However, two governing factors need to be mentioned that the width of screen affects the capacity whereas its length affects efficiency. Width is necessary to reduce the bed thickness to a practical maximum, and length to allow the undersize to be removed without an inordinate amount of fines in the oversize. It is also true that width does influence the efficiency and length affects the capacity but only to such a minor degree that it could always be neglected. The experience with both the quantities of width and length leads to a general rule that the length of screen should be between 2 to 3 times its width. In paddy cleaner with vibratory sieves and aspirator, the screen area of 0.2 m^2 per ton of paddy is recommended.

(*c*) *Grain bed depth:* A criteria for deciding the bed depth is that it should be one particle thick i.e. all particles regardless of their size should be in contact with the screen and none should ride down over the others. This is only possible with a very low feed rate and capacity. On the other hand a bed is actually needed to keep a particle in contact with the screen and also in contact with other particles surrounding it.

A general guideline for deciding the bed depth is that the thickness of bed should not exceed 4 times the size of screen aperture for material weighing 1600 kg/m^3 and 2½ to 3 times for a material weighing 800 kg/m^3. The following methods can be used to control the bed depth during screening:

(*i*) At constant feed rate the bed thickness can be reduced by increasing the screen slope.

(*ii*) At constant slope, the bed thickness is increased by increasing the feed rate and vice-versa.

(*d*) *Shape of the opening:* The shape of screen opening depends upon the shape of the grain to be handled. In general practice for round-shaped grain, the top screen of round holes and bottom screen of slotted holes are used. The round hole top screen drops the round grains through the smallest possible opening and scalp off materials longer than the desired grains. The slotted hole bottom screen holds up the round desired grain, but drops out split and broken grains, small weed seeds and other undesirable materials. Thus the smallest possible perforation is used for scalping, while the largest possible perforation is used for grading.

In case of elongated grains generally slotted top screen and slotted bottom screen are used. For the specific purposes of separation, it may be necessary to use a round hole top screen, but usually better cleaning is accomplished when both top and bottom slotted screens are used.

(*e*) *Percentage open (perforated) area:* The higher percentage open area increase both capacity and screening efficiency. To ensure good contact and maximum available use of plate area, holes should be spaced as close as practicable. Triangular spacing is invariably used with holes pitched at interval which are multiple of hole diameter. The advantage of this method is that for a large plate the ratio of hole area to available area is dependent upon only the pitch. But it should be recognised that there is definite limitation to increase the open area. Excessive loading in terms of strength means excessive cost of replacement which might be more than the saving resulting from increased capacity.

(*f*) *Aperture size:* The size of screen openings depend upon the size and shape of the grain. For the round openings, the size of scalping screen is decided slightly higher than the equivalent diameter of the grain to pass through the aperture whereas size of grading screen is kept small enough to retain the smallest desirable grain. The size of aperture in a four screen cleaner for various agricultural grains is listed in Table 2.1. The size of aperture of scalping and grading screen in a two screen grain cleaner for some of the food grains is presented in Table 2.2.

Table 2.1

Size of aperture in a four screen cleaner for some of the agricultural grains

Grain	*Screen Size, mm (from top screen)*			
	1	2	3	4
Wheat, plump	6.30	2.0 × 20	5.6, 3.6 × 20	2.2 × 20
Wheat, thin	5.6	1.8 × 20	5.0, 3.2 × 20	2.0 × 20
Maize, cleaning only	12.7	4.8	12	5.6
Paddy, long grain	5.6, 4.8	2.4, 2.6	5.0, 3.2 × 20	1.4 × 20
Paddy, short grain	5.6	2.4, 2.6	5.0, 3.2 × 20	1.8 × 20
Rye	4.8	1.4 × 20	4.8, 2.7 × 20	1.6 × 20
Safflower	7.0	3.2	4.4 × 20	2.4 × 20
Sesame	2.4	6.4 × 6.7	1.3 × 20	1.6, 2.4 × 9.5
Kidney (Red) Beans	12.0	5.2 × 20	11.0	5.5 × 20
Mung Beans (Green gram)	5.5	3.0	5.1	3.2 × 20
Chick Peas (Bengal gram)	12.0	4.4 × 20	10.3	4.8 × 20
Pigeon Peas	9.5	3.6 × 20	9.1	4.0 × 20
Lentils	7.1	4.0, 5.0	2.8 × 20	4.0, 5.0

Table 2.2

Size of aperture in a two screen cleaner for some of the agricultural grain/seeds

Grain	*Top Sieve (scalper) mm*	*Bottom sieve (grader) mm*
Wheat (bold)	5.0	2.0 × 20
Wheat (thin)	5.0	1.8 × 20
Maize, cleaning only	11, 12	6.3
Rice	4.8	5.6 × 5.6
Sorghum	5.0	2.0 × 20
Rye	4.8	1.6 × 20
Mustard (Pusa bold)	3.1	1.4
Linseed (T-59)	1.6 × 20	2.0
Safflower (TSE-1)	6.5	2.0 × 20
Sunfflower (Mordon)	8.0	3.0
Pigeon pea and Bengal gram (bold)	8.0	3.2 × 20
Pigeon pea and Bengal gram (small)	6.5	3.2 × 20

(*g*) *Stroke length:* The stroke length of the screen also affects the separation efficiency. The optimum value of stroke length for wheat grain has been recommended as 17 *mm* to give maximum efficiency.

(*h*) *Angle of inclination:* Each screen is adjustable to different degrees of inclination with the horizontal. The common range in pitch adjustment is from 4 to 12°. High capacity grain separators may have a greater adjustable pitch range to move grain over the screens rapidly, but in a precision seed cleaner 4 to 12° of pitch gives adequate cleaning capacity. It is also function of many other variables such as material to be handled, width, length of screen etc. If the angle is too high, material moves too fast and efficiency drops. As the angle of inclination is increased, the horizontal passage of opening is decreased. An angle more than 23° reduces the effective opening so much that the thinner bed produced by more rapid run off is rarely advantageous. While there is steep loss of efficiency at steep angles, there is sharper drop in capacity at low angles. The screens should be independently adjustable for pitch, since each screen may require a different pitch for optimum results. The scalper screen should be set at a steep pitch for the fast removal of trash and weed-seed while grading screen should be set at a flat pitch to hold the grain on the screen for a longer period. The steep pitch should be given to make chaffy grain move down the screen and should be flattened while cleaning round grain to prevent them from gaining too much momentum and bouncing over the screen without properly graded. The wiremesh screens should always be steep since their surface retards seed movement enough to give a thorough sifting.

(*i*) *Vibration amplitude and frequency:* The speed and amplitude of vibration should be designed to convey the material properly and to prevent the blinding of screen. These are more or less dependent upon the size and weight of the material being handled and are related to the angle of installation and type of screen surface. The objective is to properly stratify the feed over the screen for an efficient operation. The fast shake causes the seed to turn and tumble and present all their sides to the screen openings. Fast shake speeds are, therefore, very effective in cleaning chaffy materials. However, when the shake speed is too fast, rounds or heavy seeds tend to bounce over the screen and are not sifted properly.

Slow shake speeds are used to obtain accurate separation. However, the grain will be dead on the screen if the shake is too slow. This clogs the screen perforations and causes incomplete cleaning.

The frequency of oscillation has a very significant effect on the grain losses in combines. An increase in frequency of vibration to a certain maximum limit gives better effectiveness, but decrease in effectiveness sets in with higher frequencies.

(*j*) *Number of screens:* It is generally agreed that the most efficient screening results when a series of single deck screens are used. This is because lower decks of multiple-deck screens are not fed so that their entire area is used and because each separation requires a different combination of angle, speed and amplitude of vibration for maximum performance.

(*k*) *Rate of forward travel:* The rate of forward travel is completely an empirical factor that does not lend itself to generalization because it depends upon both machines and material characteristics. This indicates the travel of material in *m/min*. If the length of screen (in *m*) is multiplied by 60 and divided by the product rate travel in *m/min*, then the result is the retention time in seconds. This is useful in determining the capacity.

(*l*) *Air velocity:* The use of air in an air-screen cleaner is for removing the lighter particles based on difference in density of the material. The air is applied either before falling the grains on the screen or during the separation over the screen through the openings. the terminal velocity of wheat, maize, oats, rye and soybean are presented in Table 2.3. The terminal velocity for various components of wheat mixture is also presented in Table 2.4.

For paddy cleaner with vibratory screen and aspirator, 8 m^3/min air flow rate per ton of paddy upto 8 t/h and 5 m^3/min for capacity more than 10 t/h are recommended.

Table 2.3

Range of terminal velocity for wheat, maize, rye, oats and soybean

Grain	*Terminal Velocity m/s*
Wheat	5.8 –9.1
Maize	8.0–12.8
Rye	6.1 –8.2
Oats	5.2 –8.0
Soybean	9.1–18.3

Table 2.4

Terminal velocity for various components of wheat- chaff mixture

Particle	*Terminal velocity range, (50% and above separation level) m/s*
Wheat grain	9.4–12.5
Chaff	5.0–8.0
Nodules	5.8–8.8

Improving efficiency of screen cleaner

The efficiency of screen cleaners may be improved by providing small accessories as follows:

(a) Screen brushes

The cleaning efficiency of a screen cleaner is directly proportional to the percentage open perforations. To keep the perforations open, the brushes should be provided under each screen. These brushes travel to and fro under the screens and thereby sweep the screens and remove the clogging of perforations. The brushes should be properly adjusted and operated all the time to avoid incomplete separation.

(b) Oil-Cloth cover

Sometimes long grains and stalks stand on end and pass through the round apertures which generally drop out round grains. This may be minimized by fastening a plastic sheet or oil cloth, smooth side down, over the top screen so that it lies as close as possible to the screen. Due to weight of the oil cloth, the long grain or stalk remain flat and do not pass through the round holes.

(c) Screen knockers

To keep the screens clean, tappers or hammer like screen knockers may be installed over the scalping screens. The screen knockers tap the screens to jar loose, material wedged in the perforations.

(d) Rubber balls

Generally two or more screens are used in the cleaners. Between the two screens, the rubber balls are held in separate compartments. As the screens operate, the balls strike the screen surfaces and make the apertures free from the materials plugging them.

Disk separator

The disk separator separates materials on the basis of difference in length of various constituents. The separator has pockets or indentations on its surfaces. When the machine is operated, the smaller sized materials are caught in the pockets while the larger ones are rejected. It is used especially for removing dissimilar material like wheat, rye, mustard, barley from oats.

The indent disk separator consists of a number of disks in series fitted on a shaft inside a close housing which revolves on a horizontal axis. The pockets are slightly undercut on each disk as shown in Fig. 2.16. As the disks revolve through a mixture of grains, the pockets pick up short grains and drop them in a trough at the side

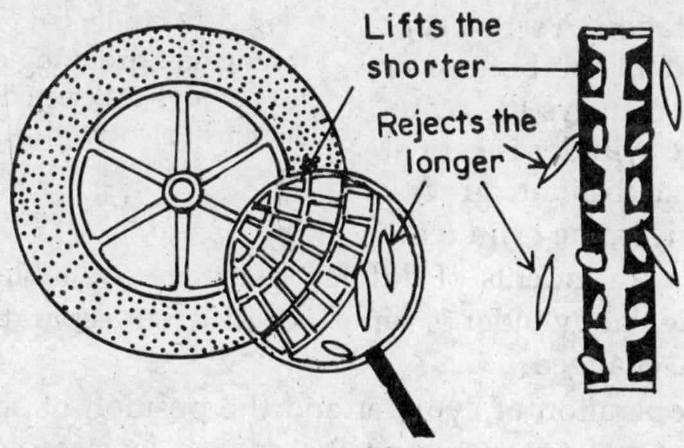

Fig. 2.16 : Disk separator

of the machine. The desirable or undesirable materials not lifted by the disks are conveyed through the disk spokes to the end of the machine and pass out through the tailing opening.

A number of distinct separations or grading of the grain of varying length can be made in a single machine by installing combination of disks having pockets of different sizes. The mixture first passes through the disks with small pockets and then disks with pockets progressively larger from inlet to discharge. If only one separation has to be performed, combination of disks of same size of pockets is used in the machine to increase its handling capacity. The removable vanes are provided with the spokes of the disks which serve the purpose to move grains through the machine, agitate them and also bring them in contact with the pockets.

The disk pockets are made of three basic shapes of various sizes. The 'R' pocket derives its name from 'rice' and was designed to remove broken rice grains from whole grains. It has a flat and horizontal lifting edge and round leading edge. It rejects round grains but lifts out cross-broken or flat grains.

The 'V' pocket derives its name from 'Vetch' and was designed to pick up and remove round shaped grains. It has a round lifting edge which tends to reject tubular or elongated grains. Disks with other letter designations are designed to perform specific separations.

Indented cylinder separator

The indented cylinder separators also separate the materials on the basis of relative lengths like disk separators. It consists of a horizontal rotating cylinder which has indents on the inside surface. The indents are closely spaced and hemispherical in shape (Fig. 2.17). When the mixture of grain is fed into one end of the cylinder, short grains are picked up by the combined effect of fitting into the indents and centrifugal force. These grains are dropped into an adjustable trough inside the cylinder near the top of rotation. A screw conveyor is provided in the bottom of the trough which conveys the material. Generally, the cylinder is kept at slight inclination to facilitate gravity flow of long grains in the cylinder.

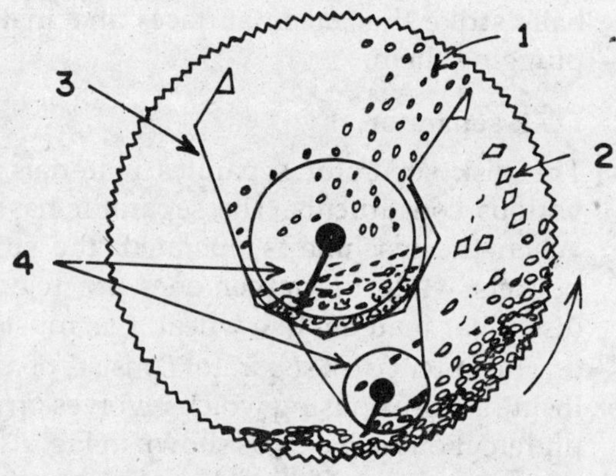

Fig. 2.17 : Indented cylinder separator
1. lifting and dropping of shorter seeds into trough 2. rejected longer seeds
3. adjustable trough 4. conveying augers

The cylinders with indents of different sizes are available, but the size of all indents in a particular cylinder is same. For different separation needs, indented cylinder has to be changed.

The speed of operation of cylinder and the position of adjustable trough are important adjustments for obtaining the desired level of separation. Since the centrifugal force helps to handle the grain in the pocket, it affects the distance travelled by grains before they fall.

The excessive speed will not allow grains to drop from the indents. Too slow speed will not lift short grains from the mixture. The position of separation edge of the adjustable trough should be such that it can catch the desired fraction of the dropping grain.

Spiral separator

The spiral separator separates the grains as per their roundness. The main component of the separator is a stationary, open screw conveyor standing on one end (Fig. 2.18). The mixture is fed at the top of the unit. The round materials of the mixture pick up speed as they slide or roll down the inclined surface until their centrifugal force becomes sufficient enough to throw them in the outer helix. While the non-round materials are caught in the inner helix and are discharged through a separate spout.

There is no moving part in the spiral separator. The rate of feeding is the only adjustable component. The feeding should be such that each grain/particle roll independently for effective separation. The main limitation of the spiral is lack of flexibility.

Separation of mustard, rape, soybean, wild peas or other round seeds can be performed from wheat, flax, oats etc. This device is less versatile as compared to other mechanical cleaners, but it is simple, inexpensive and quite useful for seed cleaning purposes.

Fig. 2. 18 : Spiral separator

Specific gravity separator

The specific gravity separator makes the separation according to difference in density or specific gravity of the materials. This separator works on two principles, (1) the characteristics of grains to flow down over an inclined surface, (2) the floatation of the particle due to upward movement of air.

The main part of the device is a triangular-shaped perforated deck. The deck is properly baffled underneath to ensure uniform distribution of air over it. The pressure or terminal velocity of the air rising through the deck is controllable very closely within a wide range. (Fig. 2.19a).

The mixture of grain is fed into the feed box. The air is blown up through the porous deck surface and bed of the grain by a fan at such a rate that the material is partially lifted from contact with the deck surface. The lightest materials are lifted to the top of the stratified mass. The heavier particles are not lifted by the air. The stratified mass moves along the direction of conveyance due to oscillating motion of the deck and is discharged at the right edge of the deck.

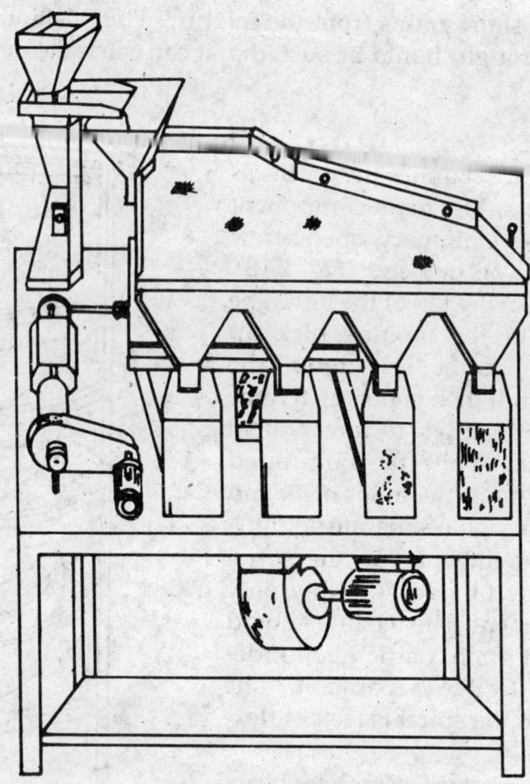

Fig. 2.19a : Specific gravity separator

Destoner or Stone separator

The stone separator is a form of specific gravity separator. It separates the grain mass into two fractions as per the difference in specific gravity.

The mixture is fed onto the centre of a perforated deck (Fig. 2.19b). The air coming through the deck from bottom stratifies the materials while the reciprocating action of the deck separates the heavy particles from the lighter particles.

The heavier materials move towards the top end of the deck whereas the lighter particles are discharged from the low end of the deck, without any middling product. The separation can be controlled by adjusting the feed rate, slope of the deck, deck vibration and the air flow rate.

Inclined draper

The separation by inclined belt draper takes place due to difference in shape and surface texture of the material. This technique of separation is used when all other methods fail.

The mixture to be separated is fed over the centre of an inclined draper belt moving in upward direction. The round and smooth grains roll or slide down the draper at faster rate than the upward motion of the belt, and these are discharged in a hopper. The flat shape or rough surfaced particles are carried to the top of the inclined draper and dropped off into another hopper. (Fig. 2.20).

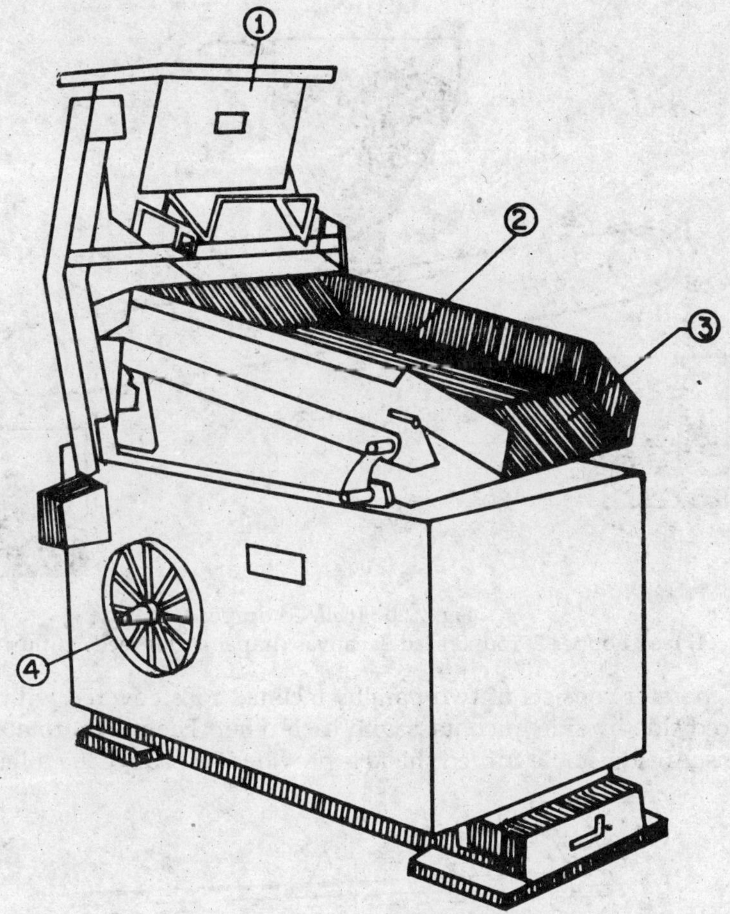

Fig. 2.19b : Stone separator
1. feed hopper 2. separating table 3. clean grain out let 4. fan

The belts of different degrees of roughness may be used as a draper for separate materials. If rolling tendencies of the grain are predominant, the rough canvas belt may be used. The smooth, plastic belt may be used in case sliding action is desired for the lower fraction. Feed rate, speed of draper and angle of inclination are other important variables for effective separation of dissimilar materials.

The feed rate is kept low enough to give opportunity to each grain for separation. The speed of the draper may be varied to simulate with the length of incline. The angle of inclination is adjusted to assure rolling or sliding of the desired lower fraction.

To increase the capacity of the separator, number of belts may be used one above another in a single machine.

Velvet roll separator

The velvet roll separator or roll mill separates grains on the basis of differences in shape and surface texture. It is a finishing machine and should be used only after cleaning and separation of grain from the chaff and trash. It is effective in separating grain with a rough seed coat or sharp angles from smooth surface grain.

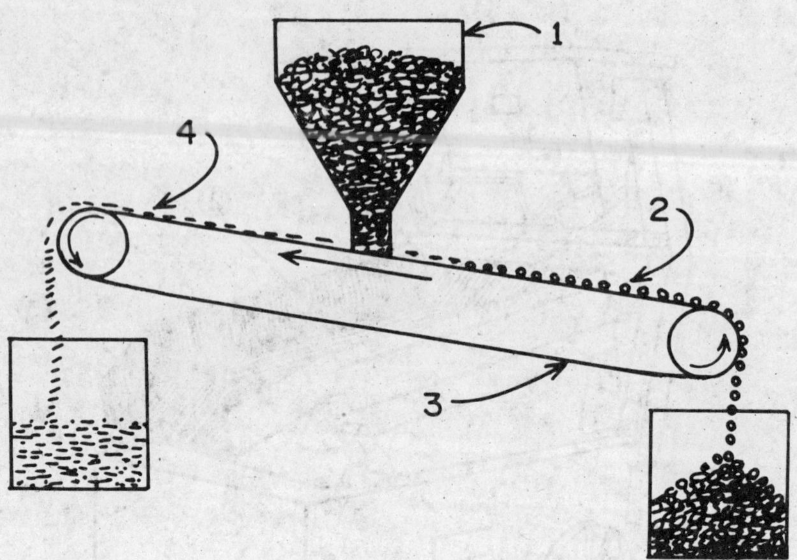

Fig. 2.20 : Inclined draper
1. feed hopper 2. round seed 3. canvas draper 4. flat seed/impurities

The separator consists of two parallel inclined rolls covered with velvet cloth and placed side by side in contact with each other. The rollers rotate in opposite directions. An adjustable curved shield is provided just above the rollers (Fig. 2.21).

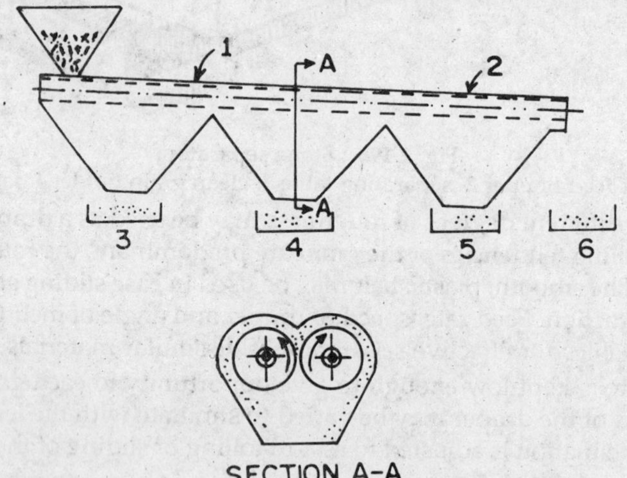

Fig. 2.21 : Schematic diagram of a velvet roll separator
1. metallic duct 2. velvet roll 3. rough seeds 4. rough and some smooth seeds 5. more smooth and some rough seeds 6. smooth seeds

The mixture to be separated is fed onto the upper end of the rollers. As the rollers rotate, the smooth grains bounce down the inclined trough and are discharged at the lower end of the machine. The rough surfaced grains or the grains having sharp or broken edges are caught in the velvet. These grains are thrown up against the shield and take a bouncing path between shield and rollers and are finally thrown over the sides.

For achieving desired separation, adjustments can be made in feed rate, speed of rollers, characteristics of cylinder roughness and roll inclination. To increase the capacity of machine, the number of roll pairs one above another can be increased.

Pneumatic and aspirator separators

The pneumatic separation is based on the difference in aerodynamic properties of various constituents of the mixture. The aerodynamic properties of a particle depends upon its shape, size, density, surface and orientation with respect to air current. Both the aspirator and the pneumatic separator use terminal velocity of the grain to separate different fractions. This refers to the velocity of air required to suspend particles in a rising air current.

In a pneumatic separator, the fan is placed at the intake end of the machine which creates higher pressure than the atmospheric pressure. The high pressure air blast separates the materials. The mixture of products is introduced into a confined rising air stream, the particles with low terminal velocities are lifted by the air current whereas the particles with higher terminal velocities than air velocity fall down. The air velocity can be adjusted by altering the speed of fan or by changing the opening of air inlet.

The aspirator has a fan at the air discharge point which creates a vacuum or negative pressure within the machine. The scalping separator is a type of aspirator separator in which rough separation is performed. The mixture of the grain is dropped into a rising air column which has a velocity slightly lower than the terminal velocity of the heavy grains. The leaves, trash and lighter particles rise with the air and are deposited in an enlarged settling chamber. The denser, plumper grains fall through the incoming air into a container.

The fractionating aspirator is another type of separator. The mixture of grain is fed into the lower end of an expanding air column, the heavy grains fall against the air flow while the lighter particles are lifted. The grains with high terminal velocity are dropped in the expanding column. The lighter fractions of grains are discharged as per the relative weight through different outlets positioned in the column. Thus the mixture is separated into various fractions (Fig. 2.22).

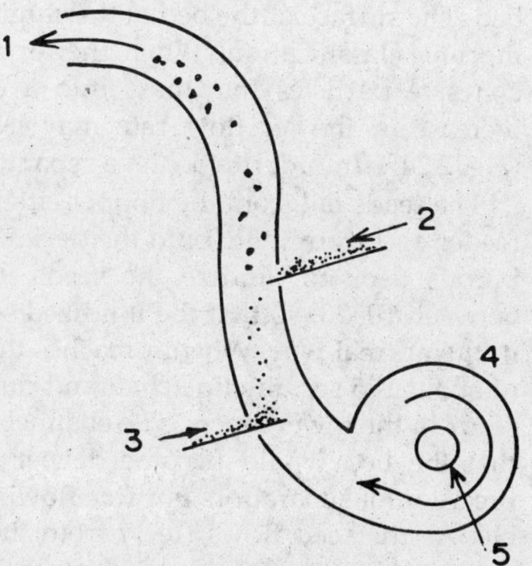

Fig. 2.22 : Diagram of a pneumatic separator 1. undersirable material removal 2. uniform feeder 3. clean grain outlet 4. centrifugal blower 5. control for air intake

Separation based on fluidisation technique

The fluidised bed cleaner/separator makes the classification of seed due to difference in density and size. This device is suitable for cleaning lighter seeds like cabbage, radish, lettuce, carrot, onion, grass seeds etc. When an air current is passed vertically upward through a perforated deck containing a bed of granular materials, various phenomenon occur (Fig. 2.23).

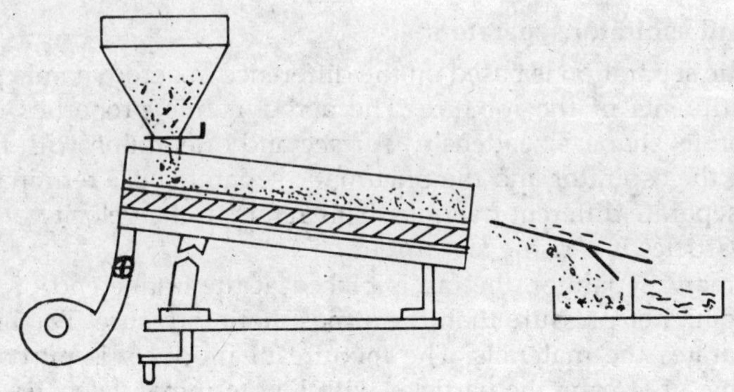

Fig. 2.23 : Diagram of a fluidised bed separator

At low air flow rate the bed remains inert, causing pressure drop in the air flowing through the bed. As the air flow rate increases, a stage is reached called the quiescent stage when the bed begins to expand to the extent of 10% volume increase [Fig. 2.24a]. The mobility of the bed depends on the shape of the particles but the external vibration is applied to produce a mobile fluid bed. At this stage the surface may be observed to bubble, small particles may be observed to rise and some general circulation of the bed may occur [Fig. 2.24b & c]. These are known as boiling and bubbling beds. Classification of the beds by density or particle size may occur at these stages. Further increase of air flow rate cause slugging of the bed. The surface of the bed at the slugging stage is turbulent and uneven preventing any classification. When the air flow rate is further increased, the bed of cohesive particles may pass into a channeling stage (Fig. 2.24e). Finally any increase in the air flow rate may lead to the formation of the spouting bed (Fig. 2.24f). In order to achieve separation the bed should be boiling or bubbling.

The seeds fall from the hopper either directly onto the deck or onto a vibratory feeder and from there onto the deck. A fan feeds a plenum chamber beneath the porous deck to fluidize the seeds immediately. The channel slopes are kept between 10–20° so that the fluidized seed flows down the slope while classifying in the normal way. When it reaches the bottom of the deck after a residence time of about 1–5 sec. the fine chaff and dust etc. are collected at the top and the good seeds in the lower layers. The channel is built in the form of a converging taper so that the bed would become deeper at the discharge end, thus increasing the precision of separation. For free flowing seeds the amount of vibration is negligible. As the seed flows freely from the end of the channel a splitter is carefully positioned to separate the rubbish and leaving the good seed to pass underneath.

Magnetic separator

The magnetic separator performs separation on the basis of surface texture and stickiness properties of the grain. Since the grains do not contain any free iron,

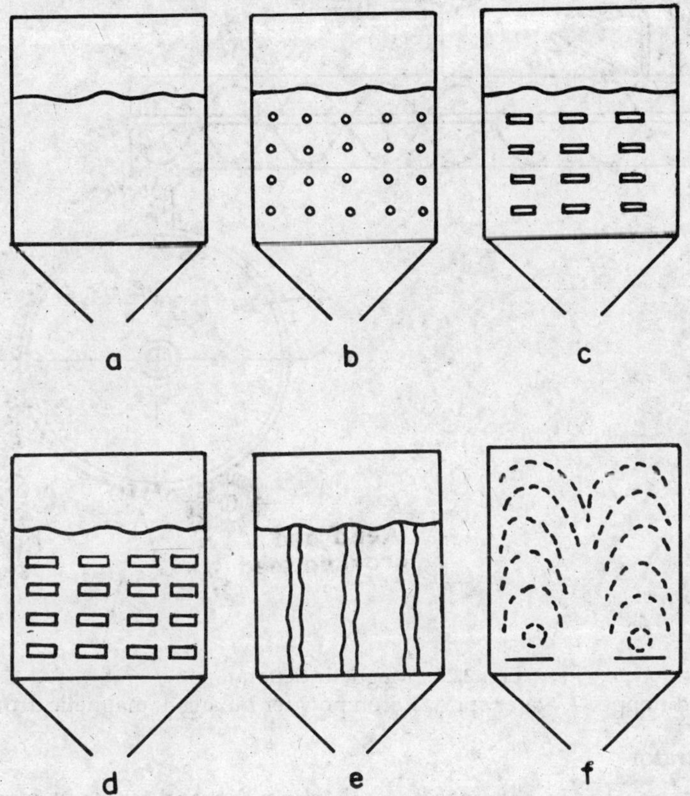

Fig. 2.24 : Types of fluidised bed

therefore, are not attracted by the magnet. A selective pretreatment of mixing finely ground iron powder to feed mass is given. The grain mixture is fed to a screw conveyor or other mixing device that tumbles and mixes the grain with a proportioned amount of water. Due to moisture, iron powder adheres to rough, cracked, broken and sticky seed coats. Moisture does not remain on smooth grains so no iron powder adheres to smooth surfaced grains.

The grain mixture is fed onto the top of a horizontal revolving magnetic drum, the smooth grains that are relatively free of powder fall along the drum simply by gravity. The materials with iron powder are attracted by the magnetic drum and stick to it and are removed by rotary brush or break in the magnetic field as shown in Fig. 2.25.

Most magnetic separators have two or three revolving magnetic drums operating in a series. The grain mixture is passed over these magnetic drums to increase the efficiency of operation. The extent of difference in seed coats, amount of water mixed, amount of iron powder and thoroughness of powder-water mixing operation affect the degree of successful separation by magnetic separators.

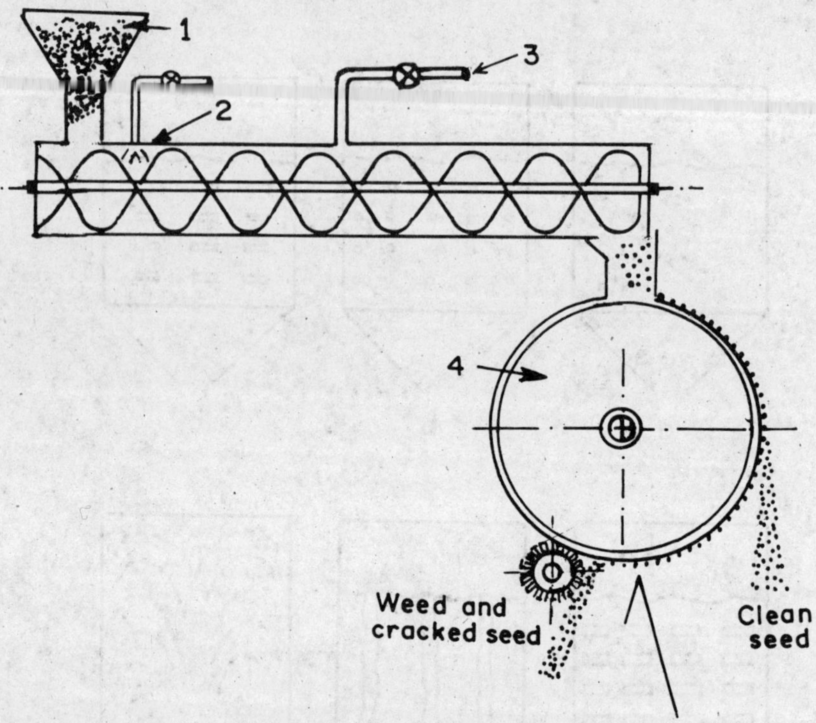

Fig. 2.25 : Magnetic Separator
1. feed hopper 2. water spray 3. iron powder mixing 4. magnetic drum

Cyclone separator

The cyclone separator is a device for collecting the end product in processing operations. It is most commonly used for collection of dust and wastes during processing of grains. It can also be used with air screen cleaners to collect light particles which could be carried by air stream. The application of cyclone separator is also made to separate out air borne material from the discharge of pneumatic conveyor. In operation of the separator, the air and material both enter the cyclone tangentially at the top of the separator where pressure drop occurs and air forms a vortex around the centre of the chamber. The whirling air being lighter gets collected at the centre and is delivered out through the top opening as shown in Fig. 2.26. The heavier materials slide down along the walls of the cyclone and is discharged at the bottom.

In a cyclone separator a particle is acted upon by two forces, the centrifugal force and the weight of the particle. The centrifugal force can be described as under.

$$c_f = \frac{W V^2}{g R} \quad \ldots 2.10$$

where, c_f = centrifugal force, kg
W = weight of particle, kg
g = acceleration due to gravity, 9.81 m/s^2
R = radius of rotation, m

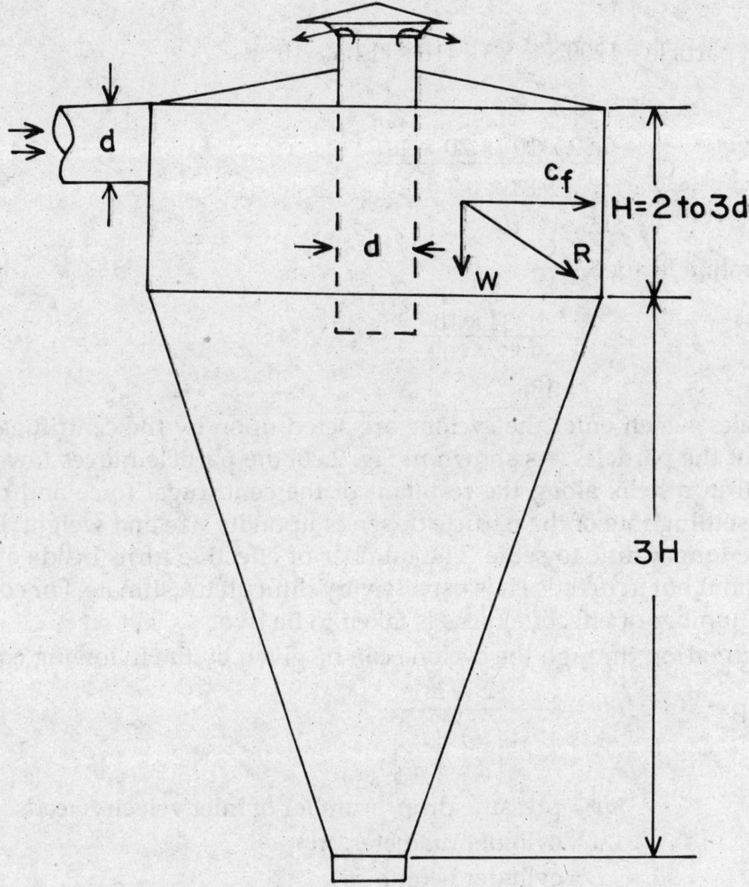

Fig. 2.26 : Cyclone separator

V = linear or tangential velocity, *m/s*

The separating force can be given as follows:

$$F = W\sqrt{\frac{V^2}{g^2 R^2} + 1} \quad \text{...2.11}$$

The performance or separation factor of cyclone can be given by the following equation:

$$\text{Performance factor, } S = \frac{c_f}{W} = \frac{V^2}{g\,R} \quad \text{...2.12}$$

It has been found that as '*S*' increases the separation becomes more effective.

Problem : Air carrying particles of density 1200 kg/m^3 and an average diameter of 25 *micron* enters a cyclone of 600 *mm* diameter at linear velocity of 20 *m/s*. Calculate the centrifugal force acting radially in the cyclone and the separation factor of the cyclone.

Solution :

$$\text{The centrifugal force, } c_f = \frac{W\,V^2}{gR}$$

$$\text{Weight of particle} = 1200 \times \frac{4}{3} \times 3.1416 \times (12.5 \times 10^{-6})^3$$

$$= 9.82 \times 10^{-6}$$

$$\therefore \quad C_f = \frac{9.82 \times 10^{-6} \times 20 \times 20}{9.81 \times 0.3}$$

$$= 1.334 \times 10^{-3}$$

The separation factor $S = \frac{C_f}{W}$

$$= \frac{1.334 \times 10^{-3}}{9.82 \times 10^{-6}}$$

$$= 136.$$

The particles which enter the cyclone are acted upon by the centrifugal force and weight of the particles. As shown in Fig. 2.26, the particle moves toward the cylinder wall in a helix along the resultant of the centrifugal force and particle weight. The settling rate of the particle depends upon its size and weight. Lighter particles take longer time to settle. The number of effective turns inside cylinder is also important but in practice it is usually very difficult to estimate. For common cyclone, the number of effective turns is taken to be two.

The pressure drop through the cyclone can be given by the following equation

Pressure drop,

$$dp = \frac{12\,W h}{K E^2 \left(\frac{L}{d}\right)^{1/3} \left(\frac{H}{d}\right)^{1/3}} \qquad \ldots 2.13$$

where,
dp = pressure drop, number of inlet velocity heads
d = cylinder diameter, m
L = cylinder height, m
H = cone height, m
W = entry width, m
h = entry height, m
E = exit duct diameter, m
K = vane constant, 0.5–0.7.

The smallest particle size which could be removed by a cyclone can be estimated by the following equation

$$Dp = \sqrt{\frac{9\,\mu\,E}{2\,\pi\,N\,V\,(\rho_p - \rho_a)\left(\frac{4\,R}{E}\right)^K}} \qquad \ldots 2.14$$

where, Dp = smallest particle diameter, m
μ = viscosity of air, $kg/m\text{–}s$
E = exit duct diameter, m.
N = number of effective turns of air stream in inner spiral
V = entrance velocity, m/s
ρ_p = density of particle, kg/m^3
ρ_a = density of air, kg/m^3
R = radius at which the spiral velocity equal to V
K = constant, 0.5–0.7.

Problem : A cyclone separator having the following specifications is used to collect particles of specific gravity 1.2.

Cyclone diameter = 180 *cm*
Air inlet diameter = 30 *cm*
Separating height = 2.5 of dia. of inlet
Helix pitch = 15°
Inlet width = 10 *cm*
Entry particle velocity = 15 *m/s*

Compute the smallest particle which can be collected. Estimate the pressure drop through the unit. (ARS 1985)

Solution : The pressure drop through the cyclone is estimated by the following equation assuming $K = 0.5$.

$$dp = \frac{12\,W\,h}{K\,E^2 \left(\frac{L}{d}\right)^{1/3} \left(\frac{H}{d}\right)^{1/3}}$$

$$= \frac{12 \times 0.1 \times 0.54}{0.5 \times 0.3 \times 0.3 \left(\frac{0.75}{1.8}\right)^{1/3} \left(\frac{1.8}{1.8}\right)^{1/3}}$$

The entry height was calculated at the centre line of cyclone considering 15° helix pitch

$$dp = \frac{0.648}{0.0336}$$

$$= 19.27.$$

The smallest particle size removed by the unit can be estimated by the following equation.

$$D_p = \sqrt{\frac{9\,\mu\,E}{2\,\pi\,N.V\,(\rho_p - \rho_a)\left(\frac{4R}{E}\right)^E}}$$

$$= \sqrt{\frac{9 \times 5 \times 10^{-5} \times 0.3}{2 \times 3.1416 \times 2 \times 15\,(1200 - 1.293)\left(\frac{4 \times 0.9}{0.3}\right)^{0.5}}}$$

μ = viscosity of air = 5×10^{-5} *kg/m–s*
ρ_a = density of air = 1.293 kg/m^3

$$Dp = \sqrt{\frac{13.5 \times 10^{-5}}{7.817 \times 10^5}}$$

$$= 1.314 \times 10^{-5}\ m$$

$= 13.14 \times 10^{-3}$ mm or 13 *microns*.

Colour separator

The colour separator separates the fruits, vegetables or grains due to difference in colour or brightness. The colour separators are generally used for larger crop seeds like peas and beans. These seeds differ in colour because of varietal differences and also due to immaturity or disease. The mud balls and discoloured or defective seeds can be removed with the help of electronic separator.

The grain mixture is fed uniformly into the optical chamber of the separator. Two photo cells are fixed at a particular angle which direct both beams to one point of the parabolic trajectory of the grains. A needle is placed on the other side which is connected to a high voltage source (Fig. 2.27). When a beam falls on a dark object through photoelectric cells, current is generated on the needle. The needle end receives a charge and imparts it to the dark seeds. The grains are then passed between two electrodes with a high potential difference between them. The seed is compared with a selected background or colour range, and is separated into two fractions according to difference in colour. Since each grain is viewed individually by this machine, the capacity is low.

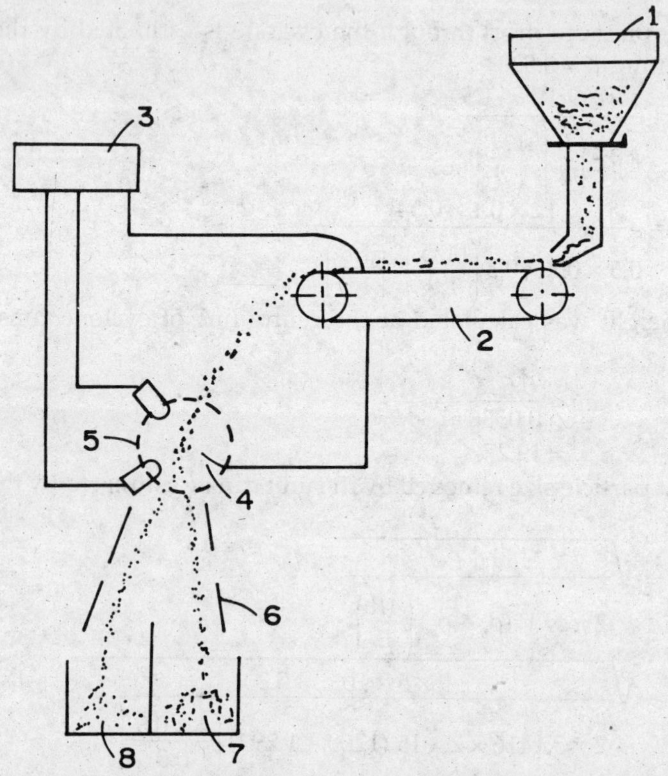

Fig. 2.27 : Schematic of a colour separator
1. hopper 2. belt conveyor 3. amplifier 4. charging needle 5. optical chamber 6. deflecting electrodes 7. foreign material 8. desired material

Bibliography

Ahuja, SS and BS Pathak, 1971. Studies on design and operational factors in mechanical separation system for wheat grain. M.Sc. Thesis, Punjab Agricultural University, Ludhiana.

Allen, GF, 1962. Screens and pneumatic classification. Industrial and Engineering Chemistry, Vol. 54(12): 39–47.

Berry, PE, 1959. Basic theory of oscillating conveyors. Journal of Agricultural Engineering Research Vol. 4 (3): 204–213.

Chakraverty, A, 1988. Post harvest technology of cereals, pulses and oilseeds. Revised Edition. Oxford and IBH Publishing Co. Pvt. Ltd. New Delhi:.

Clarke, B, 1985. Cleaning seeds by fluidisation. Journal of Agric Eng. Res. 31 : 231–242.

Fowler, BT and SC Lim, 1959. The influence of various factors on effectiveness of separation of finely divided solids by a vibrating screen. Chemical Engineering Science Vol. 10: 163–170.

Gluck, E. Samuel, 1965a. Vibrating Screens (I). Chemical Engineering. Vol. 72(4): 151–168.

Gluck, E. Samuel, 1965b. Vibratory Screens (II). Chemical Engineering. Vol. 72(6): 179–185.

Gluck, E. Samuel, 1965c. Gyratory and circular motion screen. Chemical Engineering. Vol. 72(22): 131–146.

Goldberg, SA and Warren Walter, 1959. When you specify sieve to include screen opening size. Chemical Engineering Vol. 66(3): 107–110.

Gorden Ritten house, 1943. Relation of shape to the passage of grain through sieve. Industrial and Engineering Chemistry (Analytical Edition). Vol. 15(2): 153–157.

Greg, BR, A.G. Law, SS Virdi and JS Bali, 1970. Seed Processing. Co-operatively published by Mississippi State University, National Seed Corporation and USAID.

Hall, CW and DC Davis, 1979. Seed and grain cleaning and grading equipment. In Processing Equipment for Agricultural Products (Second Edition). AVI Publishing Co. Inc. Westport, Connecticut.

Harrington, Roy, E. 1970. Thresher principle confirmed with multicrop thresher. J. Agric. Eng. Vol. 7(2): 49–60.

Harris, I J, 1965. Optimum design of sieve trays. British Chemical Eng. Vol. 10(6): 377–381.

Henderson, S M and RL Perry, 1976. Agricultural process engineering. Third Edition. The AVI Publishing Co. Inc. Westport, Connecticut.

ISI Test code for air screen seed cleaners. IS: 5718–1970. Indian Standard Institution. Manak Bhawan, New Delhi.

Kachru, RP and KM Sahay, 1988. Separation of oil seeds by pedal cum power operated air screen cleaner. J. Oilseeds Res. 5: 132–139.

Kachru, RP and KM Sahay, 1990. Development and testing of pedal cum power operated air screen grain cleaner. AMA Vol. 21(4): 29–32.

Kashyap, MM and BS Pathak, 1974–75. Air velocity requirement for winnowing of wheat grains. Paper presented at XIII meeting of ISAE.

Kaul, RN, 1967. Physical characteristics of agricultural materials affecting cleaning and separation process. Bulletin of Grain Technology, Vol. 5(2): 109–114.

Klein, Leonard M, James Henderson and Abraham D. Stoesz. 1961. Equipment for cleaning seeds. Year Book of Agriculture: 307–321.

Macaulay's, JT and JHA Lee, 1969. Influence of oscillating frequency on separation of wheat on a sieve in an air stream. Trans. ASAE. Vol 12(50): 648–654.

McCabe, WL, JC Smith and P Harriott, 1985. Unit operation of chemical engineering. 4th Ed. McGraw Hill, New York.

Perry, RH and CH. Chilton, 1973. Chemical engineering handbook, McGraw Hill Kogakusha Ltd.

Singh, Amar, Y Singh and AP. Bhatnagar, 1980. Development of rotary screen grain pre cleaner. J. Agric. Engg. Vol. 17(2): 105–107.

Singh, J and Y Singh, 1980: Development and evaluation of processing equipment. Technical Bulletin No.CIAE/PHTS/81/24, CIAE, Bhopal.

Singh, KK, 1984. Screen performance of grain pre- cleaners for wheat. M.Tech. Thesis, Punjab Agricultural University, Ludhiana.

Singh, KK and MM Kashyap, 1986. Screen performance of rotary screen grain pre-cleaner for wheat. I E (I) Journal-Agric. Vol, 67: 31–34.

Singh, KK and MM Kashyap, 1987. Performance of flat screen grain pre cleaner for wheat. I E (I) Journal-Agric. Vol.67: 65–68.

Uhl, JB and BJ. Lamp, 1966. Pneumatic separation of grains and straw mixtures. Trans. ASAE Vol. 9(2): 249.

Uichanco, VA and RM Lantin, 1972–73. Determination of optimum design parameters of rotary screen cleaners. The Philippines Agriculturist. Vol. 56 (9 and 10): 291–308.

Uppal, RSD and MM Kashyap, 1974. Particle trajectories in horizontal wind tunnel and their application to pneumatic separation. M.Sc. Thesis, Punjab Agricultural University, Ludhiana.

Walter, L. Badger and T. Julius Banchero, 1955. Introduction to chemical engineering size separation, McGraw Hill Book Co, New York: 618–627.

Weber, Moran and FM Raymond, 1938. A precise method of sieving analysis. J. Industrial and Engg. Chemistry (Analytical Edition) Vol. 30(4): 10–18.

Woft Ernst R, 1954. Screening principles and their application. J. Industrial and Engg. Chemistry Vol.46(9): 1758–1762.

Chapter 3

Drying

Drying of agricultural products is an important unit operation under post harvest phase. It refers to removal of moisture from grains and other products to a predetermined level, whereas dehydration means removal of moisture to very low levels usually to bone dry condition. Drying is a thermo-physical and physico-chemical operation by which the excess moisture from a product is removed. Drying makes the food grains and other products suitable for safe storage and protects them against attack of insects, molds and other micro-organisms during storage. It has been found by several experiments that harvesting of crops at higher moisture content and subsequent drying to safe moisture levels leads towards saving of grains to the tune of 6 to 7 per cent. For example, if paddy is harvested at 20–22% moisture content and later on dried to 14% moisture content, it has a potential of increasing rice yield by 10% in comparison to harvesting paddy at 14% moisture content. Therefore, in modern agriculture, the importance of timely drying is immense.

Moisture and its removal

During drying, the moisture from solids gets vaporised and diffused in dilute environment. Removal of moisture takes place by high pressure or high temperature conditions. Drying removes the volatile materials present inside solid matters. In some products, drying is carried to bring down the moisture content to a predetermined limit.

Some amount of moisture present in the pore spaces of grains. Excess amount of moisture remains in bound form with absorbing materials of the grains. Mutual reaction takes place between absorbing materials and moisture molecules, finally the moisture chemically bounds with the chemical constituents of the food grains. Removal of moisture from wet materials by vaporisation depends on the rate of heat and mass transfer, which is related with the following two basic phenomenon, (1) vaporisation of moisture from surface of material, and (2) movement of moisture from internal parts of material to its surface. Movement of moisture takes place because of diffusion, cell contraction and vapour pressure gradient.

In Fig. 3.1 the phenomenon of moisture removal from food grains in heated air drying is shown.

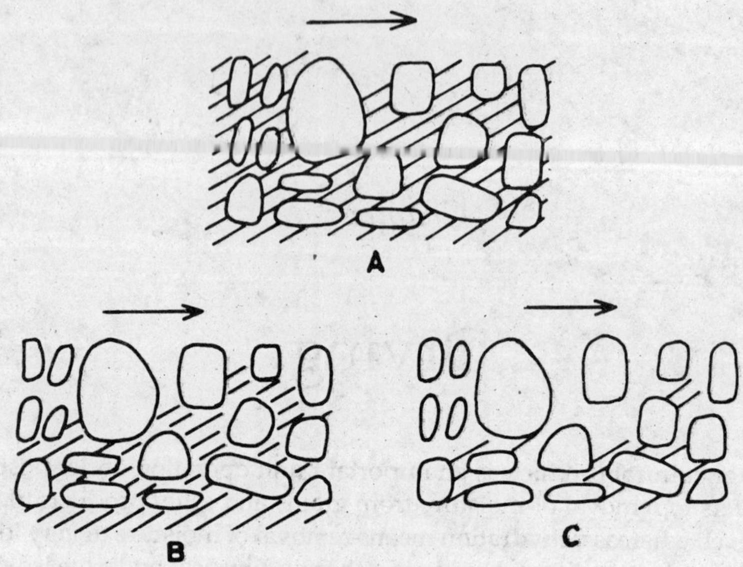

Fig. 3.1 : Removal of moisture from capillaries of grains
→Heated air flow

A : Distribution of moisture in crystalline solids in first stage
B : Distribution of moisture in solids in middle
C : Distribution of mositure in solids at the end of drying

Utilities of drying

Drying is thermo-physical and physico-chemical action and its dynamic principles are governed by heat and mass transfer laws inside and outside of the products. Drying is the most effective unit operation for protection of food grains and safe storage of agricultural products. Proper drying makes the material stable and the attack of insects and micro-organisms during storage period can be kept at negligible level. Various other advantages of drying are given below.

(1) Drying permits early harvest of crops. This leads to reduction of losses by shattering. This also leads to permit time for preparation of land for sowing of following crop.

(2) Drying helps in proper planning of harvesting season. If adequate drying facilities are available then crop can be harvested as per the availability of labourers.

(3) Drying of agricultural products to optimum moisture content results in safe storage of products over a longer period. Grains are stored in large quantities for long period in Government warehouses and industries for their subsequent use as and when required.

(4) Storage of products after drying makes products available during off seasons. selling of grains in off season would fetch additional income to growers as prices remain higher during these periods.

(5) When grains are stored, due to increase in the temperature and moisture content, heat of respiration also increases. The heat of respiration can be removed by aeration and drying. Thus the temperature and moisture con-

tent of the products can be brought down and the viability of seeds could be maintained.

(6) Some agricultural by-products and waste products can also be converted into useful products by proper drying.

Thermal properties

Information of thermal properties of products leads towards prediction of heat transfer rate for the product. For design of heating and refrigeration system for food materials, information on following thermal properties are necessary.

1. specific heat
2. thermal conductivity
3. enthalpy
4. thermal diffusion
5. surface heat transfer coefficient

In Table 3.1, the *SI* and calorie units of the above properties are given.

Table 3.1
Thermal properties units

Thermal properties	*S.I. units*		*Calorie units*
Specific heat	1.00 $kJ/kg-{}^\circ C$	=	0.239 $cal/g-{}^\circ C$
Enthalpy	1.00 kJ/kg	=	0.239 cal/g
Thermal conductivity	1.00 $W/m-{}^\circ C$	=	0.860 $kcal/m-h-{}^\circ C$
Surface heat transfer coefficient	1.00 $W/m^2-{}^\circ C$	=	0.860 $kcal/m^2-h-{}^\circ C$
Thermal diffusivity	1.00 m^2/s	=	1.00 m^2/s

Specific heat

The specific heat of a product tells us about the amount of heat required to change the temperature of the product. The specific heat of a product is not dependent on its mass density. To predict the specific heat of a mixture, we should know the values of specific heats of separate materials of the mixture. Specific heat of any product can be determined by the following expression.

$$Q = m\,c_p\,(T_2 - T_1) \qquad \ldots 3.1$$

where, Q = amount of heat, *cal*.
m = mass of product, g
c_p = specific heat, $cal/g-{}^\circ C$
T_1 and T_2 = initial and final temperatures of the product, °C

The specific heat of a wet material is the sum of specific heat of dry matter and specific heat of moisture content. Specific heat of a wet material can be determined by the following equation

$$c_p = \left(\frac{M}{100}\right) c_w + \frac{(100-M)}{100}\, c_{bd} \qquad \ldots 3.2$$

where, c_w = Specific heat of water
c_{bd} = specific heat of dry matter (grain)
M = moisture content of grain, % (d.b.)

(Specific heat of dry grains ranges between 0.35 to 0.4 $kcal/kg-{}^\circ C$)

Note: The above equation is valid for the grains having moisture content more than 8%.

To determine the specific heat of the foods, a caloriemeter or a simple thermal vacuum bottle is used.

Enthalpy

The heat content of a material is its enthalpy. It can also be said that the energy level of a material is enthalpy. The absolute value of enthalpy is very difficult to define but its zero value is usually defined at 0°C or some other convenient temperature. Enthalpy is generally used to quantify energy in steam.

$$Q = m\,(h_2 - h_1) \qquad ...3.3$$

where, Q = amount of heat, *cal.*

m = mass, g

h_2 and h_1 = enthalpy at temperatures T_2 and T_1 respectively, *cal/g*

Thus it can be said that the amount of heat energy required to heat a material from temperature T_1 to temperature T_2 is equal to $m\,(h_2 - h_1)$.

Thermal conductivity

The ability of heat conductance of a material can be judged by its thermal conductivity value. In agricultural products, thermal conductivity depends on their composition. But the value of thermal conductivity is also affected by percent void spaces, shape, size and arrangement of void spaces and other factors which restrict the heat flow. The thermal conductivity of a material can be defined as the amount of heat conducted across unit area and unit thickness of material in unit time for unit change in temperature. Therefore, thermal conductivity depends on the mass density of material. Thermal conductivity of single grain is 0.3 to 0.6 *kcal/m – h –* °C whereas thermal conductivity of a heap of grains ranges between 0.1 to 0.15 *kcal/m – h –* °C. The reason for this is the pore space between the grains. These pore spaces are occupied by air whose thermal conductivity is quite lower (0.02 *kcal/m – h –* °C) as compared to grains. Thermal conductivity can be determined by the following equation,

$$Q = \frac{A\,K\,(T_1 - T_2)}{t} \qquad ...3.4$$

where, Q = amount of heat

K = thermal conductivity

t = thickness of material layer

A = area of material perpendicular to direction of heat flow

T_1 and T_2 = temperatures at surfaces of the material

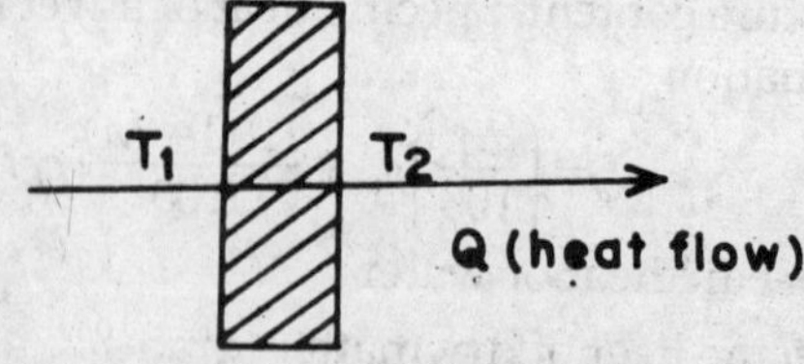

Thermal diffusivity

Thermal diffusivity is used to determine the heat transfer rates in solid agricultural products of any shape. This property of material is related to the ability of material to conduct heat and its ability to store heat energy. It can be determined by the following expression,

$$\text{Thermal diffusivity} = \frac{\text{thermal conductivity}}{\text{specific heat} \times \text{mass density}}$$

Surface heat transfer coefficient

Surface heat transfer coefficient is used to quantify rate of heat convection towards or away from the surface of a product. It is not a thermal property of any object. The surface heat transfer coefficient is needed to quantify heat transfer in most food heating and cooling applications.

The surface heat transfer coefficient is defined as the proportionality constant that relates the heat flux to or from a surface, to the temperature difference between the surface, and the fluid stream moving past the surface.

Moisture content and its measurement

The moisture content of food grains and other agricultural products plays an important role in maintaining the desirable quality of the product. Changes in moisture content of agricultural material occur during their harvesting, processing and marketing. The change in moisture content during successive post harvest stages is dependent upon the initial moisture content of product and atmospheric conditions. The information of moisture content is necessary because it tells us whether the product is suitable for safe storage or for any other processing job. For some period, if the moisture content of food grains increases beyond a fixed critical value, under such conditions chances of metabolic reactions inside the grain become higher. Due to such reactions, viability of seed reduces. Drying of agricultural products become necessary at higher moisture contents.

Moisture content representation

The amount of moisture in a product is given on the basis of the weight of water present in the product and is usually expressed in percent. Moisture content is designated by two methods, (1) Wet basis (wb) and (2) dry basis (db).

Wet basis : The moisture content in this method is represented by the following expression,

$$\text{Per cent moisture content} = \frac{\text{weight of water in product}}{\text{weight of product sample}} \times 100$$

Dry basis : In this method of representing moisture content; it is given on the basis of dry weight of product. The dry basis moisture content is determined by the following expression

$$\text{Per cent moisture content} = \frac{\text{weight of water in product}}{\text{weight of dry matter of product sample}} \times 100$$

The value of dry basis moisture content is more than the wet basis moisture content. The relationship between dry basis and wet basis moisture content is given by the following expression

$$\text{Moisture content (dry basis), \%} = \frac{\begin{array}{c}\text{moisture content (wet–basis)}\\ \text{in decimals}\end{array}}{\begin{array}{c}1 - \text{moisture content (wet–basis)}\\ \text{in decimals}\end{array}} \times 100$$

If the moisture content is expressed in wet basis terms, and the moisture content of 100 *kg* grains by drying has been brought down from 18% to 13%, it is incorrect to say that 5 *kg* of water has been removed from the grains. In wet basis moisture representation, the denominator of the equation changes in relation to the amount of moisture removed.

At 18% moisture content, the sample of grain contains 18 *kg* (100 × 0.18) of water and 82 *kg* of dry matter. The same amount of grain is dried to 13% moisture content, the final weight of water is calculated by the following relationship,

$$0.13 = \frac{\text{weight of water}}{\text{weight of water} + 82}$$

or,
$$\text{weight of water} = \frac{(0.13)\,(82)}{0.87}$$
$$= 12.25 \text{ kg}$$

Thus, the actual amount of water removed from 100 kg of wet grain will be 18 – 12.25 = 5.75 *kg*.

Problem : 500 kg of paddy at 22% moisture content (wb) is dried to 14% moisture content (wb) for milling. Calculate the amount of moisture removed in drying.

Solution:

1st method, using wet basis moisture content data

Amount of initial water content = 0.22 × 500

= 110 *kg*

Dry matter in paddy = 500 – 110 = 390 *kg*

Final moisture content in 14% or 0.14

$$0.14 = \frac{\text{weight of water } (W_w)}{\text{weight of water} + \text{dry matter}} = \frac{W_w}{W_w + 390}$$

or
$$W_w = \frac{0.14 \times 390}{0.86}$$
$$= 63.488 \; kg$$

Thus amount of water removed = 110 – 63.49

= 46.51 kg

2nd method, amount of water removed by analysis on dry basis, changing the initial and final moisture content to dry basis, initial mc (db) = $\frac{0.22}{1 - 0.22} \times 100 = 28.20\%$

$$\text{final } mc \text{ (db)} = \frac{0.14}{1 - 0.14} \times 100 = 16.27\%$$

Amount of water removed is equal to the product of initial dry matter and the difference in dry basis moisture content

$$390 \times \frac{28.20 - 16.27}{100} = 46.51 \; kg$$

Moisture content determination methods

There are several methods for determination of moisture content of agricultural products. For determination of moisture content of a particular product, the choice of method depends on many factors, they are (1) the form in which water is present in the product, (2) the relative amount of water present, (3) the rapidity of determination, (4) accuracy of method, (5) product's nature whether easily oxidised or decomposed, and (6) the cost of equipment used.

Moisture content is determined mainly by two methods, (1) direct, also called as primary and (2) indirect, also called as secondary methods. The accuracy of moisture content determination by direct methods is high, hence these methods are used by research workers. The moisture content values obtained by direct methods are used to calibrate all the indirect type moisture measuring devices. The direct methods of moisture determination are time consuming therefore, such methods are not very much useful for determination of moisture in field, mandi or ware houses. Indirect methods are faster and mostly employ the electrical properties of the grains.

The determination of moisture is empirical method because the various moisture determination methods measure more or less the water present in the product. Thus the experimental conditions or method govern to some extent, the amount of moisture obtained.

Direct methods

(i) Air oven method : When the moisture content of grains is upto 13%, 2–3 grams representative ground samples of grains are placed in an air- oven. The temperature of the oven is set at 130°C and the samples are kept in oven for 1–2 hours. Afterwards, the samples are taken out and placed in a desiccator to cool down. The drop in the weight of grain is measured based on its initial weight.

In other method, 25 to 30 grams of unground representative samples of grains are taken and placed in an air-oven at 100°C temperature. The samples are kept in it for 72 to 96 hours. Afterwards, the samples are taken out from oven placed in a desiccator to cool down to room temperature. Moisture content of samples is measured based on drop in weight from initial weight of sample.

(ii) Vacuum-oven method : In this method, 2–3 grams of representative sample of ground material is placed in a vacuum-oven (25 *mm* vacuum) and dried at 100°C for 72–96 hours. Afterwards, the samples are weighed. The drop in the weight of grain is measured based on its initial weight.

The equipment generally used for determination of moisture content by air-oven method are given below.

(a) Moisture dishes : For moisture measurement, the moisture dishes should be made of heavy gauge aluminium sheet metal. The moisture dishes may not be easily damaged by dents. The dishes should have tight covers. Same numbers should be marked on the dish and its cover. Before use, the moisture dishes should be dried in an oven for one hour thereafter these should be weighed.

(b) Desiccator : The desiccator should be properly air tight and have active absorbent inside.

(c) Oven : Convective types of dryers are mostly used for moisture determination. These are sufficiently heat resistant. These ovens maintain predetermined temperature. The oven should have arrangements for proper air circulation, transferable perforated trays and a suitable thermometer sensitive enough to show 0.5°C temperature difference. These ovens are equipped with proper temperature control system. The oven should be run for few hours prior to their use for moisture determination.

(d) Balance : An analytical balance should be used to determine the moisture content of product correctly. The balance should be sensitive enough to weigh upto 0.01 grams.

The temperature and time combination for various grains/seeds are given in Table 3.2

Table 3.2

Oven temperature and heating period for moisture content determination

Seed	*Oven temperature, °C*	*Heating period, hour*
Wheat	130	19
Sorghum	130	18
Maize	130	72
Mustard	130	04
Sunflower	130	01
Safflower	130	03
Beans	103	72
Barley	103	20

(iii) Brown-Duvel fractional distillation method : The moisture content of products is measured by fractional distillation method. The method is recognised as an official method for determination of moisture content. Hundred gram whole grains

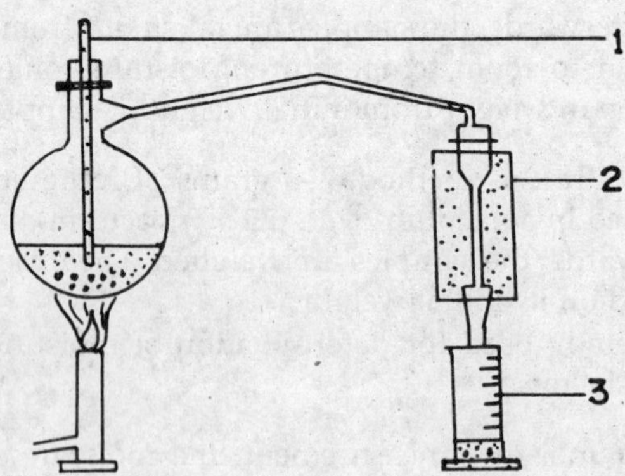

Fig. 3.2 : Brown-Duvel fractional distillation apparatus
1. thermometer 2. hot water bath 3. graduated cylinder

alongwith 150 *ml.* of mineral oil is taken in a flask. The sample is boiled as shown in Fig. 3.2. Moisture from the sample is thus evaporated, collected and condensed in a graduated cylinder. The millilitre of moisture collected shows the percentage moisture content. The determined moisture content is on wet basis. In this method the temperature of mineral oil in flask should reach to 200°C within 26 minutes. The time required for moisture determination is about 30 minutes. If the temperature of mineral oil reaches to desired temperature within time, the moisture content determination is found to be fairly accurate. When samples of grains are taken from bulk, a variation in moisture content to the tune of 1 to 1.5% is possible.

(iv) Infra-red method : In this method, grain moisture content is directly measured by evaporation of the water from a sample of grain with an infra-red heating lamp. A commercial infra-red moisture meter is also available in market. The instrument consists of a balance, a pan counter balanced by a fixed weight and a variable length of weighing chain. An infra-red lamp is mounted on an arm above the pan with a provision to change its height. A scale calibrated in percentage moisture content is incorporated in the stem of the instrument as shown in Fig 3.3. At the end of the test, when the balance is zeroed, a direct reading of moisture content is obtained. The sample of grains may be unground, but when ground sample is used, the time needed to evaporate the water is reduced.

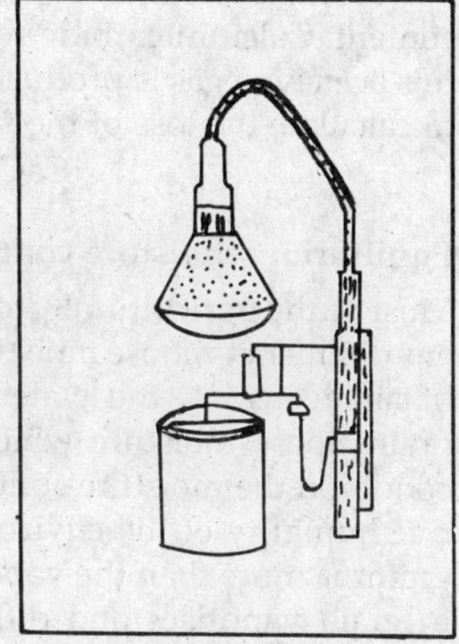

Fig. 3.3 : Infra-red moisture meter

Indirect methods

(i) Electrical resistance method : The electrical conductivity or resistance of a product depends upon its moisture content. This principle is employed in resistance measuring devices. The universal moisture meter measures the electrical resistance of the grain at a given compaction. The electrical resistance apart from compaction is also affected by grain temperature and impurities present in the sample. Such moisture meters are calibrated for grain/seed types, degree of compaction and temperature. Universal moisture meter gives fairly accurate readings of moisture content on wet basis.

(ii) Dielectric method : Such devices measure the dielectric constant of grains. The grains are filled in chamber. The sides of the chamber is formed by the plates of a condenser between which a high frequency current is passed to measure the capacitance of the sample. The capacitance varies as per the water present in sample, the degree of compaction and the grain temperature.

The electrical properties of grain are temperature dependent. Since the measurement of grain temperature is difficult, therefore, ambient air temperature is often used. This may result a source of error because of the difference between air and grain temperatures.

(iii) Chemical method : The removal of water by strong desiccants ($CaCl_2$) is caused by the vapour pressure gradients. The moisture moves from the wet grain to the drying agent, since the vapour pressure of the grain is higher than that of the

desiccant. The hydration of salt is accompanied by evolution of heat. This heat of evolution helps in driving the water out of the grains. Calcium chloride, when heated to redness reacts with superheated steam to form HCl and calcium hydroxide.

$$CaCl_2 + 2\,H_2\,O \rightarrow Ca\,(OH)_2 + 2\,HCl$$

In some other method, calcium chloride is used to measure the grain moisture content. Calcium carbide reacts with water present in the sample. As a result of this acetelyne gas is produced. This acetelyne gas can be measured volumetrically to calculate the loss of moisture. The chemical action can be given as under,

$$CaC_2 + 2\,H_2\,O \rightarrow Ca\,(OH)_2 + C_2\,H_2$$

Equilibrium moisture content (EMC)

Most of the agricultural products, specially the food grains absorb moisture from environment or loose moisture. When the ambient temperature rises and humidity of air decreases, the water present in food grains vaporises, consequently the grains loose moisture which results in drying. Thus, we find that at particular condition the moisture content of grains depends upon the temperature and relative humidity of the environment. If the vapour pressure of the water present in grains is more than the vapour pressure of water vapours in air, the water present in grain vaporises and diffuses in the atmosphere. Alternatively, if the vapour pressure of water present in grain is less than the atmospheric vapour pressure, grain will absorb moisture from atmosphere. This property of gaining or loosing of moisture as per the atmospheric conditions is known as hygroscopicity.

The moisture content attained by a grain with respect to a set of atmospheric temperature and relative humidity is called the equilibrium moisture content (EMC) of the grain. In such condition, the grain moisture is in equilibrium with surrounding air.

Methods for determination of EMC

The methods for determination of EMC of agricultural products can be categorised into two; (1) static method and (2) dynamic method. In the static method, grains are left in humid and still air until they attain equilibrium, while in the dynamic method humid air is agitated or moved by mechanical means and the grains attain equilibrium condition.

Static method : In static methods, to bring the atmospheric air to desired relative humidity levels different concentrations of sulphuric and hydrochloric acids are used. Static methods are generally too much time consuming, and to bring the grain to equilibrium condition by acids, 3–4 weeks are required. Thus in case of higher humidity and high temperature conditions, chances of attack of molds are high. Decomposition and change in grain structure is also possible. It is essential to maintain required humidity and temperature conditions of air throughout the test period. Temperature of air is generally maintained by an air-oven, whereas the relative humidity of air is maintained by acid solution. Various concentrations of sulphuric and hydrochloric acids give relative humidity of air between 0 to 100 percent. The vapour pressure above the acid solution depends upon the concentration of chemical solution and the temperature. In Table 3.3, values of relative humidity attainable by different concentration of sulphuric acid are given.

Table 3.3

Relative humidity values on the different concentration of sulphuric acid solution in water

Temperature, °C	*Acid by weight basis, %*			
	20	*40*	*60*	*80*
10	87.4	56.6	15.8	3.8
20	87.4	56.7	16.3	4.7
30	87.4	56.6	17.0	5.7
40	87.4	57.5	17.8	6.9
50	88.8	58.0	18.8	8.2

Handling of concentrated acid solution is not easy. Determination of EMC with salt solutions is relatively safer, but to maintain a wide range of relative humidity values a number of salts are needed. Considering this point, use of salts for EMC determination is cumbersome as compared to use of acids. Hall (1957) and Lal (1969) have collected information on use of salts for EMC determination. In Table 3.4 the values of relative humidities with respect to salt solutions and various temperature conditions are given.

Dynamic methods

The following methods are being followed in determination of EMC by dynamic means.

(i) Desorption method : In this method, the property of dry air to absorb moisture from moist grains is employed. Moist grains are put in an airtight container. When the air comes in equilibrium to grain its relative humidity is measured by an electric hygrometer or by a hair hygrometer. Since the container has little quantity of air, it reaches in equilibrium with grain in short period.

(ii) Isotenoscopic method : This method also employs absorption of moisture by dry air to determine grain EMC. But in this method arrangement is available to mea-

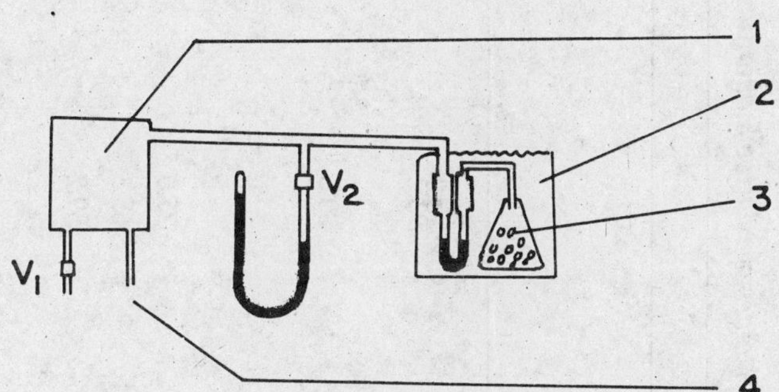

Fig. 3.4 : Schematic diagram of an isotenoscope
1. vacuum storage jar 2. constant temperature water bath
3. sample flask 4. to vacuum pump

sure directly the vapour pressure exerted by the moist grains. In Fig. 3.4 schematic diagram of isotenoscope is shown. In a conical flask, grain sample is kept.

Table 3.4

Values of relative humidity attainable by use of saturated solution of some salts at different temperatures

Salt	*Temperature, °C*						*Amount of salt needed to dissolve in 100 g water, g (as per temperature)*
	20	*25*	*30*	*35*	*40*	*50*	
	Average relative humidity						
Potassium Sulphate	97.2	96.9	96.6	96.4	96.2	95.8	11–17
Potassium Nitrate	93.2	92.0	90.7	89.3	87.9	85.0	32–75
Patassium Chloride	—	84.7	84.5	83.0	—	—	34–43
Ammonium Sulphate	80.6	—	80.0	—	79.6	79.1	—
Ammonium Chloride	—	77.0	78.2	76.6	—	—	—
Sodium Chloride	75.5	75.8	75.6	75.5	75.4	74.7	36–37
Sodium Nitrate	65.3	64.3	63.3	62.0	61.7	—	90.2–118.0
Sodium Bromide	59.2	57.8	56.3	54.6	—	—	—
Patassium Carbonate	—	43.8	43.5	43.4	—	—	—
Maagnessium Chloride	33.6	33.2	32.8	32.5	32.1	31.4	408–600
Potassium Acetate	23.2	22.6	22.0	21.0	—	—	—
Lithium Chloride	12.4	12.0	11.8	11.7	11.6	11.4	79–98
Vacuum Oven	0.0	0.0	0.0	0.0	0.0	0.0	—

Isotenoscope is a U tube filled with the liquid of negligible vapour pressure. The arms of the tube has an enlarged section above the level of liquid to prevent drawing of the liquid out of the tube while evacuating or readmitting air to the flask. The isotenoscope is connected to a vacuum pump through a vacuum storage jar. Atmospheric pressure can be brought back into this jar by means of a valve 'V_1'. The 'V_2' is a shut off valve connecting closed end of mercury manometer to the vacuum system. In operation, valve 'V_1' is closed while all air is evacuated from the flask, the vacuum storage jar, and from the system. Under this condition, vapour pressure builds up in the flask which forces the liquid in the two arms of the isotenoscope to dissimilar level. The level of the liquid is then equalised by bleeding a small amount of air into the vacuum storage jar. This equalization pressure is continued until vapour pressure built up in the flask has reached a maximum for the temperature of water bath. Valve 'V_2' is then closed and the absolute pressure indicated in the manometer is read. The isotenoscope is removed from the flask and the flask is closed by a properly weighed stopper. The weight of flask with sample is recorded to determine sample moisture content.

EMC models

For determination of EMC of agricultural products many theoretical, semi-theoretical and experimental models have been proposed. But, none of the theoretical EMC equation is found to predict or to provide the EMC values correctly in the entire range of temperature and relative humidity values. Some of EMC models are described below.

(i) Kelvin equation : Kelvin in 1871 has given the model of moisture adsorption by solid material. For evolving the model, the phenomenon of capillary condensation in pores of solid materials was considered. The relationship between the vapour of water present in capillaries and saturated vapour pressure at same temperature is the basis of capillary condensation theory. The kelvin equation is as under

$$ln\left(\frac{P_v}{P_{vs}}\right)=\frac{2\sigma\, V\cos\alpha}{rRT_a} \qquad \ldots 3.5$$

where, P_v = vapour pressure of grain

p_{vs} = saturated vapour pressure at temperature in equilibrium with the system

σ = moisture surface tension

α = angle between moisture and capillary wall

V = Volume of moisture

r = radius of cylindrical capillary

R = universal gas constant

T_a = absolute temperature

The utility of above equation for grain EMC determination is limited in conditions of relative humidities above 95% when the action of capillary condensation takes place.

(ii) Harkins-Jura equation : This model is based on the theory of existence of a potential field above surfaces of solid materials. In this concept, the work required to adsorb or desorb a water molecule is the sum of work required to overcome

vapour molecule to come on surface and the work necessary for condensation. Considering the above potential field theory, Harkins-Jura have proposed in 1944 the following equation

$$ln\left(\frac{P_v}{P_{vs}}\right) = d - \frac{e}{V^2} \qquad ...3.6$$

where, d and e = product constant depend on temperature

The Harkins-Jura equation does not predict satisfactorily accurate EMC values when the relative humidity is more than 30%.

(iii) Chung-Pfost equation : Chung-Pfost has proposed an equation for determination of EMC on the basis of potential field theory. The equation is given below

$$ln\left(\frac{P_v}{P_{vs}}\right) = -\frac{A}{RT}\exp(-BM) \qquad ...3.7$$

where, R = universal gas constant
T = absolute temperature, °K
A and B = constant dependent upon grain temperature
M = moisture content, % (db)

The above equation provides fairly accurate EMC values of grains between 20 to 90% relative humidity values.

(iv) Henderson equation : Handerson in 1952 has proposed an equation for EMC determination. This EMC equation is very much popular and based on the Gibbs' adsorption equation. The following equation for the EMC curves (Fig. 3.5) has been derived by Henderson.

$$1 - rh = e^{-CTM_e^{\,n}} \qquad ...3.8$$

where, rh = relative humidity, decimal
T = absolute temperature, °K

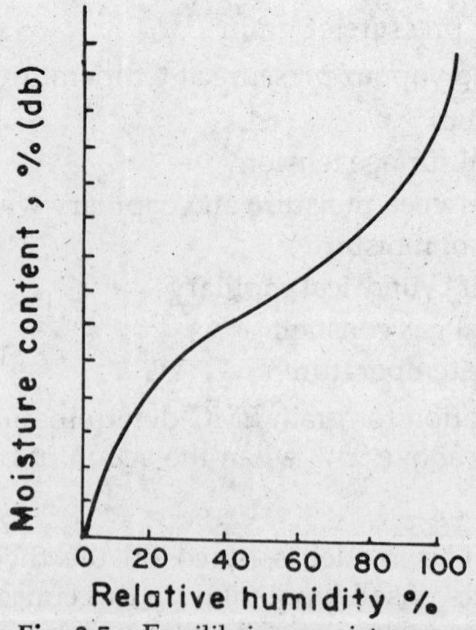

Fig. 3.5. : Equilibrium moisture curve

M_e = EMC, % (db)

C and n = constant, dependent on crop type and temperature.

Out of the above described theoretical, semi-theoretical or empirical equation, no single relationship can predict grain's *EMC* values in the full range of relative humidities and temperatures generally encountered.

Importance of EMC

EMC is of particular importance for drying and storage of agricultural materials. The usefulness of EMC are:-

(i) EMC gives us the idea whether the material will gain or loose moisture at a particular atmospheric condition.
(ii) It also gives an idea about rate of moisture removal.
(iii) EMC helps to determine drying characteristics.
(iv) With the help of EMC, it can be predicted that to which moisture level product can be dried with heated air.

Problem : Calculate the equilibrium moisture content of brinjal seed at relative humidity of 10% and temperature of 50°C using Henderson's equation. Given that constants c is 6.5×10^{-6} and n is 1.8.

Solution : Henderson's equation is

$$1 - rh = e^{-CTMe^n}$$

Substituting the given values of rh, T and c and n in the above equation, we get

$$1 - 0.1 = e^{-[6.5 \times 10^{-6} \times (50+273) Me^{1.8}]}$$

or, $$0.9 = e^{-[6.5 \times 10^{-6} \times 323 Me^{1.8}]}$$

or, $$e^{-0.105} = e^{-[2.1 \times 10^{-3} Me^{1.8}]} \qquad \text{(Since } e^{-0.105} = 0.9\text{)}$$

or, $$0.105 = 2.1 \times 10^{-3}\, Me^{1.8}$$

or, $$Me^{1.8} = 50$$

$$Me = 8.78\% \text{ (db)}$$

Problem : Determine the values of constants c and n from the Henderson's equation for the following data obtained under two different conditions of EMC studies of sunflower seed.

Condition	Relative humidity, %	Temperature °C	EMC, % (db)
1	50	40	10
2	70	50	13

Solution : Henderson's equation is $1 - rh = e^{-CTMe^n}$

Substituting the values of rh, T, Me in the equation

$$1 - 0.5 = e^{-[C \times (40+273) \times 10^n]}$$

or, $$0.5 = e^{-[C \times 313 \times 10^n]}$$

or, $$e^{-0.693} = e^{-[C \times 313 \times 10^n]} \qquad \text{(Since } e^{-0.693} = 0.5\text{)}$$

or, $$0.693 = c \times 313 \times 10^n \qquad \text{...(i)}$$

Substituting the values of rh, T and Me of condition 2, we get

$$1 - 0.7 = e^{-[C \times (50 + 273) \times 13^n]}$$

or, $$0.3 = e^{-[C \times 323 \times 13^n]}$$

or, $$e^{-1.204} = e^{-[C \times 323 \times 13^n]}$$

or, $$1.204 = c \times 323 \times 13^n \quad \text{...(ii)}$$

Dividing equation (ii) by equation (i), we get

$$\frac{1.204}{0.693} = \frac{C \times 323 \times 13^n}{C \times 313 \times 10^n}$$

or, $$\left(\frac{13}{10}\right)^n = 1.68$$

$\therefore$ $$n = 1.98$$

Substituting the value of n in equation (i)

$$0.693 = c \times 313 \times 101.98$$

or, $$c = 2.32 \times 10^{-5}$$

Heat of vaporisation

The heat of vaporisation of food grains is the amount of energy necessary for vaporising the available moisture in grains. The EMC curves make available the required data for calculation of heat of vaporisation necessary for moisture evaporation. The amount of energy needed for grain moisture evaporation is dependent on its temperature and moisture content. Lower levels of temperatures and moisture contents result in higher amount of heat or vaporisation. One of the equations developed for estimation of heat of vaporisation is given below.

$$\frac{ln\, p_v}{ln\, p_{vs}} = \frac{h_{v'}}{h_v} + q \quad \text{...3.9}$$

where, $h_{v'}$ and h_v = respectively heat of vaporisation of grain moisture and free water

q = constant dependent upon product and temperature

In Table 3.5 heat of vaporisation values of some food grains at different moisture content and temperature levels are given.

Table 3.5

Heat of vaporisation values for some grains at various moisture content, % (db) and temperature levels

Food grains	*Moisture Content, % (db)*	*Temperature, °C* 10	21	38	65
Wheat	10	676.7	670.6	658.9	640.6
	15	646.2	640.0	629.5	611.7
	20	606.1	600.6	590.0	573.4
Maize	10	761.7	755.0	741.7	721.2
	15	717.8	711.2	698.9	679.5
	20	661.7	655.6	644.5	626.2
Sorghum	10	674.5	668.4	656.7	638.4
	15	641.2	635.6	624.5	607.3
	20	609.5	603.9	593.4	576.7
Water		591.7	586.1	576.1	560.0

Hysteresis effect

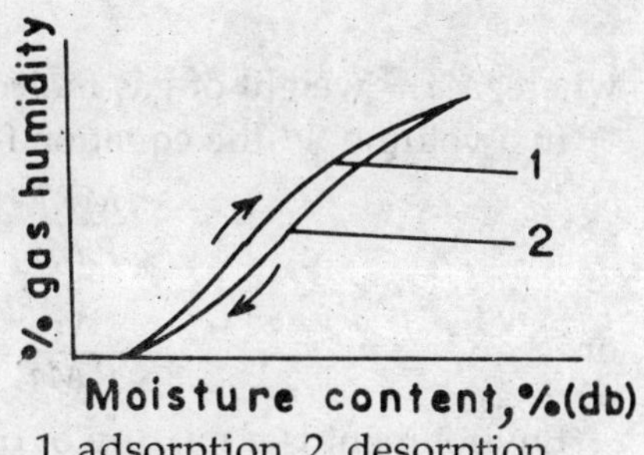

1. adsorption, 2. desorption

Fig. 3.6 : Hysteresis effect

When some agricultural products in the process of loosing moisture attains equilibrium moisture content with the surroundings, the EMC is known as desorption EMC. But when a dry product gains moisture from the surroundings and attains EMC, that value of EMC is said to be adsorption EMC. At same relative humidity and temperature level there is a meaningful difference between the desorption and adsorption EMC values. The desorption EMC values are higher than the adsorption EMC values. The difference between desorption and adsorption curves is known as hysteresis effect (Fig 3.6).

Air-Vapour mixture and Psychrometric chart

In most drying systems, air is used as a medium for heat transfer e.g. sink for water vapour, medium for oxygen transportation for ignition, and medium for vapour transport. In engineering application, air is supposed to be a mixture of nitrogen and oxygen, but air also contains other elements like helium, nion, hydrogen, argon, kripton etc.

Gas laws

The air pressure, volume, weight and other thermal properties are bound by certain mathematical equations or laws. These gas laws act satisfactorily during moderate or normal processing conditions.

Ideal gas law : There are two popular forms of the ideal gas law;

$$PV = MRT \qquad \text{...3.10}$$

and

$$\frac{P_1V_1}{T_1} = \frac{P_2V_2}{T_2} \qquad \text{...3.11}$$

where, P = absolute pressure
V = gas volume
R = universal gas constant
T = absolute temperature

Amagat law : This says that at a particular condition of temperature and pressure, the volume of a gas mixture is sum of the volume of each gas component at same temperature and pressure

$$V = V_1 + V_2 + V_3 + \ldots \qquad \text{...3.12}$$

Daltan's law : This law says that each component of a gas mixture produces same amount of pressure as produced by the component of equal volume to that of gas mixture

$$P = P_1 + P_2 + P_3 + \ldots \qquad \text{...3.13}$$

The weight of gas mixture is equal to the sum of each component's weight

$$M = M_1 + M_2 + M_3 + \ldots \qquad \text{...3.14}$$

Therefore, from gas law $V = \dfrac{MRT}{P\,Ma}$ ...3.15

where, Ma = weight of gas molecule

In a volume 'V' the equation for gas mixture

$$\frac{MRT}{P\,Ma} = \frac{M_1 RT}{P_1\,Ma_1} = \frac{M_2\,RT}{P_2\,Ma_2} = \ldots \quad \ldots 3.16$$

or,

$$\frac{M}{P\,Ma} = \frac{M_1}{P_1\,Ma_1} = \frac{M_2}{P_2\,Ma_2} \quad \ldots 3.17$$

Environment is a mixture of dry air and water vapour. The air never becomes completely dry. The physical and thermal properties of air are represented by a graph.

Psychrometric chart

For analysis of agricultural product drying problems, the thermodynamic properties of dry air-water vapour mixture are needed. Psychrometric chart is a graphical representation of thermodynamic properties of air. In this chart the absolute

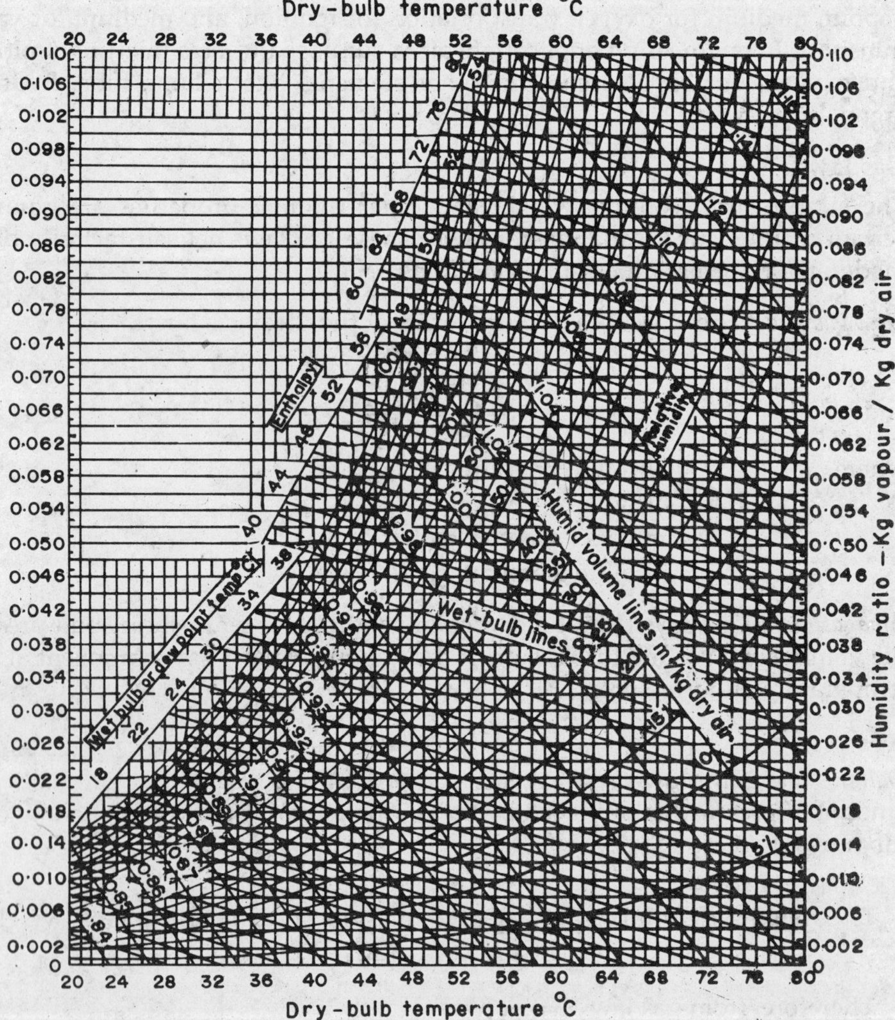

Fig. 3.7 : Psychrometric chart

humidity, percentage humidity, humid volume, enthalpy, and adiabatic saturation lines are drawn against air temperatures (Fig 3.7). Psychrometric chart gives the following thermodynamic properties of moist air at one atmosphere, (1) wet-bulb temperature, (2) dry-bulb temperature, (3) dew-point temperature, (4) humidity ratio, (5) relative humidity, (6) specific volume and (7) enthalpy. If any two of the above properties are known, a point on the chart can be located and the other properties can be determined by reading the values of required lines representing the properties and passing through the point.

Dry-bulb temperature : This is the temperature of the air, determined by a thermometer with a dry bulb.

Wet-bulb temperature: This is temperature of air which is measured by a thermometer with the bulb covered by a piece of wet cloth. Wet-bulb temperature is less than the dry-bulb temperature. A sling psychrometer (Fig. 3.8) is used to measure the dry and wet-bulb temperatures at some instant. The sling psychrometer contains two separate thermometers, the bulb of one thermometer is covered with wet cloth.

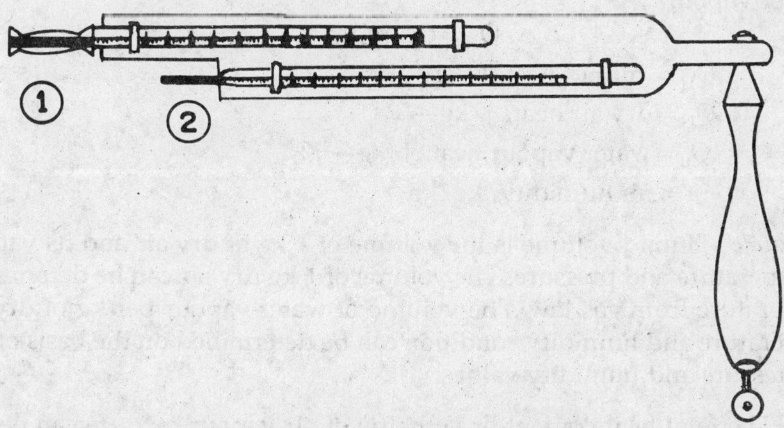

Fig. 3.8 : Sling psychrometer
1. Wet bulb 2. dry bulb

Evaporation of water from the wet bulb soon reaches to a constant condition when the psychrometer is moved in air. Transfer of heat takes place between bulb and atmosphere as a result the temperature comes down. Lowering of temperature is because of evaporation of water. The rate of evaporation can be calculated by the following equation

$$w = f_v A\,(p_{vs} - p_v) \qquad \ldots 3.18$$

where,
w = evaporation rate, kg/s
f_v = mass transfer factor, $kg/m^2 - s$
A = area of bulb, m^2
p_{vs} = saturated air humidity, $kg/kg\ (db)$
p_v = air humidity, $kg/kg\ (db)$

Therefore heat, $q = w \times$ heat for water evaporation

$$= f_v A\,(p_{vs} - p_v)\,\Delta H \qquad \ldots 3.19$$

ΔH = latent heat of evaporation, *kcal/kg*

Dew-point temperature : This is the temperature at which the saturation state of a mixture of air and water vapour reaches during a cooling process at constant pressure.

Absolute humidity : The ratio of kg of water vapour and kg of dry air is the absolute humidity.

Percentage humidity : This is the ratio of absolute air humidity with respect to the absolute air humidity at saturation point at same dry-bulb temperature.

Relative humidity : The ratio of actual vapour pressure in air-vapour mixture to the saturated water vapour pressure in air at same temperature is known as the relative humidity of air.

Saturated pressure : The water vapour in air obeys Daltan's law, therefore, it creates pressure independent of dry air. The water vapour pressure values of saturated water vapour are obtainable from steam tables.

Humid heat : Humid heat is the specific heat of a mixture of 1 kg of dry air and Y kg of water vapour

$$Q_H = Q_D + Q_v Y \qquad ...3.20$$

where, Q_H = humid heat, J/kg—°K

Q_D = dry air heat, J/kg—°K

Q_v = water vapour heat, J/kg—°K

Y = air humidity, kg/kg

Humid volume : Humid volume is the volume of 1 kg of dry air and its vapour at given temperature and pressure. The volume of 1 kg dry air can be determined at any temperature from gas law. The volume of water-vapour per kg of dry air at any temperature and humidity condition can be determined on the basis of water vapour pressure and humidity values.

Enthalpy : The total heat content or enthalpy of air vapour mixture can be calculated by the following equation

$$I_g = 0.24\,t + H\,h_g \qquad ...3.21$$

$$= 0.24\,t + (597.3 + 0.45\,t)\,H \qquad ...3.22$$

where, I_g = enthalpy, heat content of air-vapour mixture

t = temperature

H = absolute humidity

h_g = heat content of 1 kg water vapour at temperature 't' can be found out from a steam table

Adiabatic process : In an adiabatic process there is change from one condition to other condition without transfer of heat between system and atmosphere or surrounding during the entire process.

For example, the room cooler which employs evaporation cooling can be taken as an adiabatic process, assuming there is no transfer of heat from cooler body to the atmosphere. Under such condition the cooler body is considered to be perfectly insulated.

Following are the common uses of the psychrometric chart in drying problem analysis.

1. *Sensible heating and cooling* : During sensible heating and cooling process of grains there is no change in the value of humidity ratios. The heating is given to dry air or cooling is done with the help of a heat exchanger. The action of sensible heating and cooling is shown in Fig. 3.9. Due to this action, change in the values of dry and wet bulb temperature, enthalpy, specific volume and relative humidity occurs. But change in the humidity ratio, dew point temperature and vapour pressure of moist air does not occur.

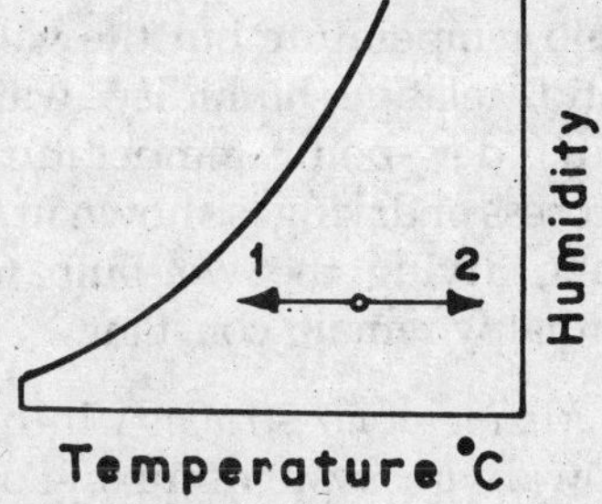

Fig. 3.9 : Sensible heating and cooling

2. *Heating with humidifying* : The heating with humidifying action of moist air is shown in Fig. 3.10. In many heated air drying systems, energy is transported to air by direct ignition of fuel. In such operation not only heat is given to drying air but in small amount water vapour is also added to the drying air. In heating with humidifying operation, increase in the enthalpy, humidity ratio, water-vapour pressure, dry and wet bulb temperature, dew-point temperature and specific volume of drying air occurs. The heating of air results in lowering of relative humidity of the air. The change in the value of relative humidity can be estimated on the basis of amount of energy and water-vapour provided to air.

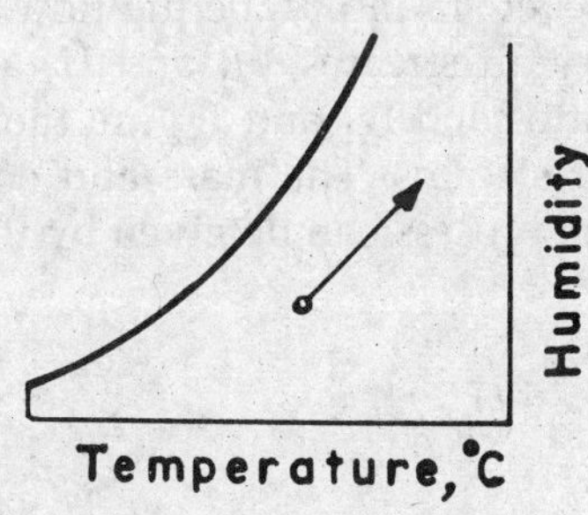

Fig. 3.10 : Heating with humidifying

3. *Cooling with dehumidifying* : For cooling of grains, the air is cooled below the dew-point temperature. Since at dew-point temperature the air is saturated with water vapour, therefore, condensation of water occurs when the temperature is further lowered. In this process decrease in the values of relative humidity, dry and wet bulb temperature, enthalpy and specific volume of air occurs. The process is shown in Fig. 3.11.

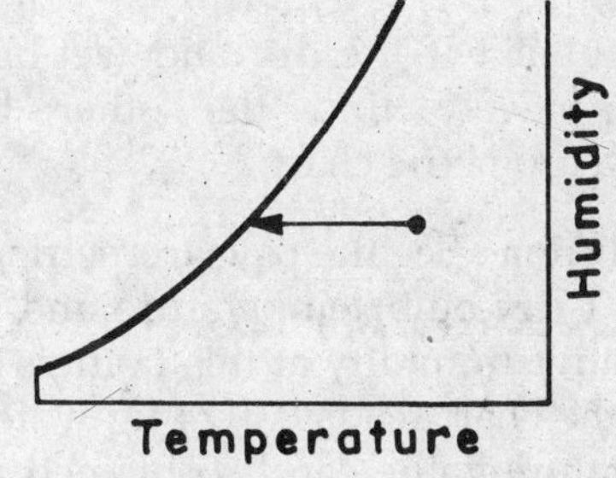

Fig. 3.11 : cooling with humidifying

4. *Drying* : The drying of agricultural products in a columnar dryer can be taken as an adiabatic process. Hence for moisture evaporation, the necessary energy is provided through drying air only. No amount of energy in the form of conduction or radiation is transmitted to product from atmosphere. As the drying air passes through the moist product, its sensible heat is converted into the latent heat.

Therefore, increase in the water- vapour content of air takes place. In the adiabatic drying process there is decrease in the value of dry-bulb temperature but the values of humidity ratio, relative humidity, water-vapour pressure, dew-point temperature increase. The process of drying is shown in Fig. 3.12. In adiabatic drying the wet-bulb temperature and enthalpy remain constant.

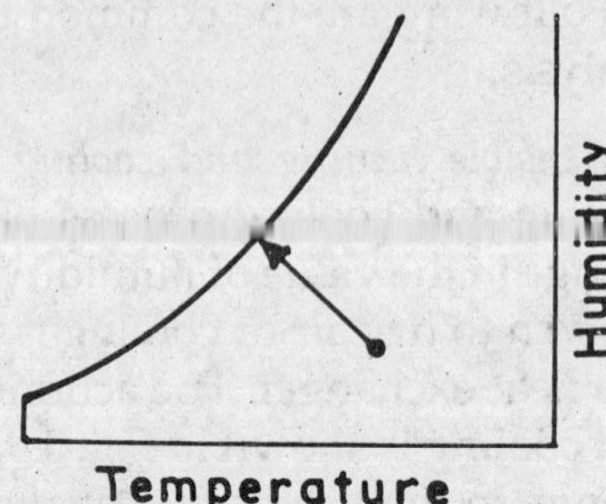

Fig. 3.12 : Drying process

5. *Mixing of air streams :* In many continuous flow drying systems mixing of two air streams of different conditions takes place. There is difference in the air flow rates, temperatures and humidity ratios of the two streams. The condition of mixed air can be found out from the psychrometric chart. Mixing of air streams process is shown in Fig. 3.13.

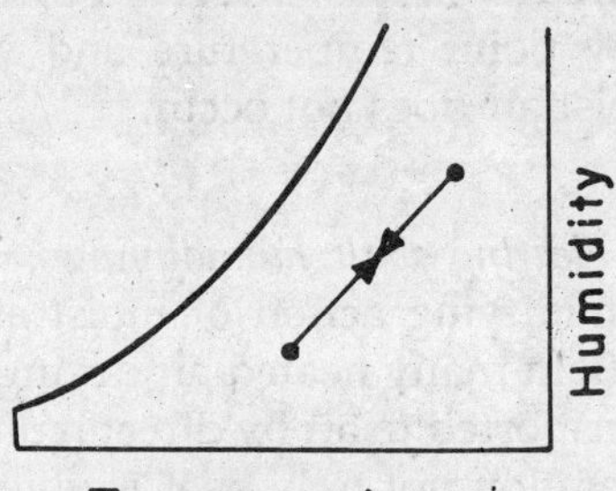

Fig. 3.13 : Mixing of two air streams

Let m_1 and m_2 be the flow rates of the two dry air streams; H_1 and H_2 are the humidity ratios and Ig_1 and Ig_2 are the enthalpy of the air streams. The mass and energy balance of the process can be given by the following equations

$$m_1 + m_2 = m_3 \qquad ...3.23$$

$$m_1H_1 + m_2H_2 = m_3 H_3 \qquad ...3.24$$

and

$$m_1Ig_1 + m_2Ig_2 = m_3Ig_3 \qquad ...3.25$$

or,

$$H_3 = \frac{m_1 H_1 + m_2 H_2}{m_1 + m_2} \qquad ..3.26$$

and

$$Ig_3 = \frac{m_1 Ig_1 + m_2Ig_2}{m_1 + m_2} \qquad ...3.27$$

Problem : If the dry and wet-bulb temperatures of moist air are 35° and 26°C respectively find the other thermodynamic properties of air from the psychrometric chart.

Solution : See the psychrometric chart Fig. 3.7. Locate the point of intersection of 35°C dry-bulb temperature and 26°C wet-bulb temperature lines. The value of relative humidity at this point is 50%. A constant humidity ratio line at right end of the chart reads 0.0174 kg water-vapour per kg of dry air. The dew-point temperature value can be obtained at the left end of the chart, which is 23°C. The enthalpy of air is 19.2 k cal/kg of dry air. The humid volume of air is 0.897 m^3/kg dry air.

Problem : Paddy is being dried in a bin. The dry and wet- bulb temperatures of ambient air are 30° and 22°C respectively. The relative humidity of exit air is 80%. If the heated air temperature is 50°C and its flow rate is 30 m^3/min, calculate the amount of sensible heat required to raise the temperature of air. Also determine the water vapour removed per hour in the drying operation.

Solution : First determine the condition of air by psychrometric chart. At 30°C dry-bulb and 22°C wet-bulb temperatures the relative humidity of air is 50%. The humid volume is 0.877 m^3/kg or dry air. Therefore, the mass flow rate of air is:

$$\frac{30 \times 60}{0.877} = 2052.4 \text{ kg air/hr}$$

This air is being heated to 50°C at constant humidity ratio line at 50°C, this point is '2'. The enthalpy of air at points 1 and 2 are 15.4 and 20.3 k cal/kg dry air respectively. Therefore, the amount of heat required in drying:

$$2052.4 \times (20.3 - 15.4) = 10056.7 \text{ kcal/hr}$$

When the air leaves the bin, the relative humidity of air becomes 80%. Assuming drying is taking place in adiabatic condition, i.e., along wet-bulb temperature line. The condition of exit air is at point '3' where wet-bulb temperature line meets line of 80% relative humidity. At points 2 and 3 the humidity ratios are 0.013 and 0.0218 kg water vapour/kg dry-air respectively.

Increase in humidity ratio of air = 0.0218 – 0.013

= 0.0088 kg water–vapour per kg of dry air

∴ when 30m^3/min air flows it picks up moisture at the rate of

$$\frac{30 \times 60}{0.877} \times 0.0088 = 18.06 \text{ kg water vapour/hr.}$$

Drying theory

Drying is a convection process in which moisture from a product is removed. Dehydration is a process in which products are dried to very low moisture levels. It is necessary to understand the nature of air and composition of products to know the drying process.

The atmospheric air surrounding the grains is a mixture of different gases. It contains mainly oxygen 20.95%, nitrogen 78.09%, carbondioxide 0.03% and argon 0.93%. Air also contains water-vapour. Water vapour creates pressure which is dependent on the amount present in air.

The composition of food grains is mainly dry matter i.e., starch and moisture. The water content of agricultural products is given in terms of moisture content. At the time of harvesting moisture content of products is high. By nature, food grains are hygroscopic. They gain or loose moisture as per the atmospheric conditions. The moisture migration into or from a product is dependent on the difference of vapour pressure between atmosphere and product. If the vapour pressure of grain is greater than atmospheric vapour pressure, transfer of moisture from grain to atmosphere takes place. But when the atmospheric vapour pressure is greater than grain vapour pressure, grain absorbs moisture from atmosphere.

Moisture remains inside any hygroscopic product in two forms; (1) bound and (2) unbound. In bound moisture, the water-vapour pressure of grain is lower than the vapour pressure of free water surface. The highest bound moisture content is reached when the hygroscopic product is kept under 100% relative humidity condition. The unbound moisture content is more than the bound moisture content.

Methods of drying

There are many methods to develop water-vapour pressure difference between product and atmosphere. For drying of any product, the water vapour pressure of product can be increased or the air water-vapour pressure is decreased, or both can be achieved simultaneously. For increasing the water vapour pressure of grains, the methods mostly followed are; (1) sun or solar drying, (2) conduction drying, (3) infra-red drying. The methods usually employed to lower the water vapour pressure of air are; (1) heated air drying, (2) desiccated air drying, (3) refrigerated air drying.

Drying rate periods

The convection drying process can be divided into three periods. The drying rate is a function of product moisture content. Drying period rates are shown in Fig. 3.14a. The period 'C' shown in Fig 3.14b is called as constant drying rate. In

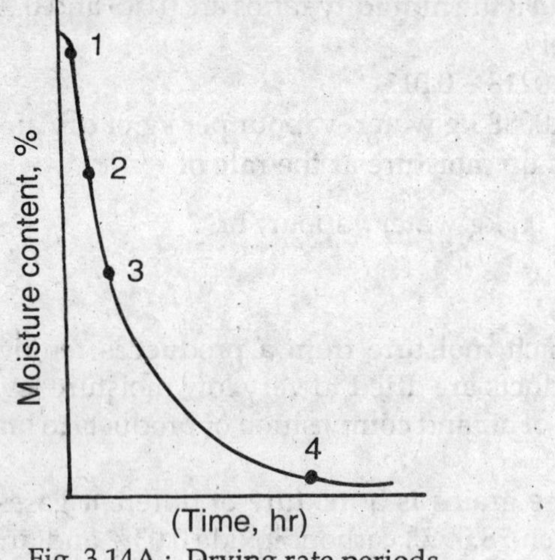

Fig. 3.14A : Drying rate periods
1–2c; 2 – 3

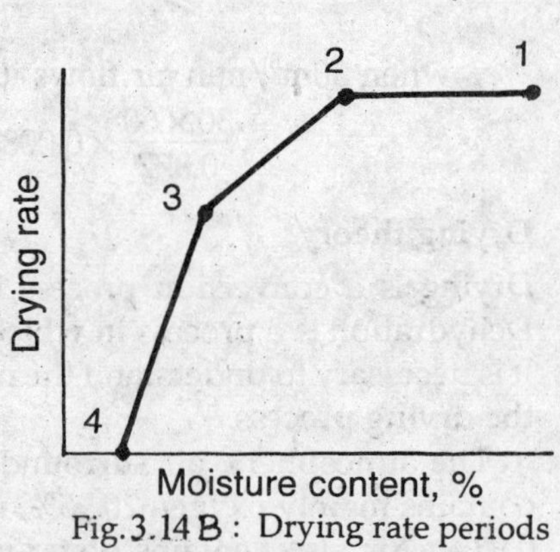

Fig. 3.14 B : Drying rate periods
F_1; 3 – 4 F_2

This period the moist surface of product behaves like a free water surface. This period continues till the rate of evaporation of water from the surface is equal to the rate of moisture migration from interior of product to its surface. The constant drying period continues till critical moisture content is reached. The critical moisture content is dependent on product and its thickness.

The period 'F_1' is known as the first falling rate period. In this period, because of decrease in wet surface area the drying rate decreases. As drying proceeds, the fraction of wet surface decreases to zero, where period F_1 ends.

The period 'F_2' is called the second falling rate period. In this period, subsurface evaporation takes place and it continues until the equilibrium moisture content is reached.

During the drying process, movement of moisture inside a porous product takes place due to; (1) capillary flow, (2) liquid diffusion, (3) surface diffusion, (4) vapour diffusion, (5) thermal diffusion and (6) hydrodynamic flow.

Removal of moisture greatly depends upon the atmospheric condition and not on type of product. During constant period rate, moisture migration rate from inside of product to its surface is equal to the rate of evaporation of water from surface. Drying of sand and washed seeds takes place in constant rate period. But for agricultural products, constant rate period is for very short period and depends on the following factors.

(*i*) Uncovered area

(*ii*) Difference in humidity levels between wet surface and air

(*iii*) Mass transfer coefficient

(*iv*) Drying air velocity

Constant rate period relationship can be given by the following equation

$$\frac{dw}{d\theta} = W_v A (p_s - p_a) = \frac{h_{fa} A (t_a - t_s)}{\lambda_s} \qquad \ldots 3.28$$

where, $\frac{dw}{d\theta}$ = constant period rate drying, kg/hr

p_a, p_s = water–vapour pressure at temperatures t_a and t_s respectively, kg/m^2

t_a, t_s = respectively air and water temperature, °C

h_{fa} = air film heat conduction, kcal/hr–m^2–°C

λ_s = latent heat of water at t_s, k cal/kg

A = water surface area, m^2

W_v = water–vapour transfer coefficient, kg/hr–m^2

After the constant rate, the drying proceeds in falling rate period. In between the constant and falling rate periods lies the critical moisture content of the product. The moisture content of product from where constant rate drying ceases and falling rate drying starts is called as the critical moisture content. Most of the agricultural products are dried in falling rate drying period, and in this condition the movement and diffusion of moisture in interior of grains controls the entire drying process. Since the constant rate is for very short period, it is of less importance for research studies. The critical moisture content of wheat is between 69 to 85% (*db*). The falling rate period is controlled by; (1) migration of moisture from interior of grains to upper surface due to water vapour diffusion and (2) removal of moisture from the surface. The falling rate period is usually divided into two, (1) unsaturated surface drying, and (2) drying, when moisture migration from interior of product becomes slow. These drying conditions are also called as first and second falling rate drying periods respectively. Many mechanisms have been proposed to describe the drying in capillary porous products. They are, (1) liquid movement due to difference in moisture concentration (liquid diffusion), (2) vapour movement due to moisture concentration differences (vapour diffusion), (3) liquid movement due to moisture diffusion on the pore surfaces (surface diffusion), (4) liquid movement due to surface forces (capillary flow), (5) vapour movement due to temperature differences (thermal diffusion) and (6) water and vapour movement due to total pressure differences (hydrodynamic flow).

Many investigations on drying process have been conducted based on the theories of moisture migration/diffusion through capillaries of grains. Luikov and other research workers (1966) have developed the following mathematical models for the capillary porous products in partial differential equation form:

$$\frac{\partial M}{\partial \theta} = \nabla^2 K_{11} M + \nabla^2 K_{12} t + \nabla^2 K_{13} P \qquad \ldots 3.29$$

$$\frac{\partial t}{\partial \theta} = \nabla^2 K_{21} M + \nabla^2 K_{22} t + \nabla^2 K_{23} P \qquad \ldots 3.30$$

$$\frac{\partial P}{\partial \theta} = \nabla^2 K_{31} M + \nabla^2 K_{32} + + \nabla^2 K_{33} P \quad \text{...3.31}$$

The K_{11}, K_{22} and K_{33} are the phenomenological coefficients, whereas the other K-values represent the coupling coefficients. Very limited number of the phenomenological coefficients are known, therefore, the Luikov equation become in simplified from:

$$\frac{\partial M}{\partial \theta} = \nabla^2 K_{11} M \quad \text{...3.32}$$

$$\frac{\partial t}{\partial \theta} = \nabla^2 K_{22} t \quad \text{...3.33}$$

$M =$ moisture content, % (db); $\theta =$ time, hr. and $t =$ temperature, °C

The moisture migration within a grain kernel is mainly because of liquid and/or vapour diffusion, the transfer coefficient K_{11} of equation 3.33 is replaced by diffusion coefficient D_v, and we get the following equation

$$\frac{\partial M}{\partial \theta} = D_v \left[\frac{\partial^2 M}{\partial r^2} + \frac{c}{r} \frac{\partial M}{\partial r} \right] \quad \text{...3.34}$$

where, $c = 0$ for planar symmetry

$c = 1$ for cylindrical body

and, $c = 2$ for spheres

If the boundary conditions are taken:

$M(r, 0) = M_0$ (initial moisture content) and $M(r_0, \theta) = M_e$ [EMC], then the equation becomes,

$$\frac{M - M_e}{M_0 - M_e} = \frac{8}{\pi^2} \sum_{n=0}^{\infty} \frac{1}{(2n+1)^2} \exp\left[-\frac{(2n+1)^2 \pi^2}{4} X^2 \right] \quad \text{...3.35}$$

for planar symmetry.

$$\frac{M - M_e}{M_0 - M_e} = \frac{6}{\pi^2} \sum_{n=0}^{\infty} \frac{1}{n^2} \exp\left[-\frac{n^2 \pi^2}{9} X^2 \right] \quad \text{...3.36}$$

for sphere.

$$\frac{M - M_e}{M_o - M_e} = \sum_{n=1}^{\infty} \frac{4}{\lambda_n^2} \exp\left[-\frac{\lambda_n}{4} X^2 \right] \quad \text{...3.37}$$

for infinite cylinder.

Where, $\lambda_n =$ zero order Bessel function's roots.

A simple drying equation was proposed on the basis of Newton's equation as,

$$\frac{dt}{d\theta} = -K(t - t_e) \quad \text{...3.38}$$

In the equation 3.38 by replacing the temperature unit 't' by moisture content unit 'M', we get the following equation

$$\frac{dM}{d\theta} = -K(M - M_e) \quad \text{...3.39}$$

where, $M =$ moisture content, % (db)

$\theta =$ time, hour

$M_e = EMC$, % (db)

$K =$ drying constant

or, $$\frac{dM}{M - M_e} = -K\,d\theta \qquad ...3.40$$

By integrating between limits $M(r, 0) = M$, initial moisture content and $M(r_0, \theta) = M_e$ (EMC) we get;

$$\frac{M - M_e}{M_0 - M_e} = \exp\ (-k\,\theta) \qquad ...3.41$$

or, $$\theta = \frac{1}{k}\ ln\ \frac{M - M_e}{M_0 - M_e} \qquad ...3.42$$

$\dfrac{M - M_e}{M_0 - M_e}$ is called as moisture ratio

The equation 3.41 can be equated as per equations 3.35 and 3.36, if we assume that the solids-liquid diffusion takes place due to the difference in saturation levels;

$$\frac{\partial c}{\partial \theta} = D_v \frac{\partial^2 c}{\partial x^2} \qquad ...3.43$$

where, D_v = diffusivity, m²/hr
c = saturation
θ = time, hour
x = distance from the centre of product, m

By showing the saturation term with moisture constant term

$$\frac{\partial M}{\partial \theta} = D_v \frac{\partial^2 M}{\partial x^2} \qquad ...3.44$$

For a spherical body, we get the following equation

$$\frac{M - M_e}{M_0 - M_e} = \frac{6}{\pi^2}\left[\exp\left(-D_v\theta\frac{\pi^2}{r^2}\right) + \frac{1}{9}\exp\left(-9\,D_v\theta\frac{\pi^2}{r^2}\right) + \ldots\right] \qquad ...3.45$$

where, r = radius of sphere, m

If the moisture ratio versus time is plotted on a semi-log paper, curves as shown in Fig 3.15 is received. The curve portion of the graph is because of second, third and other higher terms.

In a drying process, with increase in time beyond the first term of equation 3.45 other terms approach to zero. Hence after leaving the second, third and onward terms the following equation is obtained.

$$\frac{M - M_e}{M_0 - M_e} = B\exp(-k\theta) \qquad ...3.46$$

where, $K = D_v \dfrac{\pi^2}{r^2}$ (for spheres)

and $B = \dfrac{6}{\pi^2}$ (shape factor)

Single Kernel drying

Diffusion mass transfer is similar to heat conduction within the solid. The following equation is applicable for diffusion mass transfer;

$$\frac{\partial G}{\partial \theta} = -D_v \rho A\left(\frac{\partial M}{\partial x}\right) \qquad ...3.47$$

where, G = mass of water, kg

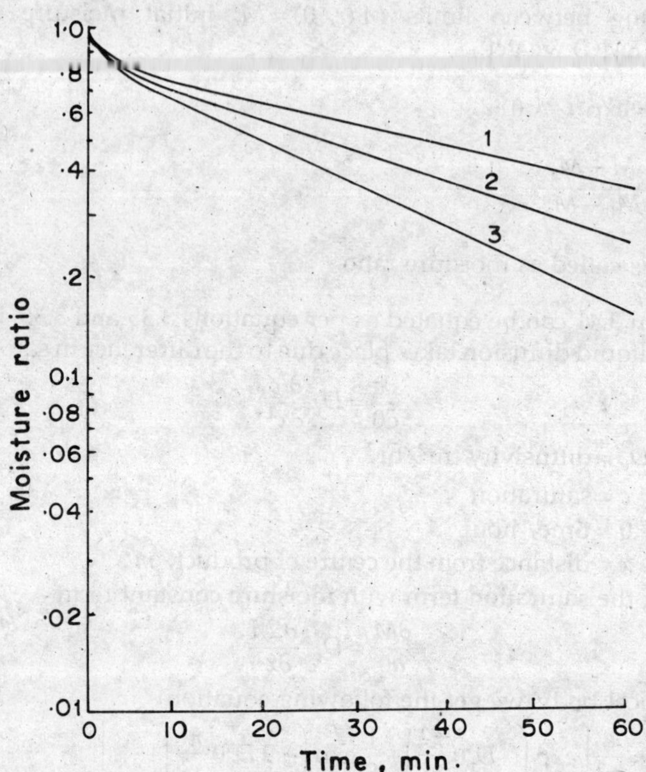

Fig. 3.15 : Moisture ratio vs. time curve
1–50°C, 2–60°C, 3–70°C

θ = time, hour
D_v = mass diffusivity, m^2/hr
ρ = dry solid density, kg/m^3
A = surface area, m^2
M = moisture content, in decimal, kg/kg
x = distance from centre of the mass, m

The moisture movement on the surface of solid is dependent on the moisture concentration gradient between the drying air and the surface of the solid. The following equation is applicable,

$$-D_v \rho \left(\frac{\partial M}{\partial x}\right) = S\,(M_s - M_e)\,\rho \qquad \ldots 3.48$$

and,

$$\frac{\partial M}{\partial \theta} = D_v \left(\frac{\partial^2 M}{\partial x^2}\right) \qquad \ldots 3.49$$

where, S = surface conductance, m/hr
M_s = moisture content at surface, % (db)
M_e = EMC, % (db) at the relative humidity of drying air

Heat transfer in grain drying

There are four types of heat flows between the kernel and the surroundings in grain drying process. The types of heat flows are, (1) latent heat due to moisture evaporation (q_l), (2) sensible heat due to convection heat transfer (q_s), (3) sensible heat due to radiation heat transfer (q_r) and (4) sensible heat due to conduction heat transfer (q_c). These heat flows are shown in Fig. 3.16.

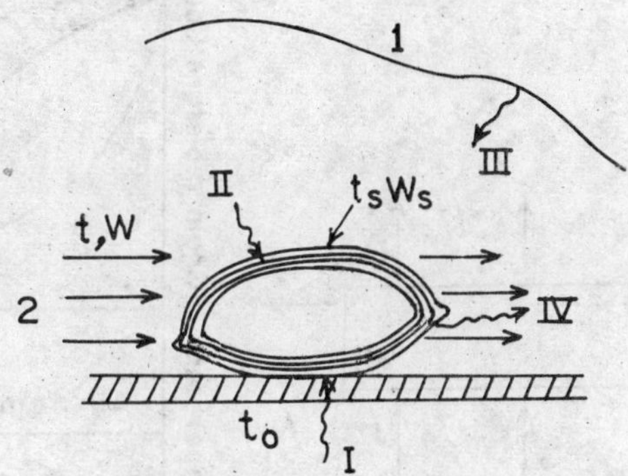

Fig. 3.16 : Single kernel drying
1. atmospheric surface 2. air flow
I conduction heat, II convection heat, III radiation heat, IV latent heat

The above heat flows are represented by,

$$q_l = q_s + q_r + q_c \qquad \text{...3.50}$$

Heat balance for a unit area of solid surface gives the following equation

$$h_d (w_s - w) h_{fg} = h_c (t - t_s) + h_r (t_r - t_s) + U (t_o - t_s) \qquad \text{...3.51}$$

where, w_s = humidity ratio of saturated air at given surface
w = humidity ratio of drying air
t_s = temperature of grain surface
t_r = temperature of surrounding surface
t_o = atmospheric temperature
h_d = mass transfer coefficient
h_r = radiation heat transfer coefficient
U = overall heat transfer coefficient
h_{fg} = latent heat at the temperature of water vaporisation

Assuming the effects of radiation and conduction as negligible in convection drying, the equation becomes

$$(w_s - w) h_{fg} = \left(\frac{h_c}{h_d c_p}\right) c_p (t - t_s)$$
$$= L_c c_p (t - t_s) \qquad \text{...3.52}$$

where, L_c = Lewis number
c_p = Specific heat

In adiabatic process the Lewis number is a unity and t_s equals the wet-bulb temperature.

Types of air flow in mechanical drying systems

There are generally three types of air and grain flow in mechanical drying systems, (1) cross flow, (2) concurrent flow and (3) counter flow. In Figs. 3.17 to 3.19 the nature of the three types of air and grain flow are shown.

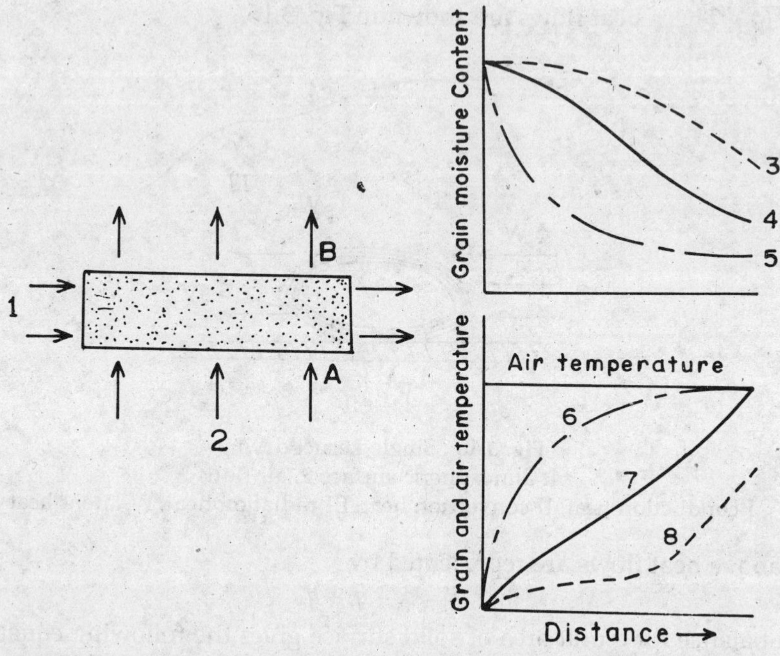

Fig. 3.17 : Moisture and temperature changes in cross flow dryers
1. grain flow 2. air flow 3. moisture at 'A' 4. mean moisture 5. moisture at 'B'
6. grain temperature at 'A' 7. grain's mean temperature 8. grain temperature at 'B'

In cross flow dryers, the flow of grain and drying air takes place at 90°, because of this the layer of grains which come in contact with heated air dries early. The temperature of grain layer nearest to flow of heated air approaches near the drying air temperature. But, the grain layer which is at farther point remains comparatively cooler, and never approaches near the temperature of drying air, therefore, it is necessary to mix the grains for uniform drying.

In concurrent flow dryers, the direction of flow of grains and heated air is parallel to each other. The cool and wet grains meet the heated air first. Therefore, during evaporation of moisture the sensible heat for drying is adsorbed as latent heat, as a result cooling of air takes place and this restricts the elevation of grain temperature to a level of drying air temperature. Thus, in concurrent flow dryers the temperature of heated air can be kept at higher levels than safe temperature limit.

In counter flow dryers, the direction of flow of grains and heated air is opposite to each other, thus the hottest grains come in contact with heated air. Since, at this point drying is almost complete, therefore, very little amount of latent heat is adsorbed for evaporation of moisture. The temperature of grain approaches the

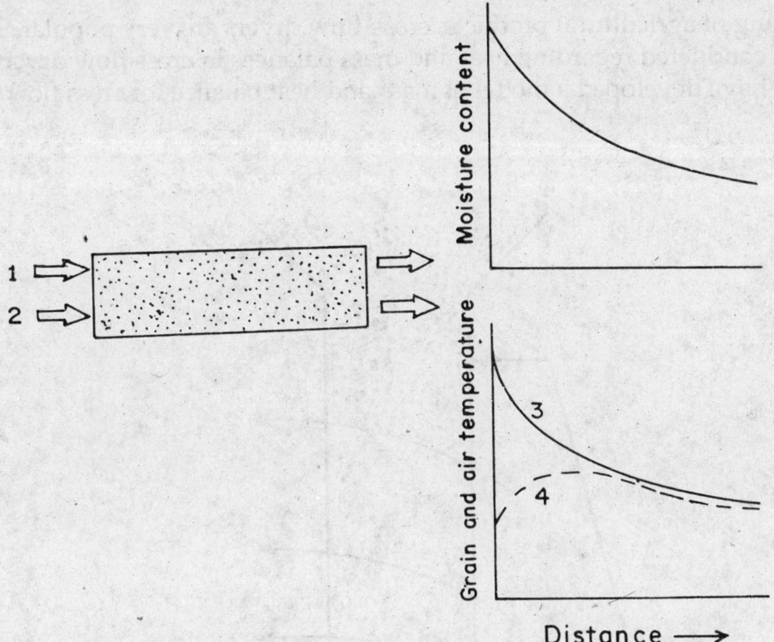

Fig. 3.18 : Moisture and temperature changes in concurrent flow dryers
1. grain flow 2. air flow 3. air temperature 4. grain temperature

temperature of incoming drying air, hence the drying air temperature should not be kept more than safe temperature for the product.

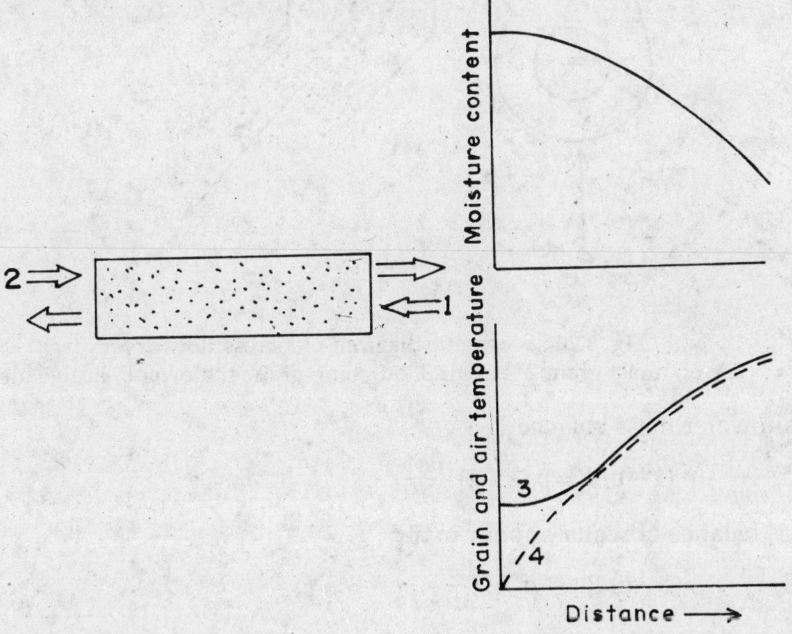

Fig. 3.19 : Moisture and temperature changes in counter-flow dryers
1. grain flow 2. air flow 3. air temperature 4. grain temperature

For drying of agricultural products, cross-flow dryers are very popular. Studies have been conducted regarding heat and mass balances in cross flow dryers. Platt etal (1991) have developed a model of mass and heat balance for cross flow dryers (Fig. 3.20).

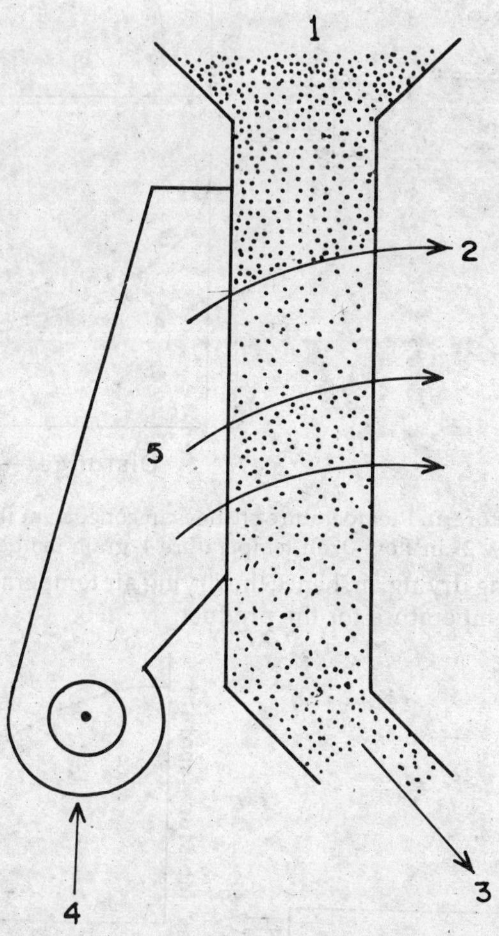

Fig. 3.20 : Schematic diagram of a cross flow dryer
1. wet/moist grain 2. exit air 3. outgoing grain 4. blower 5. heated air

Grain water mass balance

$$-G_p \frac{\partial M}{\partial y} - \rho\,(1-\varepsilon)\,R = \rho\,(1-\varepsilon)\,\frac{\partial M}{\partial t} \qquad \ldots 3.53$$

Mass balance of water-vapour in air

$$-G_a \frac{\partial w}{\partial x} + \rho\,(1-\varepsilon)\,R = -\rho_a\,\varepsilon\,\frac{\partial w}{\partial t} \qquad \ldots 3.54$$

Grain energy balance

$$-G_p\,c_{pg} \frac{\partial \theta}{\partial y} + h\,(T-\theta) - \rho\,(1-\varepsilon)\,R\,(H_w - h_w) = \rho\,(1-\varepsilon)\,c_{pg}\,\frac{\partial \theta}{\partial t} \qquad \ldots 3.55$$

Energy balance of air

$$-G_a c_{pa} \frac{\partial T}{\partial x} - h(T - \theta) = \rho_a \varepsilon c_{pa} \frac{\partial T}{\partial t} \quad \text{...3.56}$$

In above equations the values of c_{pg} and c_{pa} in the entire range of M, w, T and θ are assumed to be constant.

where, G_p = mass flow of dry grain, kg/s-m^2

G_a = mass flow of dry air, kg/s–m^2

M = moisture content, % (*db*)

t = time, *s*

R – drying rate, kg water/kg dry grain—*s*

ε = pore space, m^3 pores/m^3 total volume

ρ = density of dry grain, kg/m^3

ρ_a = density of dry air, kg/m^3

w = absolute humidity of air in dryer, kg water/kg dry air

c_{pg} = specific heat of wet grain, J/kg dry grain—°C

T = air temperature, °C

θ = grain temperature °C

C_{pa} = specific heat of humid air, J/kg dry air — °C

H_w = enthalpy of water–vapour, J/kg

h_w = enthalpy of water, J/kg

Mechanical drying types

Drying of agricultural products is broadly carried in thin and deep layers.

Thin-layer drying

The thin-layer drying process shows the condition of nearly complete exposure of grains to heated air. The thickness of grains in thin-layer drying is normally upto 15 cm. The drying action can be represented on the basis of Newton's law (equation 3.38) by replacing moisture content in place of temperature

$$\frac{dM}{d\theta} = -K\,(M - M_e) \quad \text{...3.57}$$

$$\frac{M - M_e}{M_o - M_e} = e^{-K\theta} \quad \text{...3.58}$$

where, M = moisture content at any time θ, % (*db*)

M_e = *EMC*, % (*db*)

M_o = initial moisture content, % (*db*)

K = drying constant

θ = time, hour

The drying rate in thin-layer drying is equal to the ratio of water vapour pressure driving force and resistance to drying. Therefore, the drying rate can be shown by the following equation

$$\frac{dM}{d\theta} = \frac{(p_g - p_a)}{(1/K_g a_m)} = K_g a_m (p_g - p_a) \quad \text{...3.59}$$

where, M = moisture content, % (*db*)

θ = time, hour

K_g = mass transfer coefficient, kg water/hr—m^2

a_m = effective area, m^2

p_g = grain vapour pressure, H_g mercury

p_a = air vapour pressure, H_g mercury

Deep bed drying

In deep bed dryers, the drying takes place in a drying zone and the layer of grains is more than 15 cm. A typical deep bed dryer is shown in Fig. 3.21. At heated or drying air entry point, drying of products starts. A drying zone is formed above the bottom layer which comes in contact of heated air. Bulk of the drying takes place in the drying zone and this moves along the direction of drying air. In deep bed system at a particular air flow rate, if the layer of product is more then the

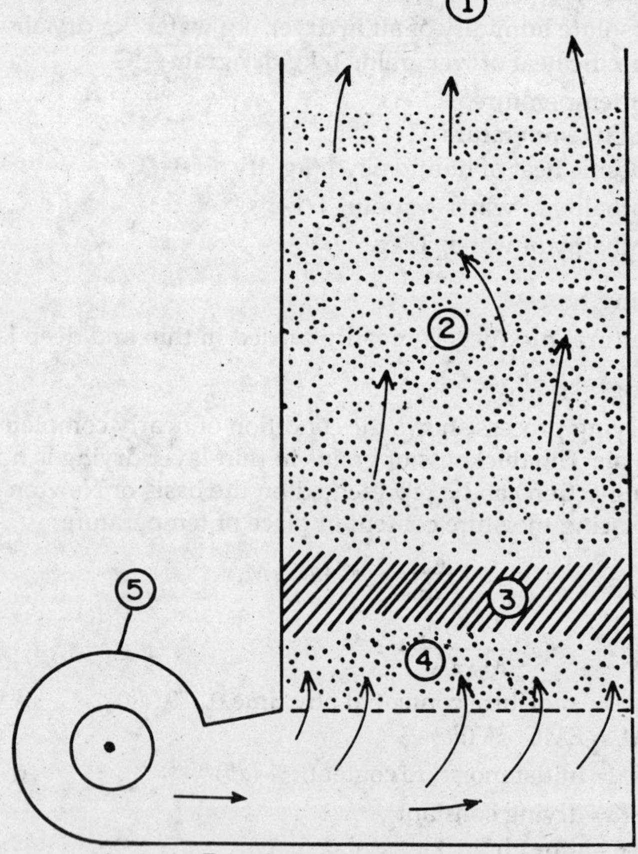

Fig. 3.21 : Deep bed dryer's schematic diagram
1. exit air 2. wet grain 3. drying zone 4. dry grain 5. blower

layer which comes in contact with heated air first, may overdry. Therefore, it has been recommended that if the drying air is 43°C, the thickness of grain layer may be limited to 45 cm. For analysis of deep bed drying method, it is assumed that the deep bed is formed by several thin layers. In each thin layer, the incoming and outgoing air humidity and temperature change with time and drying stages. This method is used to determine the drying period and the amount of water-vapour removed from the products. Additional amount of moisture may be removed from dry layer till it attains the equilibrium moisture content.

Deep bed drying system has been analysed by Hukill. The Hukill analysis is suitable for correlating the drying period, product moisture and product depth factors for the solution of drying problems.

Drying time : The time unit is the time of one-half response of grain fully exposed to the air entering the drying system.

Grain moisture : The grain moisture is given by the moisture ratio, $\frac{M - M_e}{M_o - M_e}$

Grain depth : A depth factor has been defined as "depth containing enough grain that if all the theoretical available heat could be used, it would all dry to equilibrium" in a period of one-half response time.

The amount of grain in each depth is calculated by using the following heat balance

$$\frac{dQ}{d\theta} \times c_p \times \Delta t = \Delta M \times \frac{dw}{d\theta} \times h_{fg} \qquad \text{...3.60}$$

where, $\frac{dQ}{d\theta}$ = air flow rate, kg/hr

c_p = specific heat of air, k cal/kg—°C

Δt = maximum temperature drop of the drying air, °C

$\Delta M = M_o - M_e$, % (db)

$\frac{dw}{d\theta}$ = moisture removal from the grain volume under consideration, kg/hr

h_{fg} = latent heat of drying, k cal/kg

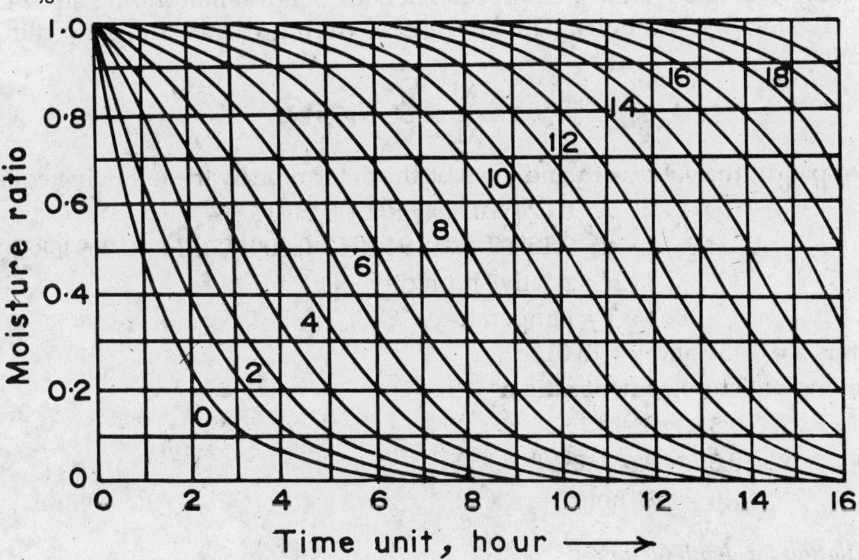

Fig. 3.22 : Depth factors for analysis of deep bed drying systems
0–18 are depth factors

The amount of grain in each depth factor $= \frac{dw}{d\theta} \times \theta_{1/2}$. For analysis of deep bed drying problems the depth factors are shown in Fig. 3.22.

Problem : A dryer whose floor area is 130 m^2 can hold 60 tonne paddy at a depth of 0.8 *m*. The ambient air is at 28°C and 80% relative humidity. The flow rate of air is 600 m^3/min. The air is being heated by a heater consuming 25 litre of mineral oil per hour. The calorific value of oil is 28000 kJ/l. If the initial moisture content of wet paddy is 21% (*wb*) then determine the moisture content of paddy at the bottom and top layers after 12 hours of drying. Also determine the average moisture content of paddy. Given that the specific heat of paddy 2800 kJ/kg—°C. Assume the *EMC* of paddy at 45°C and 32% rh as 8.6% (*db*).

Solution : From the psychrometric chart, the specific volume of air at 28°C and 80% relative humidity is 0.88 m^3/kg.

The mass flow rate of ambient air would be

$$\frac{600}{0.88} = 681.8 \text{ kg/min}$$

Amount of heat energy available from the mineral oil $= 28000 \times 25/60$

$= 11666.6$ kJ/min

Since the specific heat of air is 1.0 kJ/kg—°C, the temperature rise by the heater

$$= \frac{11666.6}{681.8} = 17.11°\text{C}$$

∴ The temperature of air entering the dryer $= 28 + 17.11$

$= 45.11$ °C

From the psychrometric chart, at first point and along the constant absolute humidity line at 45°C temperature the relative humidity of air is 32%. Now taking a thin layer just above the plenum chamber, the condition of drying air is 45.11°C and 32% relative humidity. The analysis of drying can be done by thin layer equation

$$MR = \frac{M - M_e}{M_o - M_e} = e^{(-x\,\theta)^y}$$

For paddy, the values of x and y can be the determine by the following equations

$$x = 0.026 - 0.0045\, RH + 0.01215\, T$$

and

$$y = 0.13362 + 0.0194\, RH - 0.000177\, RH + 0.009468\, T$$

RH = relative humidity, %

T = temperature, °C

Thus $x = 0.431$ and $y = 1.001$

Taking moisture content ratio as 0.5

$$0.5 = e^{-0.431\,\theta^{1.001}}$$

or, $\quad ln\ 0.5 = -0.431\ \theta^{1.001}$

or, $\quad \theta = 1.61$ hour

Calculation for depth factors

Initial moisture content of paddy is 21% (*wb*) or 26.6% (*db*). The *EMC* of paddy at 45°C and 32% rh = 8.6% (*db*)

Therefore the moisture content at half response time $= \dfrac{26.6 + 8.6}{2} = 17.6\%$

The latent heat for half response time

$$(0.266 - 0.176)\,DM \times 2800 = 252\,DM$$

$DM =$ dry matter in each depth factor

Amount of sensible heat per unit time,

$1.61\,(45.11 - 28) \times 60 \times 681.8 = 1126896.8$ kJ

Since the sensible heat and latent heat are equal

$$252\,DM = 1126896.8$$

or,

$$DM = 4471.8 \text{ kg}$$

Dry matter weight in dryer $= 60000\,(1 - 0.21)$

$= 47400$ kg

$\therefore$ depth factor $= \dfrac{47400}{4471.8} = 10.6$

Thickness of each depth factor $= \dfrac{0.8}{10.6} = 0.075$ m

time unit $= 1.61$ hr.

depth factor $= 0.075$ m

$\therefore$ time unit for 12 hours of drying $= \dfrac{12}{1.61} = 7.45$

Now from Fig. 3.22 at 7.45 time unit, we can find the values of moisture ratios at various depth factors. The values of moisture ratios at each depth factors and the calculated values of moisture contents are given in Table 3.6.

Table 3.6

Moisture ratios and corresponding moisture contents

Depth factor	*Thickness, m*	*Moisture ratio*	*Moisture Contant, % (db)*
0–1	0.075	0.01	8.78
2	0.150	0.03	9.14
3	0.225	0.05	9.50
4	0.300	0.08	10.04
5	0.375	0.16	11.48
6	0.450	0.27	13.46
7	0.525	0.42	16.16
8	0.600	0.58	19.04
9	0.675	0.75	22.10
10	0.750	0.85	23.90
11	0.825	0.93	25.34

The values of moisture contents of the above table are changed to wet basis. We have;

(*i*) Moisture content of paddy at bottom layer = 8.07%

(*ii*) Moisture content of paddy at top layer 20.21%

(*iii*) Average moisture content of paddy 13.3%

Problem : The diameter of a circular dryer is 3 m and paddy is kept in the dryer to a depth of 1.5 m for drying. Initial moisture content of paddy is 28% (db). Paddy is being dried by 35°C hot air. The relative humidity of air is 30% and flow rate is

1500 m^3/hr–tonne. If the half response time is 3 hours and the density of dry paddy is 580 kg/m^3, then determine the moisture content of paddy at any depth after 18 hours of drying. Also determine the amount of water vapour removed from the dryer after 18 hours. Given that the specific heat of air 0.24 kcal/kg—°C, latent heat of evaporation 661 kcal/kg.

Solution : Dry weight of paddy in the bin

$$= \frac{\pi}{4} \times 3 \times 3 \times 1.5 \times 580 = 6150 \text{ kg}$$

From the psychrometric chart, at air temperature of 35°C and 30% rh, the specific density is 0.887 m^3/kg

$$\therefore \quad \frac{dQ}{d\theta} = \frac{1500 \times 6150}{1000 \times 0.887} = 10400 \text{ kg/hr}$$

Since, $M_o = 28\% (db)$

$M_e = 9\%$ (db) [EMC of paddy at 30% rh and 35°C temperature]

$$\therefore \quad \Delta M = 28 - 9 = 19\%$$

The paddy is in equilibrium with 28°C at 95% rh. The wet- bulb temperature of air is 15°C (from psychrometric chart). The 15°C wet-bulb temperature line intersects 95% rh line at 22.3°C or 22°C dry-bulb temperature. Hence, the exit air temperature is 22°C.

$$\Delta t = 35 - 22 = 13°\text{C}$$

$$\frac{dQ}{d\theta} \times c_p \times \Delta t = \Delta M \times \frac{dw}{d\theta} \times h_{fg}$$

$$\therefore \quad \frac{dw}{d\theta} = \frac{10400 \times 0.24 \times 13}{(0.28 - 0.09) \times 661} = 258.36 \; kg/hr$$

$$\therefore \quad w = 258.36 \times 3 = 775 \text{ kg}$$

$$\text{Number of depth factors} = \frac{6150}{775} = 7.938$$

$$\text{Thickness of each depth factor} = \frac{1.5}{7.938} = 0.188 \text{ m}$$

$$\text{Time unit} = \frac{18}{3} = 6.$$

The moisture ratios at various depth factors at 6 hours unit can be found from the Fig. 3.22. From the values of moisture ratio, we can calculate the corresponding moisture content. At 0.5 m from the bottom of bin the moisture content of paddy will be—

$$\text{depth factor,} \quad \frac{0.5}{0.188} = 2.6$$

Now the value of moisture ratio at time unit '6' and at 2.6 depth factor is 0.085

$$0.085 = \frac{M - 0.09}{0.28 - 0.09}$$

or, $$M = 10.061\% \; (db)$$

The depth factor at 0.75 m height from the bottom

$$= \frac{0.75}{0.188} = 3.989$$

The values of moisture ratio at '6' time unit and at 3.989 depth factor is 0.2.

$$\therefore \quad 0.2 = \frac{M - 0.09}{0.28 - 0.09}$$

or, $\quad M = 12.8\% \ (db)$

(*ii*) Amount of moisture removed from the paddy

We can find out the values of moisture ratios at various depth factors, and by multiplying the values of moisture contents at each depth factor by the dry matter weight, we can know the amount of moisture removed. The amount of moisture removed at each depth factor is tabulated below.

Depth factor	*Moisture Content × dry matter weight*	*Amount of moisture in paddy, kg*
1	0.0936 × 775	72.54
2	0.1014 × 775	78.58
3	0.1128 × 775	87.42
4	0.1318 × 775	102.14
5	0.1584 × 775	122.76
6	0.1888 × 775	146.32
7	0.2192 × 775	169.88
8	0.2477 × 775	191.96

The amount of moisture present in the paddy after drying = 971.6 kg

Amount of moisture initially present in paddy

$$= 6150 \times 0.28 = 1722 \text{ kg}$$

∴ The amount of moisture removed in drying

$$= 1722 - 971.6 = 750.4 \text{ kg}$$

Dryer performance

The performance of any dryer depends upon the design of overall drying system, plant maintenance and its operational method. The desired performance objectives of the drying systems are;

(1) The grain quality must be preserved.
(2) The grain should be uniformly dried.
(3) The grain should dry fast enough to arrest moulding and germination.
(4) The dryer should be efficient in utilisation. The drying potential of the heated air should be maximised.

To evaluate the performance of dryer, the design specifications should be known or estimated and compared with measured indicators given below:

(1) Static pressure at the plenum chamber, this gives an indication of air delivery of the fan and can be roughly cross-checked with an air-velometer.
(2) The moisture content of grain at the top and bottom layers, and at various points in the dryer bed area to monitor drying progress and its uniformity.
(3) The temperature and relative humidity of exit air, this gives an indication of utilisation of drying potential of heated air.
(4) The blower speed.
(5) The amount of fuel used by the burner or energy consumption of the heater. This gives an indication about efficiency of fuel combustion.
(6) Dry-bulb temperature and relative humidity of the ambient air.
(7) The actual drying period.
(8) The milling quality of dried products.

If the drying system is efficiently designed, the drying of products will be faster and uniform. Drying of products will take place within desirable time and quality products will be obtainable. Overall thermal efficiency of drying system can be expressed in the following manner.

$$\text{Overall thermal efficiency} = \frac{\text{Amount of heat utilised}}{\text{Amount of heat available}}$$

The overall thermal efficiency of a drying system can be estimated by the following equation.

$$\text{Overall thermal efficiency, \%} = \frac{\text{(Amount of water vapour removed, kg)} \times \text{(Latent heat of evaporation, kcal/kg)}}{\text{(Amount of fuel, kg)} \times \text{(Net calorific value of fuel, kcal/kg)}} \times 100 \quad ...3.61$$

Heat utilisation factor : The utilisation factor of a drying system is the ratio of drop in temperature of drying air by drying process and the increase in the temperature of ambient air by heating

$$\text{Heat utilisation factor} = \frac{\text{Drop in dry bulb temperature of drying air, °C}}{\text{Increase in dry bulb temperature of ambient air, °C}} = \frac{t_2 - t_3}{t_2 - t_1} \quad ...3.62$$

where, t_1 = dry-bulb temperature of ambient air, °C

t_2 = dry–bulb temperature of heated air, °C

t_3 = dry–bulb temperature of exit air, °C

$$\text{Coefficient of performance } \frac{t_3 - t_1}{t_2 - t_1} \quad ...3.63$$

Heat utilisation factor = 1 – coefficient of performance

$$\text{Effective heat efficiency} = \frac{t_2 - t_3}{t_2 - t_{w_2}}$$

Where, t_{w_2} = wet–bulb temperature of the drying air, °C

For calculation of effective heat efficiency of the drying system, the sensible heat of heated air is taken into consideration.

Problem : In a dryer 4 tonnes sorghum (final weight) is dried from 18% (*db*) moisture content to 13 % (*db*) moisture content per hour. Calculate the heat requirement for drying and the necessary air flow rate for the drying system. Given that—

Dryer efficiency = 50%

Latent heat of evaporation = 574 k cal/kg

Specific heat of air = 0.24 k cal/kg—°C

Drying air temperature = 44°C

Ambient air condition = 25°C and 20°C dry and wet bulb temperature respectively

Solution : Dry matter weight of sorghum after drying operation = 4000 (1 – 0.13) = 3480 kg

$$\text{Initial weight of wet sorghum} = \frac{3480}{1-0.18}$$
$$= 4244 \text{ kg}$$

$$\therefore \text{ Amount of moisture to be removed } = 4244 - 4000$$
$$= 244 \text{ kg}$$

$$\text{Heat required to remove moisture} = \frac{244 \times 574}{0.5}$$
$$= 280112 \text{ k cal/hr}$$

From the psychrometric chart, at 25°C dry-bulb temperature and 20°C wet-bulb temperature, specific volume of air = 0.866 m³/kg

$$\text{Amount of air needed} = \frac{280112}{\Delta t \times c_p} = \frac{280112}{(44-25) \times 0.24}$$
$$= 61428 \text{ kg/hr.}$$

$$\therefore \text{ Volumetric air flow rate} = 0.866 \times 61428 = 53196.6 \text{ m}^3/\text{hr}$$
$$= 886.6 \text{ m}^3/\text{min.}$$

Drying methods

The agricultural product drying methods can be broadly grouped into two, (1) sun/solar drying, and (2) artificial drying with mechanical means.

Sun drying : This is a traditional method of drying of crops and grains and probably being followed by farmers since man has developed art of cultivation. Still, in India, major portion of crops is left in the field and threshing yard for drying under sun.

Sun is a very large nuclear fusion reactor which converts 40 lakh tonnes of hydrogen into helium in one second. Although earth absorbs a minute portion of sun energy, but the amount of energy received is approximately of 5.4×10^{24} Joule/year. This amount of energy is equivalent to 27,000 times the total energy produced by man.

The radiation of heat from sun is through electro magnetic waves. The different forms of radiation is due to length of these electro-magnetic waves. The wave length of heat radiation may vary from zero to infinity, but the majority of heat radiation lies between 0.1 to 100 μm. Sun's radiation corresponds closely to that of a black body at 6000°K. About half of the energy is in the visible part of the spectrum and most of the remaining energy is in shortwave infra-red region below 35000 A.U. Sun drying of agricultural crops and grains are associated with the following features.

(1) Uncontrolled and non-uniform drying results in cracks in kernels. Such grains yield large quantity of brokens during milling.
(2) The process is dependent on the availability of the sun energy, hence the crop harvested during monsoon season can not be dried.
(3) It requires a large number of unskilled labourers.
(4) About 1-2% grain is lost due to birds, insects and rodent attack.
(5) It requires no fuel or mechanical energy, drying cost is low as compared to mechanical drying.
(6) It requires very large area for drying.

Methods of sun drying

(1) *Drying of standing crops* : Grains are dried on the plant till proper moisture content is attained. Then, the crop is harvested and threshed. But considerable quantity is lost due to shattering. This drying process is slow and takes about 2-3 weeks after the grains have attained biological maturity.

(2) *Drying of grains on stalk* : This is an improvement over the previous method. The crop is harvested at higher moisture content and is left in the field or on bunds of the field till it has dried to proper moisture content. Another improved method is to dry the harvested crop on racks. Crop is bundled near the ears and hung on a rope exposing to the sun.

(3) *Drying of threshed grains* : The harvested crop at a higher moisture content is threshed and the grains are spread on the floor in 1 to 3 cm thick layer (Fig. 3.23). It is continuously stirred manually till it has attained the desirable moisture content. In this process, the grains are subjected to uncontrolled drying causing cracks in the kernels.

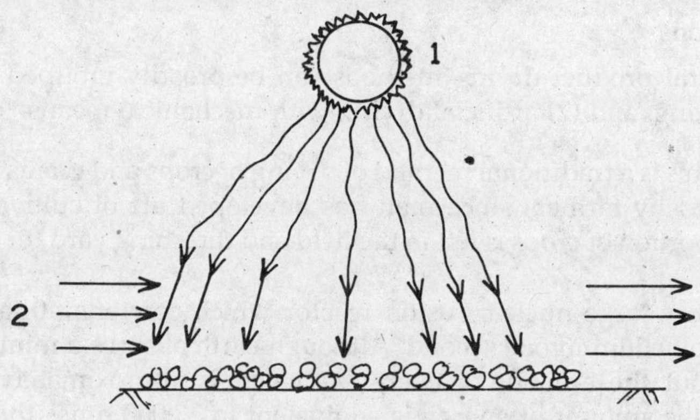

Fig. 3.23 : Sundrying of grains
1. sun 2. air

Mechanical drying

It is the process of utilising mechanical means for drying of grains by ventilating natural or heated air through the grain mass to accomplish removal of moisture from it. Mechanical drying system has the following features.

(1) Rate of drying of grains can be controlled by controlling the temperature of hot air ventilating through grain mass. Therefore, it is possible to reduce the temperature and moisture stresses developed during the drying process responsible for the cracks in the kernels to a minimum level by controlling the hot air temperature. This results in superior milling quality of grains.
(2) Grains can be dried in all seasons and also during night.
(3) The process is automatic and requires a small number of unskilled labourers, however, a trained person is required to operate the dryer.
(4) Losses due to insects, birds and rodents are avoided.
(5) The process requires fuel or electrical or mechanical power to drive the air and their accessories. Cost of drying per unit mass is comparatively higher than sundrying.

(6) The process requires little space for operation and is suitable for big mills.

(7) Mechanical drying in conjunction with early harvest improves the milling quality of grains.

The amount of moisture removed from the grain is dependent upon, (1) the moisture level and the temperature of the grain, (2) the temperature and relative humidity of the ventilating air, and (3) the quantity of air flowing around the grain.

Mechanical drying of grains is normally accomplished by passing air through grain mass. Based on the temperature of the air, drying methods are classified as, (1) unheated or natural air drying, (2) air drying with supplemented heating, and (3) heated air drying.

Natural air drying uses low humidity air. It is a slow process and is economical for in-storage drying of grains over a longer duration. Air drying with supplemented heating is used during high humidity with unheated air drying installation. A small amount of heat is added to air to decrease its humidity and thereby to increase its water holding capacity. Heated air drying is normally accomplished by convection. This system has the following units.

(1) Arrangement for heating of natural air.

(2) A blower by which heated air can be passed through the grains.

(3) A bin to hold the grains.

(4) Facility for loading and un-loading of grains.

(5) A system to control the drying process thereby to save grains from spoilage.

(6) A safety system or network for dryer operation.

In heated air drying systems, heated air is ventilated through a mass of agricultural products by forced convection method. The heat of drying air provides necessary energy for moisture removal from the products. During the process, the air creates more vapour pressure in interior parts of products than the outer surface. The water-vapour pressure gradient between inside and surface of the product is responsible for drying or water-vapour removal. The drying air also acts as a carrier for transfer of water-vapour. The amount of moisture removal by the drying air is dependent upon the psychrometric condition of the air. When heated air is ventilated through a grain mass then, (1) it vaporises moisture from the surface of grain, (2) increases the temperature of grain, and (3) carries the water-vapour outside the grain. Due to increase in grain temperature, drying becomes faster. The moisture from interior of grain moves faster towards surface due to osmotic pressure. The amount of moisture evaporation depends upon, (1) amount of moisture and temperature of grain, (2) the relative humidity and temperature of drying air, and (3) the drying air flow rate.

Mechanical drying methods

Though moisture can be removed from products even by filteration or by pressing, but, the drying process commonly means removal of moisture due to simultaneous heat and mass transfer processes. The water-vapour from wet material is carried by a medium usually drying air. There are a number of products with varying physical and chemical properties can be dried by heated air. Methods of transfer of heat to products are also numerous. Therefore, it is very difficult to classify all possible drying methods. However, some of the basic and commonly used drying methods are described below.

(1) *Contact drying* : In this method of drying, heat is supplied to wet products by conduction. For heat transfer, a heated surface like plate, cylinder, dryer wall is used (Fig. 3.24). The rate of heat transfer is dependent upon the thermal conductivity of the heated surface and also depends on the heat transfer coefficient from

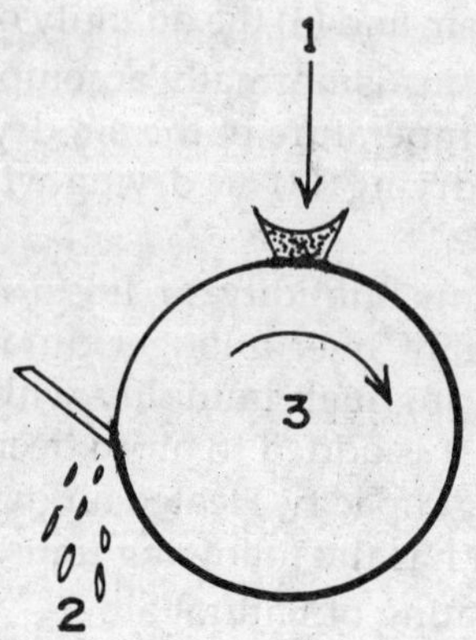

Fig. 3.24 : Contact drying
1. wet material 2. dry material 3. hot cylinder

the heating medium to the surface. The commonly used heating medium in conduction drying are steam, organic liquids, metals and other materials with high values of heat transfer coefficient. Since all the heat which passes through the material is utilised for moisture evaporation, thermal efficiency of contact drying is higher.

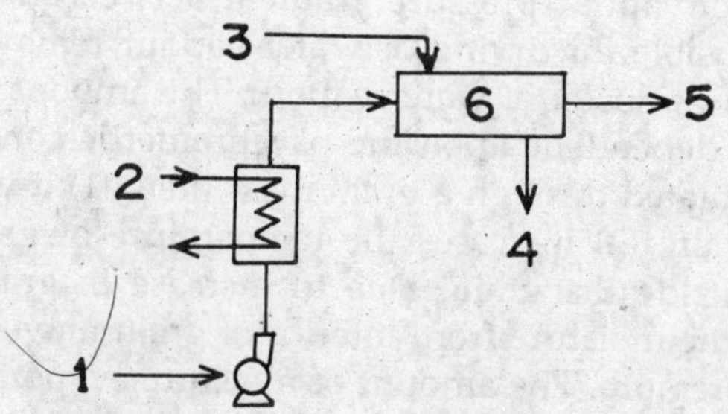

Fig. 3.25 : Convective drying
1. ambient air 2. heater 3. wet material
4. dry material 5. exit air 6. drying chamber

(2) *Convective drying* : In this method the sensible heat of heated air is transferred to the wet products by convection. Heated gaseous medium (usually air) is ventilated through a mass of wet materials and carries with it the water-vapour evaporated from the material (Fig. 3.25). To save energy, sometimes the exit air from dryer is recirculated. Sometimes, the partially dried material is recirculated for further drying (Fig. 3.26). In such situation the system is known as recirculatory dryer.

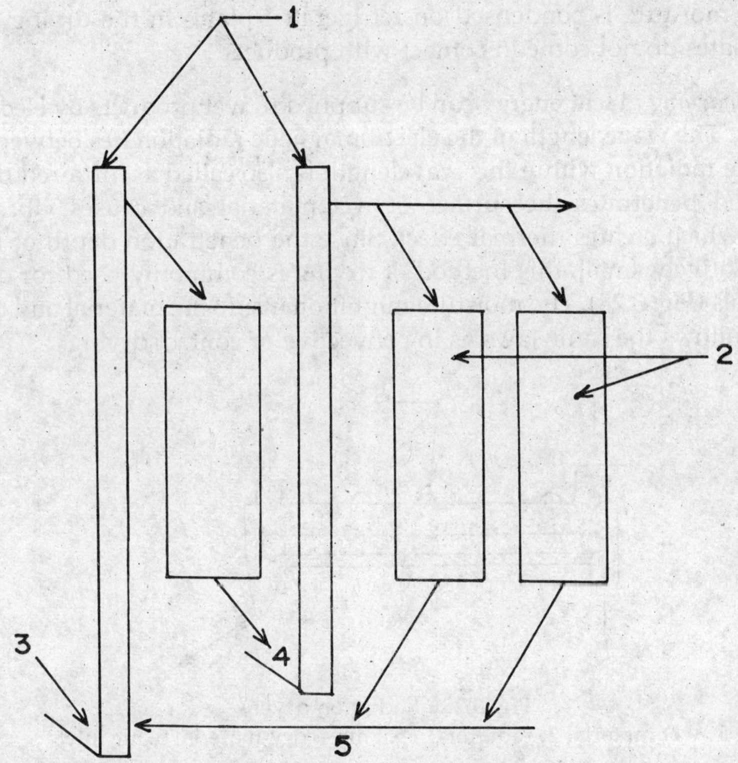

Fig. 3.26 : Recirculatory dryer's schematic diagram
1. elevator 2. tempering bins 3. wet material
4. semidried material 5. conveyor

(3) *Freeze drying :* This method of drying is based on the sublimation of frozen moisture from the wet product placed in a drying chamber. The pressure in the drying chamber is low. Heat is supplied by radiation or conduction from heated trays and the temperature of product is not raised above 0°C (Fig. 3.27). The

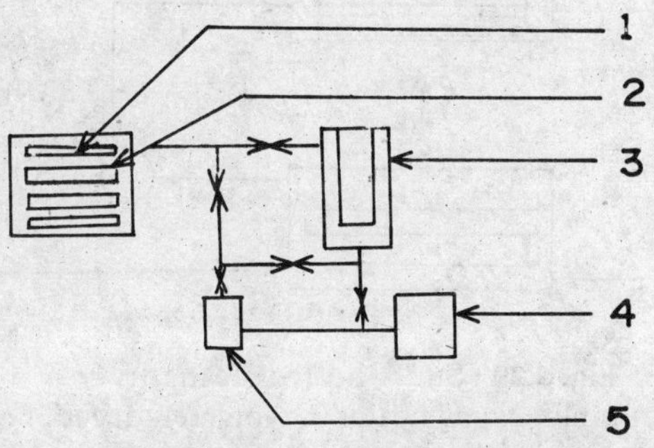

Fig. 3.27 : Freeze dryer's schematic diagram
1. dry material 2. heating surface 3. condenser
4. vacuum pump 5. diffusion pump

sublimated moisture is condensed on refrigerated plates in the drying chamber but these plates do not come in contact with products.

(4) *Radiation drying :* Heat energy can be supplied to wet products by electro-magnetic waves. The wave length of the electromagnetic radiation lies between 0.76 to 400 μ m. The radiation within this wavelength is also called as infra-red radiation. The infra-red penetrates the surface of wet material and causes vibrations of molecules, which creates thermal effect. Since the penetration depth of infra-red waves is relatively small, this method of drying is commonly used for drying of thin materials (Fig. 3.28). The moisture migration inside the material and diffusion of vapour follows the same laws as in convective or contact drying.

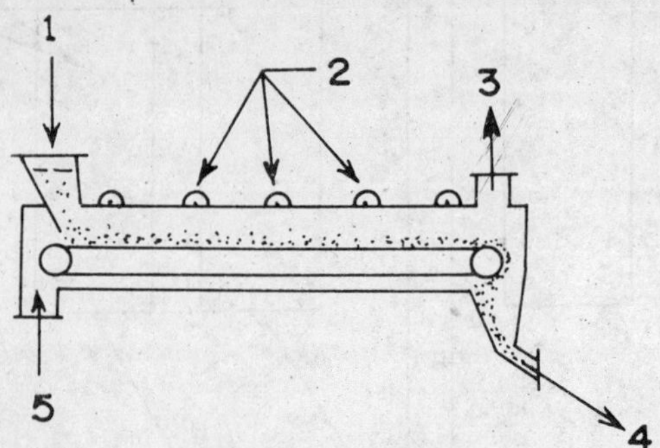

Fig. 3.28 : Radiation dryer
1. wet material 2. radiator 3. exit air 4. dry material 5. air inlet

(5) *Super heated steam drying :* In this method of drying, as shown in Fig. 3.29, first of all the dryer is filled with hot air and normal convection drying takes place. The water vapour evaporated during drying operation starts to circulate with hot- air. Over pressure is released by a pressure control valve and the air is gradually replaced by water vapour. At the end of drying, water vapour with a small amount of air is circulated.

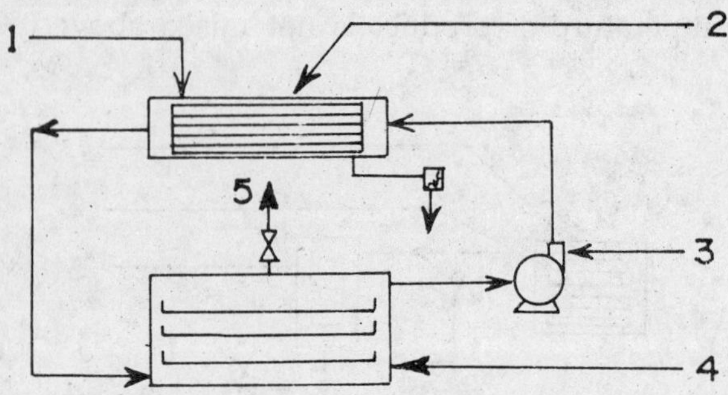

Fig. 3.29 : Super heated steam dryer
1. steam inlet 2. gas heater 3. blower 4. dryer 5. exit

(6) *Osmotic drying :* This method of drying can be successfully applied for dewatering of cellular products such as fruits, vegetables, meat etc. For the moisture removal, osmo-active substances like; 60% aqueous solution of saccharose or 25% aqueous solution of sodium chloride is used. The difference in osmotic pressure on both sides of the cell wall having properties of a semi permeable membrane

causes moisture removal from the products. The semi-permeable property of cell wall permits migration of water and low molecular molecules to the osmo-active liquid having higher osmotic pressure. Thus, thinner cell sap causes increase in water activity of the medium.

(7) *Fluidised bed drying* : In this method of drying, products are being dried under fluidised condition in a dryer. The materials are fluidised by drying air with sufficiently high velocity to cause suspension. In this drying process, higher rates of moisture migration takes place. Since every surface of product is in contact with drying air, uniform drying of products takes place. This method is normally used for the materials which have high initial moisture content and are lighter and at the same time requires to be dried quickly such as vegetable seeds.

(8) *Desiccated air drying* : In this method of drying, the ambient air is passed through a desiccated medium or source e.g. silica gel. The desiccator absorbs moisture from the air, as a result, the relative humidity of air is reduced and at the same time there is an increase in its temperature. When such air comes in contact with wet grains, transfer of moisture from wet grains to drying air takes place. The drying action occurs due to convective heat transfer. Schematic diagram of desiccated air drying is given in Fig. 3.30.

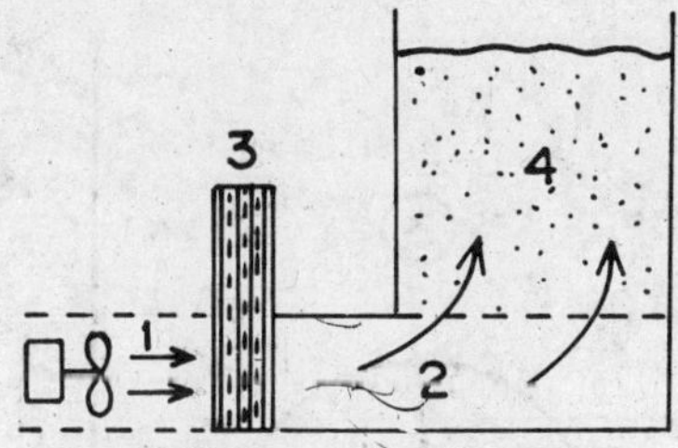

Fig. 3.30 : Schematic diagram of desiccated air drying
1. ambient air 2. desiccated air
3. desiccated source
4. grains in drying chamber

Types of dryers

Different types of dryers can be categorised as follow:

1. Dryers in which products are kept in a bin and the heated or drying air is ventilated through them.
2. Dryers in which products are kept moving. The movement of products is by gravity or by some mechanical means and the drying air is ventilated through them.
3. Dryers with free falling bed due to gravity and the drying air is ventilated through the falling film of products.
4. Dryers in which the products are dried in a state of fluidisation by strong blowing heated air.

Deep bed dryers : These batch-in-bin dryers are of large capacities to several hundred tonnes (Fig. 3.21). The most common shapes are round or rectangular. To operate deep bed dryers efficiently the following rules may be followed.

(*i*) An air flow rate of 2.94–3.92 m^3/min per tonne is recommended. The lower range of 2.94–3.43 m^3/min per tonne can be used safely in cooler and drier places, while the higher range of 3.43–3.92 m^3/min per tonne may be used in hot and humid climates. Rates above 3.92 m^3/min per tonne may result in uneven drying and is expensive in operation.

(*ii*) If the moisture content of grains is upto 18%, the layer depth of grain should be limited to 3 m. For grains whose moisture content is above 18%, the maximum depth recommended is 2.5 m. Paddy with 2.5 m deep layer may take 20 days to dry during favourable weather and upto 40 days during bad weather. In tropics where sprouting occurs in 4-8 days, deep bed dryers are not very much successful.

(*iii*) The net perforated area of the floor should be 15% of the total floor area. Air velocity of 300 m/min through the opening is preferable.

Flat bed dryers : The flat bed batch type dryer is similar to deep bed dryers except that the surface area of the dryer is more and the depth of the drying layer is less. These dryers are of usually 1–2 tonne capacity. These are designed for farm level operation (Fig. 3.31). Grains are spread 0.6 to 1.2 *m* deep over the perforated floor

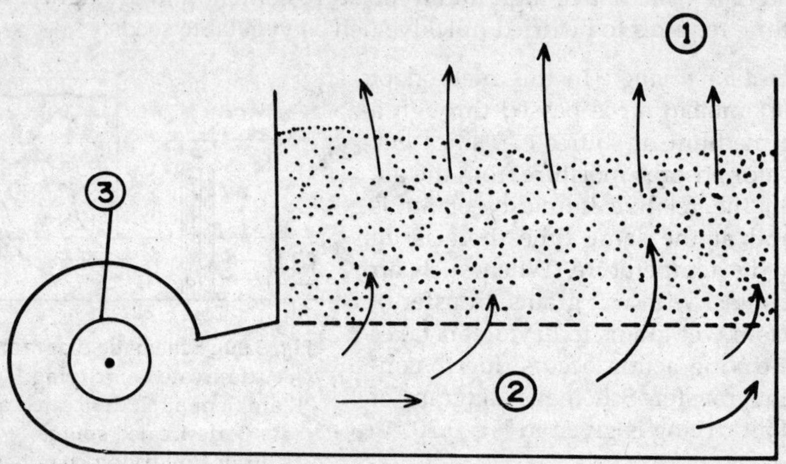

Fig. 3.31 : Flat bed dryer
1. exit air 2. plenum chamber 3. blower

and dried. The main advantages of this type of dryer are, (1) the whole batch is dried quickly and there are less chances of spoilage due to moulding, (2) there is less likelihood of over-drying of the grain, and (3) a lower air pressure is required to force the drying air.

Continuous flow dryers : These dryers are columnar type in which wet grains flow from the top to the bottom of the dryers. The rate of flow of grains through columns can be regulated by conveyors. These dryers are of two types, (1) mixing, and (2) non-mixing as shown in Fig. 3.32 and 3.33. If the grains flow in a straight path, the dryer is called a non-mixing type, and when the grains are diverted in the dryer, it is called a mixing type dryer. Drying is accomplished by forcing heated air across the falling layers of grains.

In mixing dryers, baffles are provided to cause the grains to mix during their downward flow. These dryers use low air flow rates of 50–95 m^3/min–tonne and high drying temperature of 65°C. Zig-Zag columns enclosed by screens on both sides are used primarily to achieve mixing action during drying process.

In non-mixing dryers, baffles are not provided in the columns and drying takes place between two parallel screens, 15–25 cm apart. In these dryers high air flow rates of 125–250 m^3/min–tonne can be used. It permits a faster movement of grains in columns. Drying air temperature of 54°C is used in non-mixing dryers.

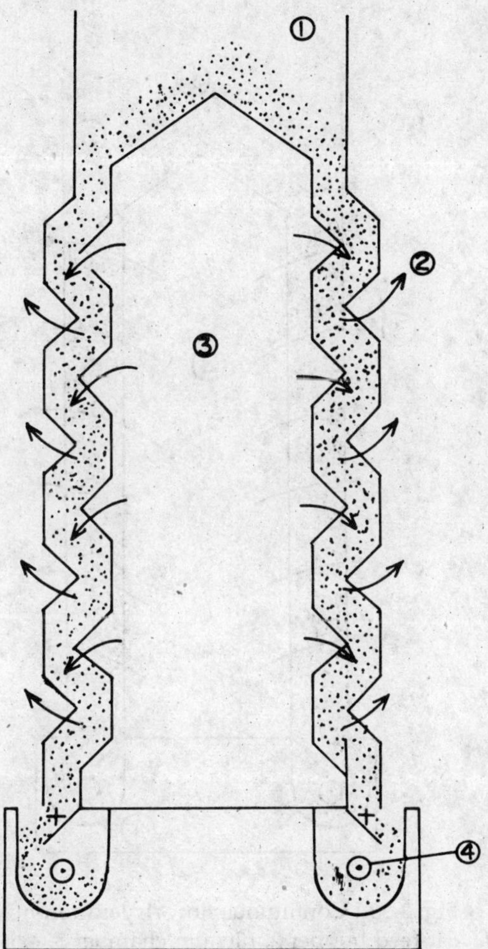

Fig. 3.32 : Continuous flow dryer (mixing type)
1. feed hopper 2. exit air 3. plenum chamber 4. dry material outlet

Recirculating dryers : In continuous flow dryers, a multipass procedure is used to avoid excessive drying stress (Fig. 3.26). During each pass, the grains are exposed to the heated air for a short time (15–30 min) and about 1–3% of the moisture content is removed. In such dryers, drying air temperature of 60–80°C is used. Drying is faster and effective because of the continuous movement of grains during the short drying times. Between drying passes, the grain is stored in a tempering bin for 4 to 24 hours to allow the moisture content at the centre and the surface of the grain to equalise.

LSU dryer : The design of this continu us dryer was developed at the Lousiana State University in the mid 1950, called the LSU dryer. This design was developed specifically for rice, to ensure gentle treatment, good grain mixing and good air-to-grain contact. It is a mixing type continuous flow dryer, in which layers of inverted V-shaped channels are installed (Fig. 3.34). The layers alternate between hot air intake and exhaust air outlets and are staggered to provide mixing. The LSU dryer is of high capacity and being used in big commercial rice mills.

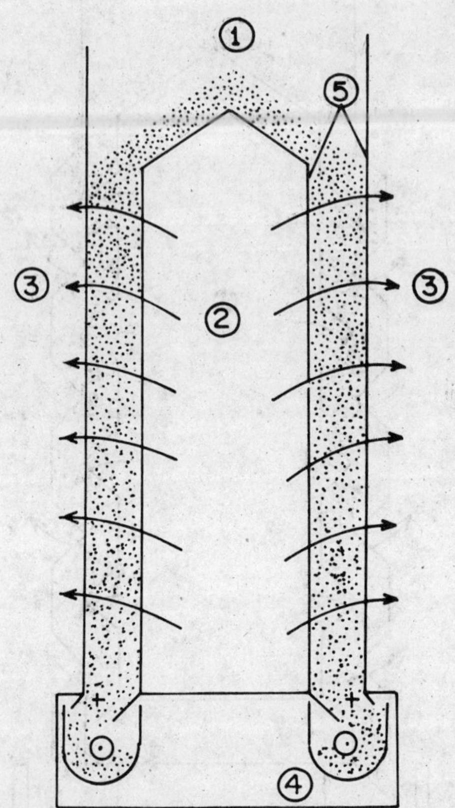

Fig. 3.33 : Continuous flow dryer (non-mixing)
1. feed hopper 2. plenum chamber 3. exit air
4. dry grain out let 5. screened grain column

Fluidised bed dryers : The fluidised bed drying technique holds an important position among modern drying methods. It is used mainly for granular materials, nevertheless, it is also applicable in the drying of solutions, pastes and liquids sprayed onto the fluidised inert bed. The principle of operation of fluidised bed dryer is to provide sufficient air pressure to fluidise a thin bed of grain/product, giving excellent air/grain contact. Above a certain pressure, related to the weight per unit area of the grain bed, the pressure drop across the bed becomes constant with volume flow rate, so that fast drying can occur. The main advantages of fluidised bed drying are;

1. Fluidity of bed facilitates continuous and easy equipment performance even on very large scale plants.
2. Lack of moving parts.
3. Very good condition for heat and mass transfer.
4. Provides good mixing i.e. uniformity of drying of the material in the bed.
5. By combining high air temperatures and suitable grain tempering, the fluidised bed offers better economy, gentle grain handling and reduced drying times.

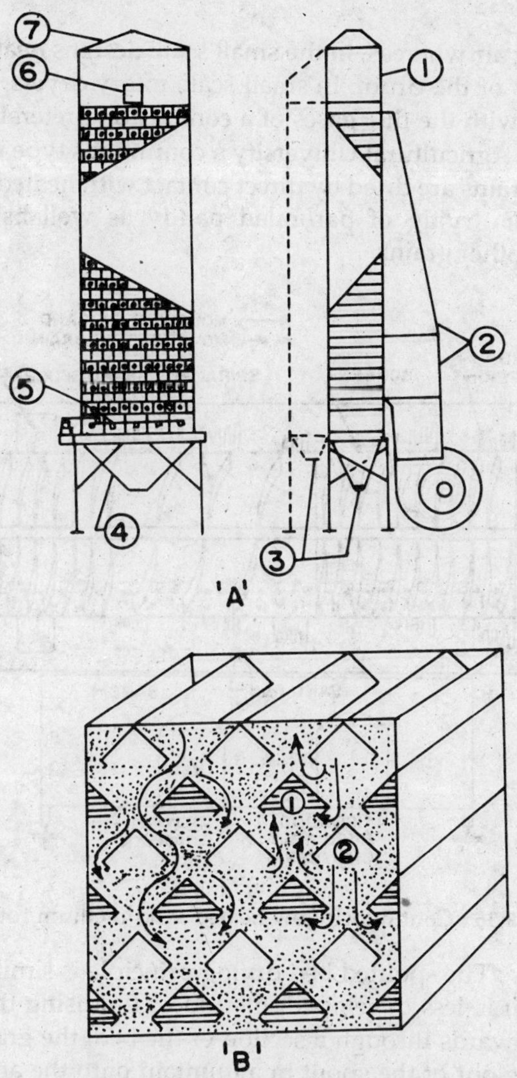

Fig. 3.34 : LSU dryer
A' 1. garner 2. duct 3. dry material outlet
4. hopper 5. continuous flow 6. door 7. roof
'B' cross-section of drying chamber
1. air exhaust 2. air intake

The drying method is based on ventilating hot air through the bed of product kept on the support grid or the gas distributor. When the air velocity becomes higher than the critical fluidisation velocity, the bed progressively expands until it reaches a state that of boiling liquid. This phenomenon is called fluidization.

Rotary dryer : In commercial rotary dryers the diameter of drum is between 1 to 3 m and the length is from 3 to 6 m. It is operated at slight inclination. The drum rotates on its axis. The grain flows downward through the rotating drum and is periodically lifted by inclined flights, then dropped, ensuring good air/grain contact. However, in commercial practice the main method of drying is convective heating by drying air whereas, in the small scale designs heating is by conduction through the walls of the drum. In small scale rotary dryers, the walls are heated by direct contact with the flue gases of a combustor, preferably biomass fired.

In Tamil Nadu Agricultural University a continuous type rotary dryer has been developed. The grains are dried by direct contact with heated sand (Fig. 3.35). The dryer is used for drying of parboiled paddy as well as for dry roasting of Bengalgram and other grains.

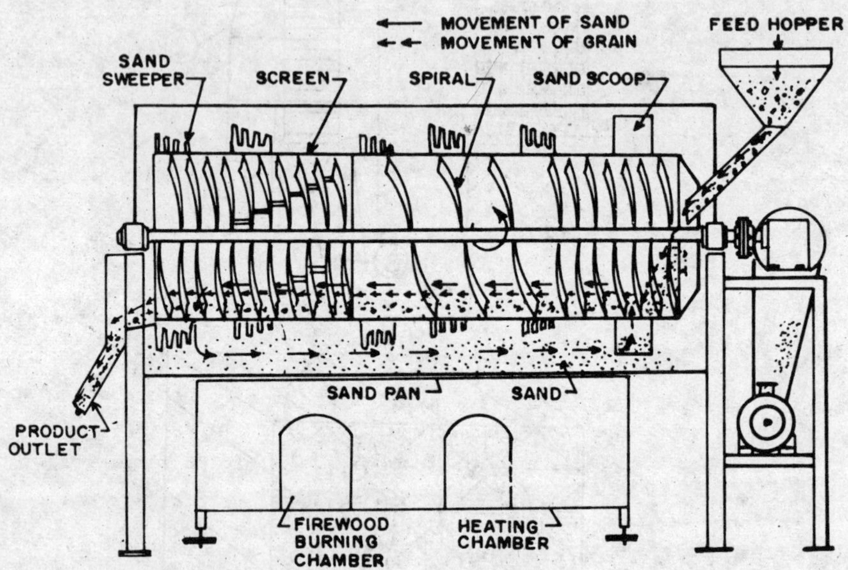

Fig. 3.35 : Continuous flow heated sand medium rotary dryer

Spouted bed dryers : The spouted bed drying principle is similar to that of fluidised bed drying, but it is less developed. Instead of fluidising the whole bed, a jet of air is spouted upwards through a section of the bed, the grain being entrained in the jet and falling out of the spout in a fountain onto the annular region of grain surrounding the spout. The whole bed of grain is continually being heated and dried. Thus the process is energy efficient and uniform in treatment. The grain is automatically tempered during drying operation.

Tray dryer : In a tray dryer, many shallow trays are kept one above the other with a gap in between, in the drying chamber. Tray dryer is generally used for drying of vegetables and similar semiperishables. The trays may or may not have perforated bottom. Perforated trays are used when the plenum chamber is at the bottom of drying chamber (Fig. 3.36). If the heated air is coming from the sides of drying chamber, the trays may not have perforated bottom. The gap in between the group of trays permits air ventilation. Products are kept in thin layers in the trays.

Tunnel dryer : It is similar to tray dryer. When the group of trays is stationary, the system is called a tray dryer but when the group of trays is moving in a tunnel, the system becomes a tunnel dryer (Fig. 3.37). The flow of heated air in a tunnel dryer may be concurrent or counter current.

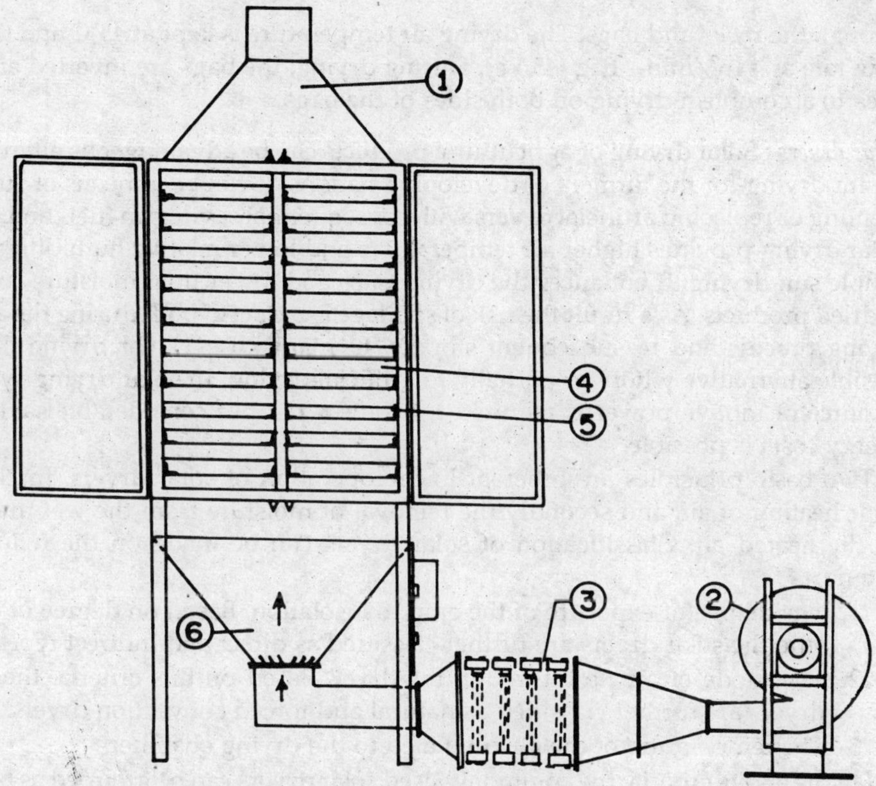

Fig. 3.36 : Diagram of a tray dryer
1. exit air 2. blower 3. heater 4. inter space between trays
5. trays 6. plenum chamber

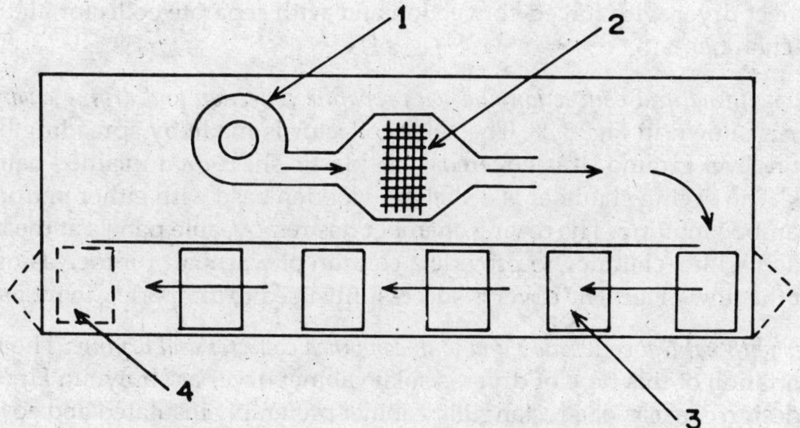

Fig. 3.37 : Schematic diagram of funnel dryer
1. blower 2. heater 3. trays 4. exit air chimney

Grain drying in bags : This method is useful to dry grains in small quantities. The method requires large number of unskilled labourers. More space is needed because bags of grain are put on perforated racks. The heated air is forced through the racks and bags. The drying air temperature is kept at 45°C and the air flow rate at 4 m^3/min—bag (45 kg). During drying, the bags are inverted at least once to accomplish drying on both sides of the bags.

Solar dryers : Solar drying of agricultural products can be advantageous alternative to sun drying for the farmers of developing nations. It can be a means of supplementing or replacing artificial dryers with consequential savings in fuels and costs. Solar drying provides higher air temperatures and lower relative humidities than simple sun drying. It enhances the drying rates and lower final moisture content of dried products. As a result the risk of spoilage is reduced, both during the actual drying process and in subsequent storage. In many cases, solar drying can be feasible alternative wholy or partially to artificial drying. In solar drying system, a source of motive power is required for some types but considerable saving in energy costs is possible.

Two basic principles are inherent in the operation of solar dryers, firstly the solar heating of air, and secondly the removal of moisture from the wet material by the heated air. Classification of solar dryers can be made on the following criteria;

1. The degree of exposure of the crop to insolation. Based on degree of exposure the solar dryers are further classified as direct and indirect types.
2. The mode of air flow through the dryer. Based on this criteria, the solar dryers are further classified as natural and forced convection dryers.
3. The temperature of the air circulated to the drying chamber.

On the above criteria, the commonly used solar dryers can be grouped as below:

1. Direct dryer with natural convection and with combined collector and drying chamber.
2. Direct dryer with natural convection and with separate collector and drying chamber.
3. Indirect dryer with forced convection and with separate collector and drying chamber.

Direct dryer with natural convection and with separate collection and drying chamber : Such dryer is shown in Fig. 3.38. The solar collector is made by spreading burnt rice husk on level ground. It is covered with plastic sheet on a inclined bamboo framework. The drying chamber is a shallow wooden base with either perforated metal or bamboo matting. The drying chamber has removeable panels at the back. It is provided with a chimney to provide a column of warm air for increasing the draught or the flow of air. The dryer is successfully used to dry paddy in monsoon.

Direct dryer with natural convection and with combined collector and drying : The most popular variation of this type of dryer is solar cabinet dryer as shown in Fig. 3.39. The basic design consists of a rectangular cabinet preferably insulated and covered with a roof of glass or clear plastic. Holes are made through the base and upper parts of the cabinet. Perforated drying trays are positioned within the cabinet. The insolation passes through the roof and is absorbed on the blackened interior surfaces which are then heated and warm the air within the cabinet. The warmed air rises by natural convection and passes out of the upper holes. Fresh air in the meantime enters through the base holes.

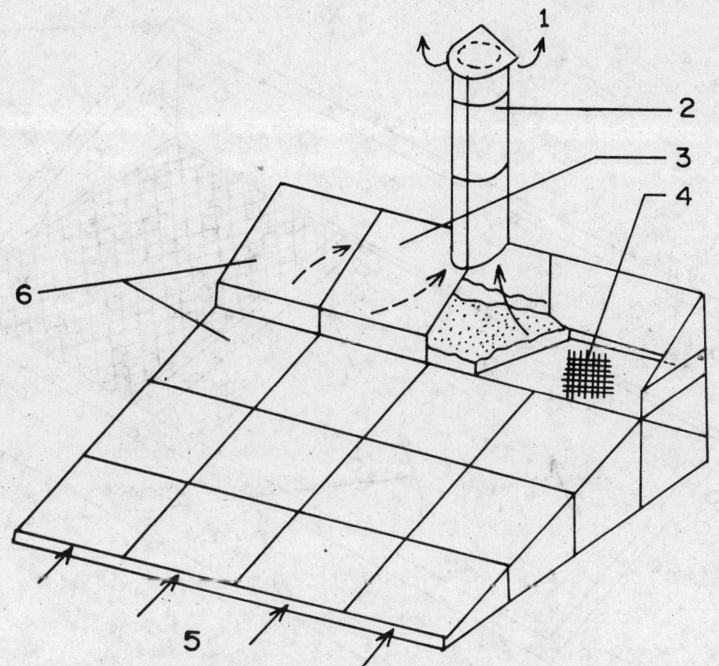

Fig. 3.38 : Direct dryer (natural convection) with separate collector and drying chamber
1. exit air 2. chimney 3. transferable panel 4. bamboo/nylon mesh 5. air intake 6. clear plastic sheet

The length of the cabinet should be three times the width to minimise the shading effect of the sides. The angle of the roof should be such as to maximise the amount of insolation received. The global latitudes and time of year decides the angle of roof.

Indirect dryer with forced convection and with separate collector and drying chamber : In this type, the roof of the dryer is commonly used as solar collector. The warm air from the roof section is sucked by a blower and delivered to the lower plenum chamber under the drying bin. The warm air is ventilated through the products kept in drying racks/chamber and it leaves the room through north facing roof.

Design of solar collector

The temperature rise of air leaving the solar collector can be computed by the following equation

$$t - t_0 = \frac{ER}{U}[1 - \exp(-N)] + (t_e - t_0)\ \exp(-N) \qquad \ldots 3.64$$

where, t = temperature of air leaving the collector, °C

t_0 = ambient air temperature, °C

t_e = temperature of air entering the collector, °C

E = fraction of incoming insolation absorbed by the collector

R = rate at which solar energy falls on unit surface area of the collector, W/m^2

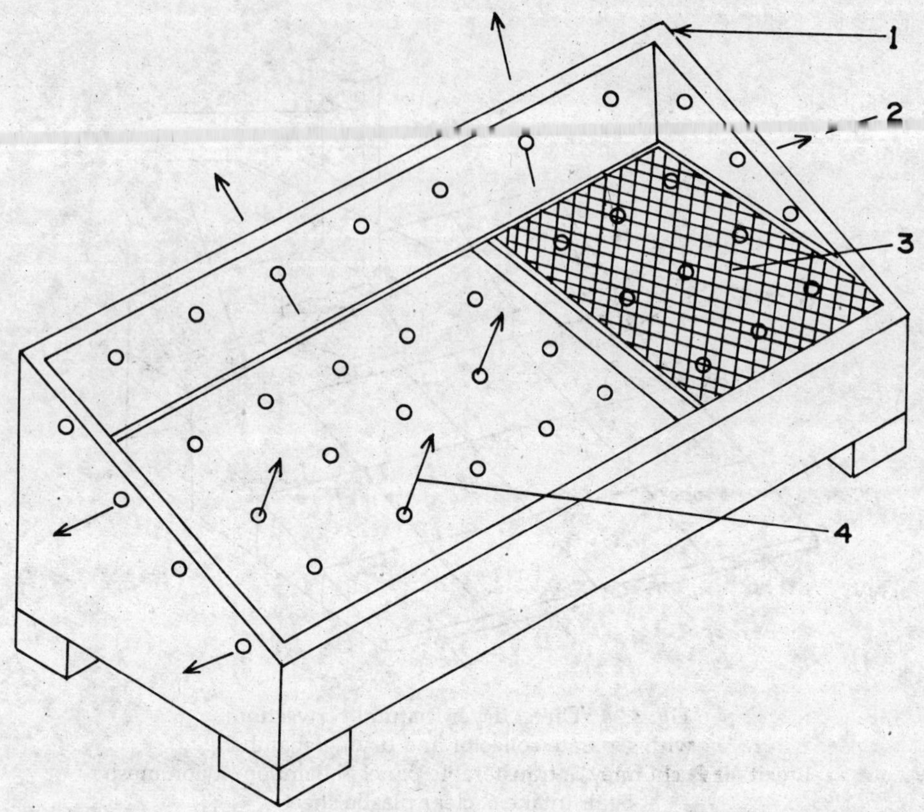

Fig. 3.39 : Solar cabinet dryer
1. frame 2. exit air 3. drying tray 4. air intake

U = overall heat transfer coefficient between the heated air with the collector and ambient air

$N = \frac{UA}{Mc_p}$

A = area of collector, m^2

M = mass flow rate of air through the collector,

c_p = specific heat of air passing through the collector,

When the air entering the collector is at the same temperature as the ambient air the equation becomes,

$$t - t_0 = \frac{ER}{U}[1 - \exp(-N)] \quad \ldots 3.65$$

For the specific cases of a metal absorber plate covered with asphalt roofing paint,

$$t - t_0 = 0.019\,R\left[1 - \exp\left(-\frac{0.041}{V}\right)\right] \quad \ldots 3.66$$

For an uncovered collector and for a collector with a single glass covering

$$t - t_0 = 0.064\,R\left[1 - \exp\left(-\frac{0.011}{V}\right)\right] \quad \ldots 3.67$$

where, V = volumetric air flow rate per unit area of collector surface, $m^3/s{-}m^2$

There are different types of solar collector configurations. These can be grouped into five modes as shown in Fig. 3.40. The bare plate collector is simplest one. The air is flowing under the uncovered absorber plate and over the casing, generally insulated. In the single front pass collector, air flows between the transparent cover and the absorber plate and casing. A complex but efficient design is the double pass collector, in this, the air flows initially between the absorber plate and the casing and then between the cover and the absorber but in the opposite direction. In the parallel pass arrangement, the air partially flowing between the cover and the absorber plate and the remaining air is flowing between the absorber plate and casing.

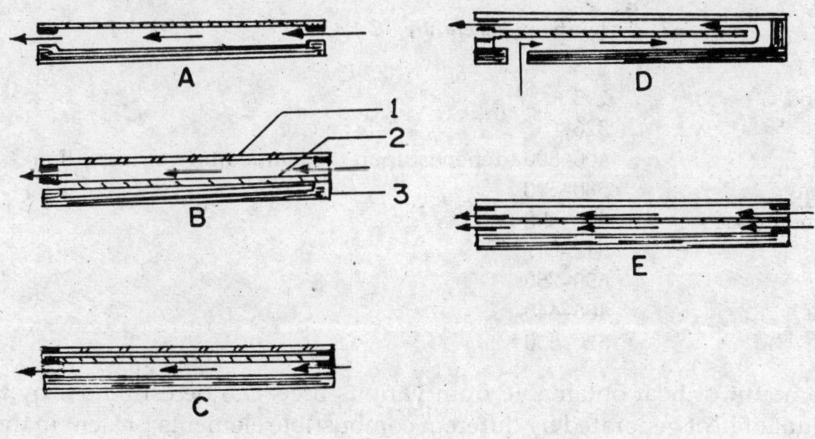

Fig. 3.40 : Various solar collector configurations
1. cover 2. collector surface 3. casing
→ air movement
'A' uncovered surface 'B' single front pass
'C' single back pass 'D' double pass 'E' parallel pass

Fuels for drying

Fuels are those substances which give heat by combustion in presence of air or oxygen. Therefore, the fuels contain one or many combustible elements like carbon, hydrogen and hydrocarbon, though iron, mangnese, sulphur and phosphorus etc. also give heat upon burning in presence of oxygen but these are not counted as fuels.

Fuels have been categorised as per the followings;

(1) the physical state of availability in nature like solid, liquid and gases, (2) source of availability like natural or man made (artificial), and (3) on the basis of calorific value.

The above classification does not provide exact quality and calorific value of fuels as their combustion with air is a cumbersome process. The amount of heat depends on the fuel/air ratio. In Table 3.7, the natural and man-made fuels have been described.

Table 3.7
Natural and man-made fuels

From	*Natural*	*Man-made*
1. Solid	Wood, peat, coal, lignite, oilshell	husk, baggase, coke, charcoal, briquette, wood chips, bark
2. Liquid	Petroleum	Oils received by distillation of petroleum, coaltar, shell oil, gasolene, alcohal etc.
3. Gas	Natural gas	Coal gas, water gas, producer gas, oilgas, hydrogen, acetylene

The ignition temperatures of some of the combustible materials are given below.

Matter	*Ignition temperature, °C*
peat	225
pine wood	295
coal	326
charcoal	400–800 (depends upon the temperature of formation)
hydrogen	580–590
carbon monoxide	644–658
ethylene	542–547
methane	650–750
acetylene	406–440
ethane	520–630

The amount of heat obtainable from various fuels can be estimated by taking the amount of heat generated by different combustible elements present in the fuel. Some of the important combustion reactions are given below.

$$H_2 + \frac{1}{2} O_2 = H_2 O \text{ (vapour)} + 57810 \text{ kcal/mole}$$

$$H_2 + \frac{1}{2} O_2 = H_2 O \text{ (liquid)} + 68360 \text{ kcal/mole}$$

$$C + \frac{1}{2} O_2 = CO + 29430 \text{ kcal/mole}$$

$$CO + \frac{1}{2} O_2 = CO_2 + 68220 \text{ kcal/mole}$$

$$C + O_2 = CO_2 + 97650 \text{ kcal/mole}$$

$$S + O_2 = SO_2 + 69800 \text{ kcal/mole}$$

$$CH_4 + 2\,O_2 = CO_2 + 2\,H_2 O + 192400 \text{ kcal/mole}$$

$$C_2 H_2 + 2\frac{1}{2} O_2 = 2\,CO_2 + H_2 O + 312400 \text{ kcal/mole}$$

The amount of heat generated/liberated by burning of fuels can be estimated by the following equation on the basis of amount of carbon and hydrogen present in the fuel.

$$\text{Heat, Calorie} = \frac{\text{carbon} \times 8137.5 + \text{hydrogen} \times 34180}{100} \qquad \ldots 3.68$$

where carbon and hydrogen are in percent.

Oxygen is also present in different amounts in fuels, therefore, air in small quantities is also required for its combustion. As a result of this, amount of heat available lowers. Since 8 grams of oxygen reacts with 1 gram of hydrogen, therefore, 8 parts of heat available from a fuel is estimated by the following expression.

$$\text{Calorie} = \frac{\text{carbon} \times 8137.5 + \left(\text{hydrogen} - \dfrac{\text{oxygen}}{8}\right) \times 34180}{100} \quad \text{...3.69}$$

In many fuels other matter like nitrogen, sulphur and water are also present along with carbon, hydrogen and oxygen. The heat capacity of a fuel with different constituents can be determined by the following equation

$$\text{Heat, calorie} = \frac{C \times 8137.5 + \left(H - \dfrac{O+N-1}{8}\right) \times 34180 + S \times 2181 - H_2O \times 600}{100} \quad \text{...3.70}$$

For drying of agricultural products, the following factors may be looked into while selecting a fuel, (1) cost of fuel, (2) heat capacity of fuel, (3) combustion efficiency of fuel, (4) cost of handling and storage of fuel, (5) disposal of fuel waste and (6) cost of maintenance of heater and furnaces.

To get complete amount of heat from a fuel, it should ignite completely. For complete ignition of any fuel, amount of air should not be less than theoretical amount. If the air amount is more than theoretical amount, some heat will be lost and it will go along flue gases and can not be utilised for drying. But in practice for ignition of fuels, higher amount of air is used than theoretical amount.

Exercise

1. 400 kg of wet pigeonpea at 23% moisture content (wb) is dried to 10% moisture content (wb) for milling. Calculate the amount of moisture evaporated in drying.
2. If the dry bulb temperature of moist air is 40°C and the humidity ratio is 0.014, find the other thermodynamic properties of air from the psychrometric chart.
3. Calculate the values of the constants *C* and *n* of Henderson's equation for cauliflower seeds with the help of following data obtained during EMC studies.

Condition	Relative humidity, %	Temperature °C	EMC % (db)
1	70	30	9
2	20	40	5

4. Sorghum is being dried in a bin type dryer. The dry and wet bulb temperatures of ambient air are 35° and 23°C, respectively. The relative humidity of exit air is 75%. If the heated air temperature is 48°C and its flow rate is 30 m^3/min., calculate the amount of sensible heat required to

raise the temperature of air. Also determine the water vapour removed per hour in the drying operation.

5. In a dryer 1 tonne of wet pulses (final weight) is dried from 27% moisture content (wb) to 10% moisture content (wb) per hour. Calculate the heat requirement for drying and the necessary air flow rate for the drying system. Given that

Dryer efficiency	:	50%
Latent-heat of evaporation	:	570 k Cal/kg
Specific heat of air	:	0.24 k Cal/kg°C
Drying air temperature	:	45°C
Ambient air conditions	:	35°C and 20°C dry and wet bulb temperatures, respectively.

6. The diameter of a circular dryer is 2.5 m and wet pulse is kept in the dryer to a depth of 1.5 m for drying. The moisture content of pulse is 28% (wb) and it is being dried by 40°C hot air. The relative humidity of air is 30% and flow rate is 1500 m^3/hr – *tonne*. If the half response time is 3 hours and the density of dry material is 700 kg/m^3, then determine the moisture content of pulses at any depth after 15 hours of drying. Also calculate the amount of water vapour removed from the dryer after 15 hours. Given that the specific heat of air as 0.24 k cal/kg °C and latent heat of evaporation 590 k Cal/kg.

Bibliography

Andales, SC, 1981. Drying of cereal grains in the Phillippines in "Food Drying" Ed. Gordon Yachiuk, IDRC-195e.

Annon, 1950. The Wealth of India. Raw materials Vol. II. CSIR. New Delhi.

A.S.A.E., 1980-81. Agricultural Engineers years book p 352.

Becker, HA and HR Sullins, 1956. A study of the desorption moisture movement in the drying of the wheat kernel. Cereal Chemistry, 32:212-216.

Broker Donald B, Fin Bakker-Arkema and CW Hall, 1974. Drying cereal grains. AVI Publishing Company. West Port Connecticut.

Chakraverty, A and DS De, 1981. Post harvest technology of cereals and pulses. Oxford and IBH Publishing Co.

Chancellor, WJ, 1968. A simple grain dryer using conducted heat. Trans. of A.S.A.E. 11:857-862.

Chung, DS and HB Pfost, 1967. Adsorption and desorption of water-vapour by cereal grains and their products. Trans. A.S.A.E. 10:552-575.

Day DL and GL Nelson, 1965. Desorption isotherms for wheat. Trans. A.S.A.E. 8:293-297.

Foster, GH, 1953. Minimum air flow requirements for drying grain with unheated air. Agric. Eng. 34(10): 681-684.

Foster, GH, 1967. Moisture changes during aeration of grain. Trans. A.S.A.E. 10:344-351.

Gershoi, AP and VF Samochetov.,1980. Grain drying and grain dryers NTIS. US Dept. of Commerce. TT-60 21164.

Gough, MC and JA Mc Farlene, 1984. Aeration of stored grains. Some psychrometric considerations. Tropical Stored Products Information 50:32-35.

Hall, CW, 1970. Drying farm crops. Lyall Book Depot. Ludhiana.
Hayens, BS, 1961. Vapour pressure determination of seed hygroscopicity. Tech Bulletin 1229, ARS USDA, Washington DC.
Henderson, SM, 1970. Equilibrium moisture content of small grain hysteresis. Trans. A.S.A.E. 13 : 762–764.
Hukill, NV., 1954. Grain drying with unheated air, Agric. Eng. 35(6) : 393–405.
Kachru, RP and KM Sahay, 1988. Status of paddy drying in India. Souvenir SJC of the associated rice mill stores. The Palghat district rice mill owners and rice merchants association.
Keswani, HB, 1982. Heat and mass transfer. Satya Prakashan New Delhi.
Lal Radhey and K.K. Agrawal, 1968. Principles and practices of rice drying. Dept. of Agric. Eng. IIT, Kharagpur.
Luikov, AV, 1966. Heat and mass transfer in capillary porous bodies. Pergman Press. London.
Maheshwari, RC., 1975. Utilisation of rice husk as fuel Unpublished Ph.D. Thesis Agric. Eng. Dept. IIT, Kharagpur.
McLean, KA, 1980. Drying and storing combinable crops. Farming Press Limited, Suffolk.
Mohsenin, NN, 1970. Physical properties of plant and animal materials. Gordon and Breach. New York.
Navarro, S and M. Calderon., 1982. Aeration of grain in subtropical climates. FAO Agric. Services Bulletin 52, FAO Rome.
Platt, D. TR Rumsey and A Palozoglu, 1991. Dynamics and control of cross flow grain dryers I. Model development and testing. Drying Tech. 9(1) : 27–60.
Pratap, V, 1979. Equilibrium moisture contents of some flours. M. Tech. Thesis. G.B.P.U.A.& T, Pantnagar.
Reyes, VG Jr.. and VK Jindal, 1988. Effect of heat treatments on the head yields of high moisture rough rice. Journal of Food. Sci. 52 (2) : 482–487.
Sahay, KM, RP Saxena and BPN Singh, 1981. Development and performance evaluation of husk fired furnace with continuous dryer for small rice mill. Paper presented in the International Conference on Agric. Eng. and Agro-Industries in Asia at AIT, Bangkok.
Shaha, AK., 1974. Combustion engineering and fuel technology. Oxford and IBH Publishing Co. 732 P.
Shedd, K., 1953. Resistance of grain and seeds to air- flow. Agric. Eng. 34(9) : 616–619.
Shove, GC and EF Olver, 1967. Temperature gradients in grain drying. Trans. A.S.A.E. 10, 152, 156.
Shukla, BD, 1990. Rice husk—a source of energy. Report on Agric. Wastes furnaces. CIAE, Bhopal.
Singh, KK, 1994. Development of a small capacity dryer for vegetables. J. of Fd. Eng. 21(1): 19–30.
Sweat Vincent E, 1986. Thermal properties of food in "Enginering properties of foods" Eds M A Rao and SSH Rizvi. Marcel Dekker Inc. New York Basel.
Thompson, JH and CK shedd 1954. Equilibrium moisture and heat of vaporisation of shelled corn and wheat. Agric. Eng. 35 : 786–788.
Trim, DS, 1982. Solar crop dryers. Paper presented at the Arab Conference on solar energy utilisation in agriculture at Amman, Jordan

Chapter 4

Storage

Storage is an interim and a repeated phase during transit of agricultural products from producer to processor and its products from processor to consumer. Besides, agricultural products need to be stored from one harvest to next thus, demanding additional carry over as safeguard against a following crop of low yield or poor quality, against speculation in price and market demand or against shortage and famines.

Most of the existing crop storage methods were known and used thousand of years ago. Man has been storing grain and its products for a longtime. It is of all time concern for man to reserve food supply to mitigate hunger at a future date. Over the long years man has learned primarily by trial and error, that dry grain can be stored for long periods and can be protected against attack of various pests.

Storage of grain and control of quality occur in three locations; on the farm, at collection points serving a number of farms, and at terminal points where grain is processed or moved forward in still larger bulks. There are costs for loading and unloading at each type of storage, for transportation between storages and for storage itself. Additional costs are involved in conditioning of grains, mainly cleaning and drying, and for the control of quality.

Storage of food grains is inevitable both in times of deficit and surplus production. This single component in the chain of post harvest handling of grains is responsible for losses of the order of 7% which has a far reaching effect on the rural economy owing to 70% of the total food grain production being stored at the farm level. All farmers store food grains for their own consumption, 64 to 96% for seed and 48 to 90% for future sale and the quantity stored for the above purpose varied from 22 to 71% of the total production.

Storage of grains and other agricultural products is necessary to provide consistent supply with the demand to provide surplus storage to carry over supplies in the years of low production and to adjust and maintain quality consistent with the intended use of the products. This is not the only one requirement of storage, but proper storage facility is essential when the agricultural commodity is harvested at one time and is used for converting it into final form at the other time, leaving wide gap between. The preservation of all seeds such as soybean, groundnut, rapeseed as well as the food grains and other commodities for a period of say 3–6 months is essential, since these are seasonal products, whereas further processing/milling continues throughout. Safe storage, therefore, helps to maintain continuity in the production.

Conventional storage methods seem to be totally inadequate for the preservation of the quality and prevention of wastage. Storage of food grains and oilseeds in the traditional structures or in jute bags does not provide longer storage period. Proper scientific methods and controlling insect infestation in these grains are of vital necessity. Safe storage of agricultural products is essential to avoid losses due to rats, insects, diseases and to maintain good quality for a longer period.

Insects/pests of stored grains

Insect pests form one of the most important factors responsible for losses in agricultural production at various stages. Living organisms and the environment interact to bring about spoilage of stored products. Living organisms may be plant, insects, pests, man, animal, bacteria, fungi etc.

The number of insect species may be nearly one million, but only one per cent of the species may perhaps be harmful to man by consuming and contaminating the food. Insects have the capacity to adopt to any environmental condition. Several species can live in stored grain having very low moisture content. They enter into any hard-to-get-to places because of their small size. Some insects can live in darkness while some other live in light.

It is estimated that 5 to 10% of the world food production is damaged by insects during storage. The estimated losses due to insects in India, have been estimated to be around 3% of the country's production. Insects feed on the germ and endosperm causing loss in weight as well as nutrients. Besides, they cause contamination with their excreta and dead bodies. The damages can be grouped into, (1) direct damages, and (2) indirect damages.

Direct damages

(*i*) Some insects consume germ, some endosperm and the others eat away both. This results in loss of weight, loss or conversion of nutrients, loss of germination power, loss in gradation and consequently fall in market value.

(*ii*) The contamination may be with the dead bodies, cast skin, excreta, obnoxious odour and/or webbings.

(*iii*) Structures and containers may also be damaged by causing tunnelling in wooden parts resulting in the weakening of the structure/container.

Indirect damages

(*i*) It may create heating and migration of moisture.

(*ii*) It may create distribution of parasites to man. Certain tape worms use stored grain insects as intermediate hosts.

(*iii*) It causes customer's resistance/repulsion which may lower the prestige.

Sources of infestation

There are mainly five sources of infestation namely, the field itself, infested gunnies, infested transport, infested godown and infested stocks.

Field : Insects may attack the crop in the field itself. Grains affected by these insects when brought to the storage centres will be subjected to their continuous attack.

The infestation may be visible or invisible. Their stocks should, therefore, be fumigated immediately on receipt.

Infested gunnies : Even if sound and new gunnies are used for packing the newly harvested grains, the insects hiding in the seams of the gunnies will attack the freshly harvested stocks. The gunnies should be fumigated before packing the freshly harvested grains.

Infested transport : The transport used for carrying the newly harvested stocks should invariably be sprayed with DDVP or malathion. If the transport like bullock cart or tractor trolley has been used for carrying infested stocks on the previous occasion, the left over insects may attack the stocks now loaded into them.

Infested godowns : Even if the stocks are moved to a freshly taken over godown, it is not free from trouble. The insects present in the cracks and crevices of the wall or that hibernate in the structures, may emerge out and attack the grains. Hence the cracks and crevices should be plugged and empty godowns should be thoroughly cleaned and fumigated.

Infested stocks : In case sound stocks are brought to a godown where infested stocks are in storage, cross infestation takes place.

Types of insects

Insects can be grouped as major, minor, incidental and parasites and predators. Major insects are those which attack the food grains directly and destroy them completely. Khapra larvae, Rhizopertha Sp. and Sitopholius Sp. are major insect pests. Minor insects are those which follow the destruction already started by the major pests but cannot directly attack the grain and cause destruction independently. Laemophleous minutes and Oryzaephilus surinanmensis are some of the minor pests commonly seen in foodgrain storage. These insects are associated with particular environmental conditions such as moisture, temperature, sanitation level etc.

There are certain undetermined species of insects which may or may not harm stored food grains. These are called incidental insects which contaminate the grain with their presence.

The insects which depend or prey on the above types of insects are called parasites and predators, respectively.

Common insects of stored grains

Some of the insects generally found in storage of foodgrains, and other agricultural commodities are described in brief.

(i) Sitophilus oryzae and S. Granarius

They are called as rice weevil and granary weevil. The adult can be easily identified by the snout like projection or probosis from the head. The larvae are white tiny irregular masses. The two species can be distinguished by the dense round punctures on the thorax in rice weevil, and less dense elongated pits in granary weevil. Both adult and larvae stages are harmful.

(ii) Rhizopertha dominica

This is called lesser grain borer and can be identified by the cylindrical body of the insects with the head bent downward and its slow movement. The larvae are coniform in shape. Both adults and larvae are harmful.

(iii) Trogoderma granariam

The general name is 'Khapra'. The adults are brownish, oval bodied fliers having a very short span of life. The larvae of this insect is harmful. The larvae are many segmented and yellowish brown in colour with movable, erectile hair with the posterior segments forming a tail or a tuft of hair.

(iv) Lasiodarma serricorne

These are commonly known as cigarette beetles or Tobacco borer beetles affecting tobacco, turmeric, ginger, chillies, pepper etc. Both larvae and adults are harmful. The adults can be identified by their yellowish brown colour and the hemispherical shape, quite different from 'Khapra', with head hidden under the domed neck shield. The freshly emerged larvae are very mobile.

(v) Tribolium castaneum and Tribolium confusion

The common names are red flower beetles and confused flour beetles. They can be noticed every-where in association with other insects. The adults are red in colour with brisk movement. The confused beetles can be distinguished from the other species examining the antennae. In the red flour beetles, the segments of the antennae are smaller at the base and abruptly larger at the apex whereas in the confused beetles, the size of the segments gradually increases from the base to the apex.

(vi) Latheticus oryzae

They are commonly known as long head flour beetles. The adults are slender, dorsoventrally flattened insects pale yellow in colour. The antennae are seven segmented, club shaped and small.

(vii) Laemophleous minutes

These are called flat grain beetles and can be distinguished from their small size, brisk movement, bright brown colour and the longer antennae. They can also be seen in association with Oryzaephilus and Tribolium.

(viii) Bruchus sp.

They are commonly known as pulse beetles. The adults have a peculiar, almost triangular shape, brownish grey in colour covered with tiny bristles.

(ix) Stegodium paniceum

These are commonly known as spice beetle, the adults have cylindrical shape and brown in colour. They can be distinguished from Lasioderma serricorne by the antennae which are long and the last three segments which are clubbed at the end.

(x) Corcyra cephalonica

This is a moth, pale greyish in colour with a wing expanse of 12–14 mm. They can be distinguished by a closed straight peak formed by the palpi pointing forward and antennae not crossed over wings while the insect is on rest. Larvae are creamy white in colour with light brownish yellow broad head. It is commonly known as rice moth.

(xi) Ephestra coutella

It is commonly known as fig moth or almond moth, greyish in colour with a wing span of about 20 mm. They can be distinguished from the transverse stripes on its outer wings. The larvae are pinkish in colour.

(xii) Plodia interpunctella

The common name is Indian meal moth. The adults have wing expanse of 14–20 mm. When at rest with closed wings it is 8–10 mm long and greyish yellow in colour. Larvae are yellowish white, some are reddish or even greenish with a brown head.

(xiii) Sitatroga cerealetta

It is commonly known as grain moth. The adult is a small buff yellowish brown in colour with a wing expanse of 10–14 mm and length of 6.9 mm. Larvae are white in colour with a yellowish head.

Rats

Rats have been found to be taking a lion's share in the devastation of food material. It has been estimated that one rat consumes 10 kg of grain in a year and contaminates more than 10 times what it actually eats, with its urine and excreta. One rat drops 25 to 150 pallets and 10–20 ml urine in a day, besides shedding its plenty of hair. Their multiplication is also so tremendous that the rat population out paces the human population by six times. Rats also spread fatal diseases like plague, jaundice, rat bite fever, typhus etc. There are 118 species of rats found in India. However, some of the common rats are house rats/black rat, house mouse, brown rat, Indian mole rat, large bandicoot rat, short tailed mole rat, Indian field mouse, Indian gerbil/antelope rat.

Rat control measures

Sanitation in houses and godowns is significantly helpful in reducing the rat menace. Rats multiply rapidly if they get food and shelter. Waste grain should be removed, grain sacks should be stored about 50 cm above the floor, and the grain should be protected with concrete, wire or sheet metal. In order to avoid the entry of rats in houses/godowns, tin plates should be fitted to the doors and windows at the bottom in such a way that once they are closed, no space is left for making entry by the rats. In construction of a rat proof godown, following things are to be given due consideration.

(*i*) The foundation is kept deep and the plinth is constructed about 1 m above the ground level in 'L' shape having collar projected from all the sides.
(*ii*) A godown should not have provision of steps.
(*iii*) Fitting of wire netting to the drain pipes.
(*iv*) If corrugated sheet or asbestos sheet is used for ceiling, the space between the wall and ceiling should also be covered with wire netting.

The chemical used to kill rats and mice are called rodenticides. Rodenticides are usually mixed with baits such as fresh meats, rice, peanut butter etc. The bait must be mixed as directed because too little may not poison.

Apart from the above methods, rat traps are also used to control rat menace.

Traditional storage structures

Some of the traditional storage structures of rural areas for storing grains are briefly described.

1. *Morai type storage structures* : In rural areas of eastern and southern regions of India, morai type storage structures are used. In morai structure generally grains like paddy, maize, and sorghum are stored. Its capacity varies from 3.5 to 18 tonnes and its shape is like an inverted turncated cone. These structures are made on a raised platform supported on wooden or masonry pillars.

In improved morai type storage structures, the circular wooden floor is supported on timber pillars. The floor planks are joined together with lap joints. This wooden floor is surrounded by 22 gauge corrugated metal cylinder of 90 cm height. The cylinder is nailed to the wooden floor. Inside the cylinder, 7.5 cm diameter ropes made of paddy straw or similar material are placed, upto the height of metal cylinder. Then, bamboo splits are placed vertically along the inner surface with no gap in between the splits. The height of bamboo splits is kept equal to the desired height of storage structure. Now grain is loaded inside the bin so formed upto a height of 90 cm or equal to metal cylinder. Now, the bamboo split becomes erect in position. Afterwards, loading of grains and winding of straw rope on splits go on simultaneously. The grains are thus loaded upto the total height of morai structure. About 1 cm thick mud plaster is done on the straw rope. The structure is covered with a conical roof with ample overhang all around.

2. *Bukhari* : Bukhari type storage structures are cylindrical in shape and are used for storage of sorghum, wheat, paddy, Bengalgram, maize etc. Bukhari structures generally have capacities between 3.5 to 18 tonnes, however, smaller capacity structures also exist. This may be made by mud alone or by mud and bamboo. The cylindrical storage structures are raised above the ground by wooden or masonry platform. The floor of the bin is made either by timber planks or by bamboo splits, plastered over with mud mixed with dung and paddy straw. The walls of the structure are made of timber or bamboo frame work and bamboo matting. Over the walls, mud-straw plaster is applied on both sides. An overhanging cone type roof is provided on the cylindrical structure. The roof is generally made of bamboo framework and straw.

In improved bukhari type structure, the basic shape remains the same but the material and method of construction have been improved to make the structure more safe and durable. The circular floor of structure is either made of wooden planks joined by lap joints or by a double layer of bamboo splits closely set at right angles to each other. Over the floor, about 5 cm thick mud plastering is provided. The walls of structure are made of two sets of strong bamboo framework. The inter-space is filled with mud. The walls on both sides are plastered with mud. The roof is conical and made of bamboo frame-work and covered with paddy straw or similar other thatching material.

The top of the conical roof is covered with 4 to 5 cm thick mud layer to provide additional protection from rains. The structure is raised on timber or masonry pillars to a height of about 1.5 m. Rat proofing cones are placed on all the four pillars to avoid rats entering the storage structure.

Kothar type storage structure : These are used to store paddy, maize, sorghum, wheat etc. Their capacity varies between 9 to 35 tonnes. The storage structure is box like made of wood and raised on pillars. Both the floor and walls are made of wooden planks whereas the thatched or tiled roof is placed over it to protect the grains from the sun or rain.

The improved Kothar structure is generally made of 5 cm thick wooden planks and beams. The walls and floor are made in such a way that no gap exists between the planks. The gabled roof on the top may be made of planks or corrugated metal sheets and should be sufficiently overhang on all sides. The storage structure is raised on timber post to a height of about 1.5 m above the ground. Rat proofing cones are provided on all posts to avoid entry of rats in the structure.

Mud Kothi : These storage structures are quite common in rural areas for storage of grains and other seeds. The capacity of such storage structures varies from 1 to 50 tonnes. These are made from mud mixed with dung and straw. These Kothies are generally rectangular in shape but cylindrical Kothi is also common in some region. There are many sizes and dimensions of Kothi made for storing grains (Fig. 4.1).

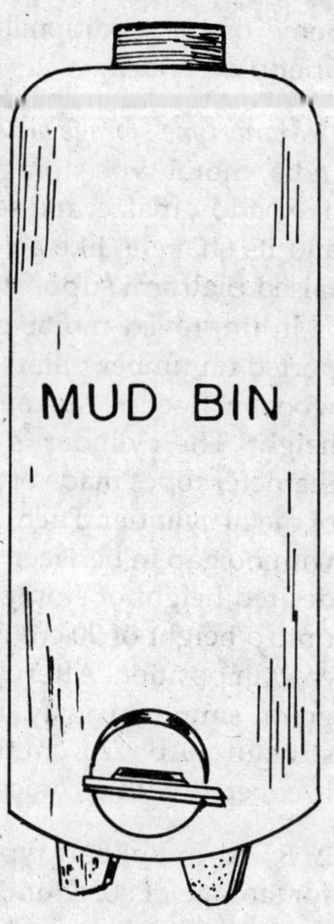

Fig. 4.1 : Mud bin

Muda : These are in use for storing grains in the rural areas of Bihar. The capacity of muda varies between 1 to 3 tonnes. It is being made of "Narai" ropes. The shape of muda is cylindrical and being made in various sizes.

Kanaj : These storage structures are very common in the rural areas of Karnataka and Maharashtra for storage of grains. The capacity of Kanaj varies between 1 to 20 tonnes. It is being made by bamboo splits. The shape of storage structure is cylindrical. The walls of storage structure is sealed with mud plaster on both sides. The roof of the structure is conical and thatched. The roof overhang on all sides.

Kuthla : These storage structures are very much common in rural areas of Bihar and Uttar Pradesh. These structures are kept inside and made of burnt mud.

Improved storage structures

Some of improved, low cost and small capacity storage structures are described in brief.

1. *Pusa bin* : Pusa bin is like other traditional storage structures made of mud. To make the storage structure moisture proof a plastic film is used in all the inner sides of the bin. It is shown in Fig. 4.2.

To construct Pusa bin first a platform of mud bricks is made. On this platform a sheet of 700 gauge plastic is spread in such a way that it overlaps the platform on all sides by atleast 6 cm. On the plastic sheet a layer of 7 cm thick Kachcha

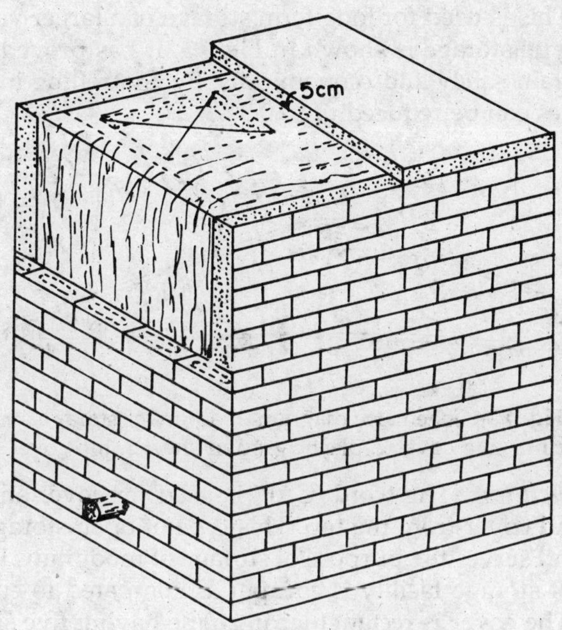

Fig. 4.2 : Pusa bin

bricks is laid. Walls are made of Kachcha bricks and these are sealed with mud plaster. When the walls are raised to proper height a wooden frame is placed on it. The upper roof is made of burnt bricks. For unloading of grains, an inclined wooden or steel pipe is fixed in such a way that grains may come out. The mouth of pipe is closed by a cover. The inside of all the four walls and roof are covered by a plastic sheet. On the top, an open space of about 50 cm × 50 cm is left for loading of grains. Leaving this open space, the roof is sealed by mud. After the bin is filled with grains, the top open space is well covered by a plastic sheet so that air may not enter the bin. Material requirement for two different capacities of Pusa bin is given below.

2 tonne Pusa bin	:	Base 140 × 100 cm; height 160 cm, plastic sheet 180 cm wide and 8.5 m long; Kachcha bricks 1150, burnt bricks 100 and a pipe for outlet.
4 tonne Pusa bin	:	Base 172 × 152 cm; height 160 cm, plastic sheet 180 cm wide, 11.6 m long, Kachcha bricks 1650, burnt bricks 210, and a pipe for outlet

2. *Brick and cement bin (grain house)* : These storage structures are very strong and effect of seasons on these are minimal. The bin is made on a platform raised at 60 cm above the ground. A ladder is provided on one side of the bin for loading of grains. A hole of 60 cm diameter is provided on the roof for loading purpose. The walls of bin are 23 cm thick with cement plaster on both sides. Roof is made of R.C.C. The base of bin is made inclined and an outlet is provided for unloading of grains. The capacity of such bin is between 1.5 to 60 tonnes. For cleaning of bin and complete unloading, a person can go inside with the help of iron rings steps provided inside the bin.

3. *Bunker Storage :* This is used for long term storage of a larger volume of grains. A typical bunker type storage is shown in Fig 4.3. It has proved successful as a means of storing grain safely and economically. By controlling insects and moisture levels, the losses can be reduced upto 0.5%.

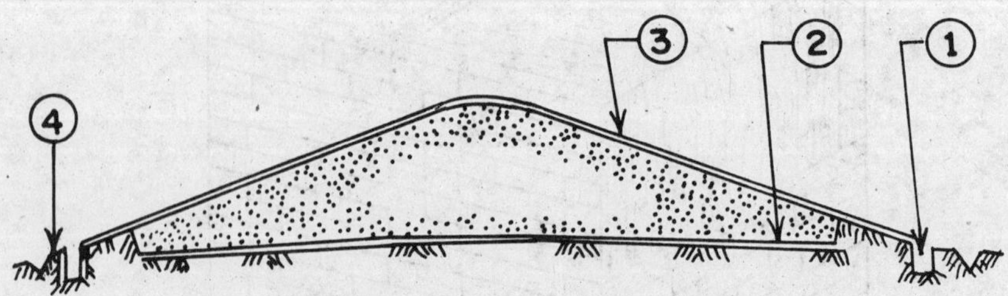

Fig. 4.3 : Schematre diagram of a bunker storage
1. trench for burying cover 2. plastic ground sheet 3. top cover 4. drain

4. *'CAP' Storage structures :'* The word 'CAP' is used for cover and plinth, plinth from the bottom and cover from the top. This type of open storage is considered as transit storage and serves the purpose of storage of foodgrains in bags for short period. This type of storage facility is cheaper as compared to conventional bag storage godowns. The cover is rectangular in shape having five sides made from polyethylene film of 1000 gauge, leaving the bottom side open. The cover is used for protecting stack of bags. Normally the stack is built over a space of 9.11×6.1 m with a height of 18 bags which gives the storage capacity of around 150 tonnes. The cover having a dimension of $9.4\,\text{m} \times 6.4\,\text{m} \times 5.5\,\text{m}$ normally weighs around 52 kg. Sometimes smaller covers are used for covering the stacks in covered varandah of conventional godowns. Such covers are called "Varandah covers". In storage of foodgrains under varandah covers, the stacks are built to a height upto 7 bags having an average capacity of 24 tonnes.

The following steps are required to be followed in the construction of 'CAP' storage:

(*i*) Select a high elevated ground and make it level.

(*ii*) Wooden sleepers are spread with one or two layers of bamboo mat cover on the top as dunnage.

(*iii*) The Gutters are provided all around the area to drain off rain water easily.

(*iv*) The stacking is done to the height upto 18 bags on the dunnage and is covered with polyethylene.

(*v*) The stacks are covered with polyethylene covers and tied with ropes to prevent from blowing off with high velocity wind.

Temperature and moisture changes in storage structures

Moisture migration in stored grain results from temperature changes. Moisture content of grain is one of the most important factors influencing its storage life. High moisture and a warm climate promote mold growth, insect growth and increase rate of respiration of grains. Moisture migration takes place in bin even though the grains are at a moisture level generally considered safe for storage.

When the winter season grains after harvest are stored in the fall, grains become warm. In this condition after passage of time the air at the surface of storage bin cools, because the outside temperature is lower than the temperature in the centre of bin. The cool air at inside surface of bin moves down along the edge of the bin, across the bottom and then up near the centre of the bin where the air and grain are warm (Fig. 4.4).

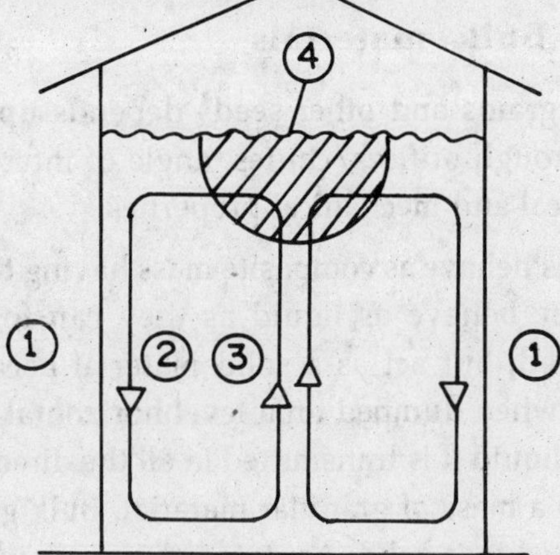

Fig. 4.4 : Loss of food grains due to temperature changes
1. outer cool temperature 2. cool grain
3. warm grain 4. spot of moisture accumulation

The air moving through the centre of the bin picks up moisture and moves across the top of the sides. At this location the surface of the grain is cold and the moisture condenses on the grain consequently the moisture content of grain is raised. Because of increased grain moisture, spoilage takes place at top of the bin.

During winter season when atmosphere is cool the temperature of grain is lowered. When such grain is stored, after sometime the temperature of air in the grain along the surface rises. The atmospheric temperature in spring and later on rises, as a result air currents move down through the centre of the bin to its bottom (Fig. 4.5). At this spot, moisture condenses on the cold spot. The air then moves to

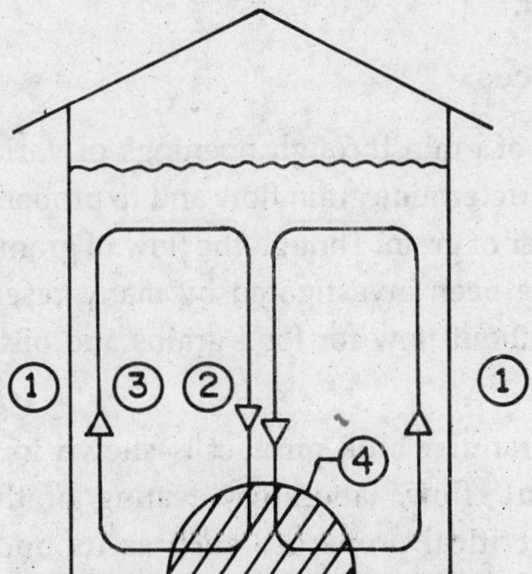

Fig. 4.5 : Loss of grains due to temperature changes in storage
1. outer warm temperature 2. cool grain 3. warm grain 4. spot of moisture accumulation

the walls and due to higher temperature goes upwards. Due to moisture accumulation at the bottom of the bin, spoilage of grain takes place.

Characteristics of bulk materials

The behaviour of food grains and other seeds depends upon their size, shape, density, ease of flow through orifices/chutes, angle of internal friction, moisture content and other physical and mechanical properties.

The granular materials behave as composite mass having the characteristics of a liquid and a solid. They behave as liquid as they can form the shape of the container in which stored, but act as a solid material because of formation of natural angle of repose when dumped on a level horizontal plane. When a static pressure is applied on a liquid it is transmitted in all the directions. To great extent this is also applicable to a mass of granular material. Bulk grains offer resistance of a sliding force as offered by a solid. As stated above, grain mass when dumped on a horizontal plane forms a heap but liquid forms a pool. The bulk granular matter also possesses cohesive strength. The cohesive strength of granular material places them between more cohesive solid and less cohesive liquids.

The grain mass is a numerous small solid particles which can also move around each other and thus becomes a flowing mass. The laws of hydro-dynamics are not applicable to the flow of granular materials, because

(*i*) Pressure is not uniformly distributed in all the directions due to development of arches and frictional forces between granular particles.

(*ii*) The solid grains have definite shape and size which govern the rate of flow. Fluids do not have these properties.

(*iii*) The rate of flow is not proportional to head except at heads smaller than container diameter.

Flow through orifices

Information on the flow of grain through openings of various sizes, shapes and orientation is required to determine grain flow and to properly size the opening for flow control during tansfer of grain. Though the flow of granular materials through horizontal openings have been investigated by many research workers, limited information is available about flow for food grains and oilseeds through vertical orifices.

The flow pattern of granular bulk product is shown in Fig. 4.6 A. To ensure reliable 'first in first out' flow, laboratory testing of the stored product to determine a number of critical properties such as its unconfined compressive strength after consolidation, internal friction angle, the wall friction angle are required. Hopper slope angles of 60° or 70° are usually necessary to achieve reliable mass flow. With normal hopper slopes of 30°–40°, pipe flow or funnel flow are more likely to occur in which the bottom layers of material remain static until the upper material has been discharged.

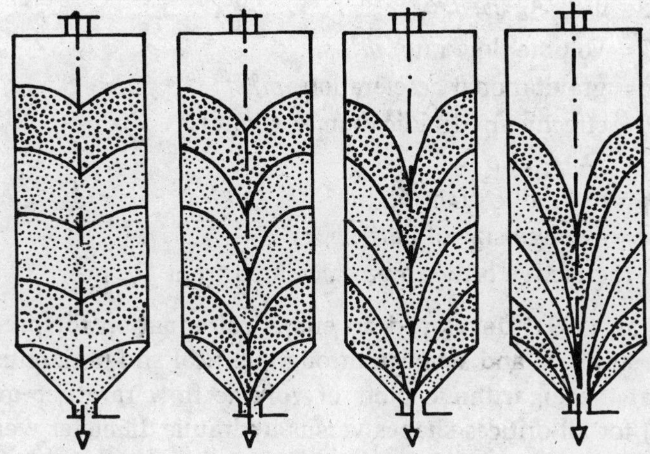

Fig. 4.6 A : Flow pattern of granular bulk product with centre discharge

Deming and Mehring applied dimensional analysis technique to study flow of solid particles through funnels and proposed the following relationship.

$$t\,D^{2.5}\,w = f\left(\frac{d}{D}\right)$$

Where, t = time, minutes for flow of 100g material

D = orifice diameter, em

d = equivalent particle diameter, em

w = bulk density, g/cc

They could not establish the actual form of the function, $f\left(\frac{d}{D}\right)$ and proposed that

$$Q = \frac{100\,D^{2.5}\,w}{\tan\phi_r\left[\left(34.6 + 37.4 + 444\sin\frac{\phi_r}{2}\right)\left(\underline{d} + 0.13 - 0.161\tan\phi_r\right)\right]}$$

where, Q = flow rate, g/min

ϕ_r = angle of repose of material

$$\underline{d} = \frac{0.8\left[d_2^5 - d_1^5\right]}{d_2^4 - d_1^4}$$

d_2, d_1 = major and minor dimensions of particle

Ewalt and Buelow (1963) reported that the flow rate of grain through an orifice is independent of the depth of grain above the opening. Beverllooetal (1961) suggested that flow rate varies with the orifice, are times the hydraulic diameter raised to 0.5 power for flow through horizontal openings.

$$Q = 0.75\ A_e \sqrt{g\ De}$$

where, Q = Volume flow rate, m^3/s

g = gravitational acceleration, m/s^2

D_e = effective hydraulic diameter, m

$= D_h - 1.4\ d$

D_h = hydraulic diameter, m

d = average size of particle, m

A_e = effective orifice area calculated from D_e, m^2

Chang et al. (1991) developed an emperical equation for predicting flow of wheat, corn, sorghum and soybean through vertical and horizontal orifices. They reported that the logarithmic plots of volume flow rate per unit orifice area ($m^3/hr/cm^2$) for all orifices shapes versus hydraulic diameter were nearly linear for all tests. They expressed the flow rate as,

$$d = KAD^n$$

where, Q = volume flow rate through the orifice, m^3/hr

A = Orifice area, cm^2

D = orifice hydraulic diameter, cm

K = coefficient, hr^{-1}

n = coefficient, dimensionless

The discharge of wheat and paddy (rough rice) through horizontal rectangular opening is studied by Bisht et al (1989), and the they proposed the following expressions of flow rates for wheat and paddy.

$$Q = 0.499\ \rho_B\ A^{1.4481} \text{ (for wheat)}$$

and $$Q = 0.578\ \rho_B\ A^{1.4481} \text{ (for paddy)}$$

where, Q = flow rate of material, kg/min

ρ_B = bulk density of material, g/cc

A = area of opening, cm^2

The above equations represent that the volume flow rates of wheat and paddy through opening of rectangular shape are proportional to their bulk density and increase rapidly as the opening area increases.

Gregory et al. (1987) simulated the conditions of grain flow through circular orifice with that of flow of sand in an old hour glass. They mentioned that the flow in the top reaction has a high degree of acceleration due to flow lines of high curvature. This section represents a transition from relatively low velocities at the outer wall to high vertical velocities as the flow converges to the main core (Fig. 4.6B). Below the transition zone and just above the orifice, the core is relatively straight and has the size and shape of the orifice. The material discharged is replaced by new material from above, but material on the outside of the core does not move. The downward force causing flow can be represented as a function of cross-sectional area of the orifice, depth of flow, bulk density of material and acceleration on due to gravity. While the upward force restricting flow can be expressed as a function of the shear stress of the material in the boundary layer exists between the core flow and the non-flowing material.

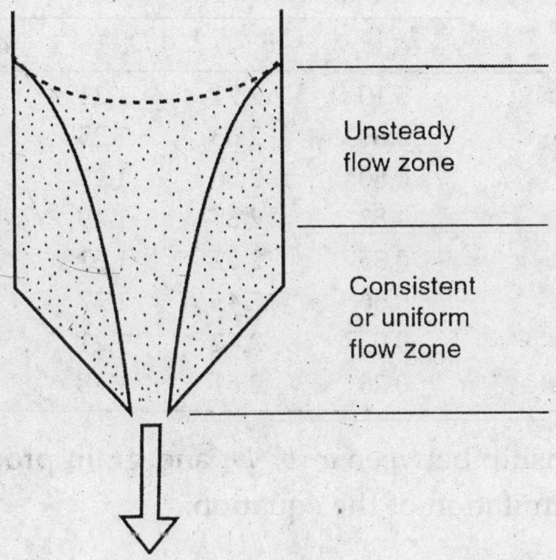

Fig. 4.6 B: Flow through circular orifice

Some studies on flow characteristics of selected Indian varieties of food grains and fertilisers through horizontal circular openings were conducted at Pantnagar.

No definite relationship between a, b, D_c and grain properties could be established. This was the limitation of the equation.

Dehpour et al (1992) studied the flow of granular fertilisers and proposed an emperical expression for discharge rate. The discharge rate was expressed as a function of bulk density of material, orifice diameter, effective diameter of particle and acceleration due to gravity. Using dimensional analysis the following expression was proposed.

$$Q = a\left(\frac{A}{D^2}\right)^b PD^{2.5}\ g^{0.5}$$

where Q = discharge rate

A = area of opening

D = average particle size

P = weight density of particle

g = acceleration due to gravity

a, b = emperical constants

The initial flow of fertilisers was started when an arch was developed at the centre of the hopper and thereafter the material slides from the surface of the hopper towards its centre. Arching is associated with convergent flow of granular material which is necessary for the bulk material to loose its strength to flow easily.

a, b, D_c = experimental constants given in Table 4.1

Table 4.1
Constants of equation 4.5

Material	45°			60°		
	a	*b*	*Dc*	*a*	*b*	*Dc*
Soybean	5.10	2.71	1.31	2.44	2.99	0.57
Wheat	3.81	2.63	1.25	1.84	2.86	1.12
Millet	0.80	2.56	1.27	0.66	2.74	0.68
Pigeon-pea	4.83	2.67	1.49	2.37	2.89	0.88
Rapeseed	0.35	2.52	1.31	0.12	2.71	0.72
Urea	0.66	2.45	1.37	0.24	2.67	0.64
Super Phosphate	0.027	2.62	1.21	0.01	2.83	0.47
Di-ammonium phosphate	0.26	2.61	1.28	0.12	2.81	0.70

No definite relationship between a, b, D_c and grain properties could be established. This was the limitation of the equation.

Particle shape and size

The shape and size of the individual particles and their internal friction affect the flowability of a bulk product. Regularly shaped particles cannot pack together because of the void space around them. This property does not restrict the free flowing movement of the particle with respect to surrounding particle. But irregularly shaped particles or a mixture of large and small particles can interlock, thus they affect the flowability or flow pattern.

Since the shape of food grains, oilseed and other agricultural granular materials are generally irregular, the size of grains is represented by their equivalent diameter. It is diameter of a sphere having the same ratio of surface to volume as the actual particle. The equivalent diameter of a particle can be determined by the following equation.

Equivalent diameter = $(a\ b\ c)^{1/3}$

where, a, b, c are the maximum, intermediate and minimum mutually perpendicular dimensions respectively.

Sometimes the sphericity of particle is also necessary to know the shape. Particle sphericity is equal to the ratio of surface area of sphere having same volume as that of particle of the actual surface area of the particle.

$$\text{Sphericity} = \frac{d_i}{d_c} = \frac{(a\ b\ c)^{1/3}}{a} \qquad \ldots 4.6$$

where, d_i = diameter of largest inner circle of the particle

d_c = diameter of least circumscribed circle of the particle

Grains and other agricultural granular materials have different size ranges of the same product. This may result in completely different handling and conveying characteristics. The size distribution of a product can be determined by mechanical sieving method, by sedimentation method and by optical microscopy. Mechanical sieving can effectively determine the size distribution of grain and other products.

The larger particles of a bulk grain generally do not obstruct the flow, but the smaller particles of the same material act as binding agents as they can be fitted among larger particles. Therefore, the flow characteristics of a bulk product are mainly influenced by the proportion of small particles.

Bulk density and true density

For the design of storage and other systems, knowledge of bulk and true density of product is necessary. The bulk density of a granular product is the density measured without the influence of any compression on the product. The bulk density depends on particle density, its shape and the manner in which the constituent particles are packed or positioned with respect to each other. The bulk density is also represented by hecto-litre weight. The weight of grain per hecto-litre is called hecto-litre weight. It is measured with the help of a kettle, pan and one wooden stroker. The capacity of kettle is 500 ml. Inside diameter is 8.5 cm and the height is 8.8 cm. Its sensivity when filled with water shall be within ± 5 ml.

To measure the bulk density, product is filled in the pan and it is held over the kettle in such a way that the opening of the pan is above the centre of the kettle and pouring occured from a height of 15 cm above the kettle. The product is poured into the kettle in regular stream. The wooden stroker is placed on the edge of the kettle and with gentle stroke, extra product is removed. The product of kettle is now weighed.

Hectolitre weight $= m \times 0.2$, kg

where, m = weight of product, g

The true density represents the actual density of particle. It is determined by liquid displacement method. The liquid generally used is toluene. Toluene is filled in a measuring cylinder and the volume is recorded. Then, 100 g grain is put in cylinder. The rise in volume of toluene in cylinder is noted. True density is the ratio of weight of grain and the difference in volume of toluene.

In case of grains, the bulk density is increased when the grains are compressed. During this process reorientation and settling of particles take place. As a result air percolates out of the bulk mass and the volume decreases. This increases the bulk density. The density of compressed mass or bulk is called the 'apparent density'. The value of apparent density is higher by 20% than the normal bulk density. In design of a silo, the apparent density of grain is taken into account. Because of compaction, the bottom layer or base layer of silo has maximum density whereas the top layer has minimum density.

Angle of repose

When a granular product is dumped through a circular opening on a level horizontal surface, the product takes a shape of inverted cone (Fig. 4.7). The angle between the horizontal and inclination of heap is called the natural angle of repose. The angle of repose is an indicator of the product's ability to flow. Each product has its own natural angle of repose. The cohesive materials have larger angle of repose, whereas the lower angle of repose represents easier flowability of the product. The flowability of products with respect to the natural angle of repose is given in Table 4.2. The size, shape, moisture content and orientation of the particles influence the angle of repose.

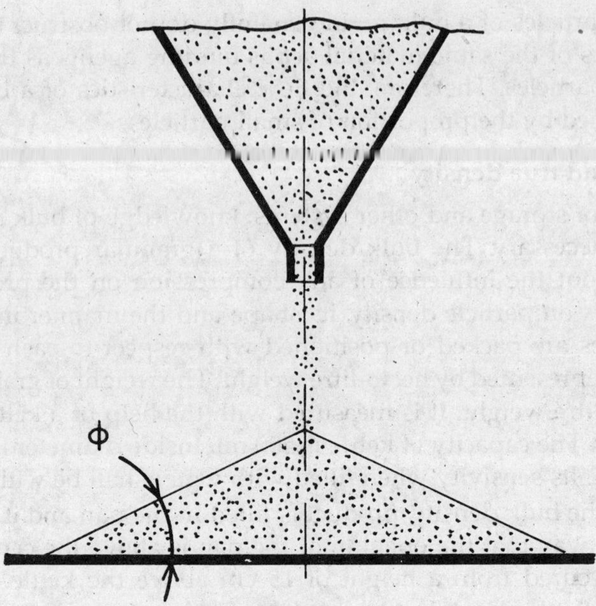

Fig. 4.7 : Angle of repose

Table 4.2

Flow characteristics with respect to angle of repose

Flow characteristics	*Angle of repose, degree*
Very free-flowing	25–30
Free-flowing	30–38
Fair flowing	38–45
Cohesive or non-easy flowing	45–55
Very cohesive	beyond 55

The natural angle of repose of a certain product is approximately equal to the maximum angle of internal friction of that product.

Internal and External Friction

In grain conveying systems, two types of friction can be identified, (1) wall friction or external friction, which is friction between grain mass of kernels and the wall of bin or a surface and (2) friction of the kernels or grains against each other known as internal friction.

Consider a sliding mass of grains kept in a container. The container has no bottom. This container is placed over a horizontal plane as shown in Fig. 4.8. If 'F' is the horizontal force or pull and N is the vertical force caused by the total weight of grain, the friction coefficient 'μ' can be given by,

$$\mu = \frac{F}{N} \qquad ...4.7$$

Coefficient of friction 'μ' between granular materials is equal to tangent of the 'angle of internal friction' for that material.

The sliding stress between the grain and the horizontal plane gives the frictional resistance per unit of surface area of the grain against the wall. This is called the

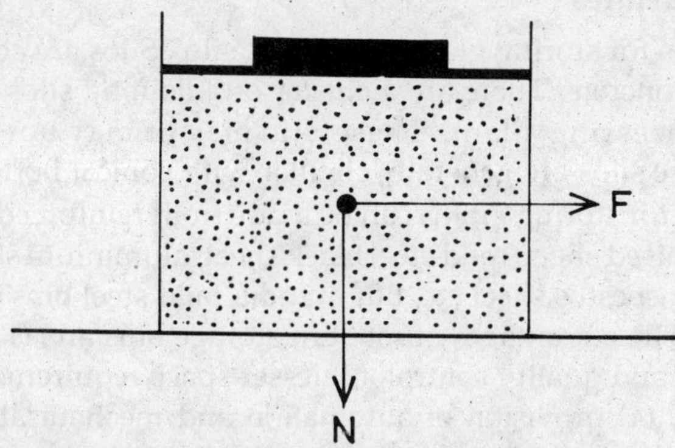

Fig. 4.8 : Sliding mass of grain on horizontal plane

angle of wall fiction or external friction. Some values of angle of friction of few silo material are given in Table 4.3.

Table 4.3
Angle of friction of various silo wall materials

Silo wall material	*Angle of friction, degree*
Oxidised sheet metal	38
Concrete	25
Smooth sheet metal	15
Stainless steel	8

In the above setup, if the horizontal plane is replaced by the grain present in the container, we can measure the internal frictional resistance of the grains against each other. The frictional resistance depends on type and roughness of the horizontal plane, weight of the product, humidity of product and the surface. The angle of internal friction of some of the grains are given in Table 4.4.

Table 4.4
Angle of internal friction of some grains

Grains	*Angle of internal friction, degree*	
	Minimum	*Maximum*
Wheat	24	26
Maize	26	29
Barley	25	29
Rice	24	26

Higher values of angle of internal friction indicate that the material is cohesive. The lower values of angle of internal friction are indicative of easy flowing products. The exact values of angle of internal friction and wall friction of products can be determined experimentally in laboratories.

Moisture content

Flow characteristics of grains are also effected by its moisture content. The moisture adhering at the surface may cause cohesiveness and restrict easy flowing capacity of the material.

Modern storage structures

The modern facilities for storing grains in bulk is 'silo'. Silos are constructed from steel or reinforced concrete. There are a cluster of adjoining silos in any modern large capacity processing plant. These silo may or may not be of different configurations. These silos are generally circular with conical bottom.

Bulk storage bins for storing grains can be made from reinforced concrete, plain or corrugated galvanised sheet, mild steel black sheet, aluminium sheet, fibre glass, brick, ferro cement, asbestos sheet etc. But in India mild steel bins and R.C.C. bins are quite common. The advantages of modern storage bins are, (1) less expensive and easier handling and quality control, (2) lesser space requirement, (3) elimination of cost of bags, (4) provision of automation and mechanisation for quicker handling and maintaining quality of stored product, (5) protection from losses due to birds and rodents.

Deep and Shallow bins

Silos/bins are classified into two groups depending upon the relative dimensions of the container. These are classified as, (1) deep bins and (2) shallow bins. To what extent of relative dimensions of depth and diameter or width, the structure

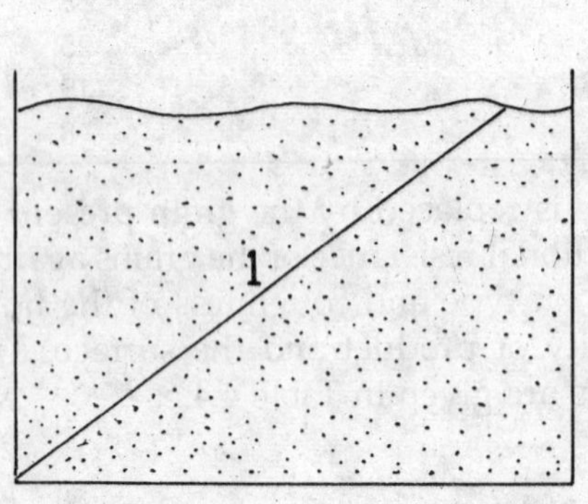

Fig. 4.9 : Shallow bin
1. plane of rupture

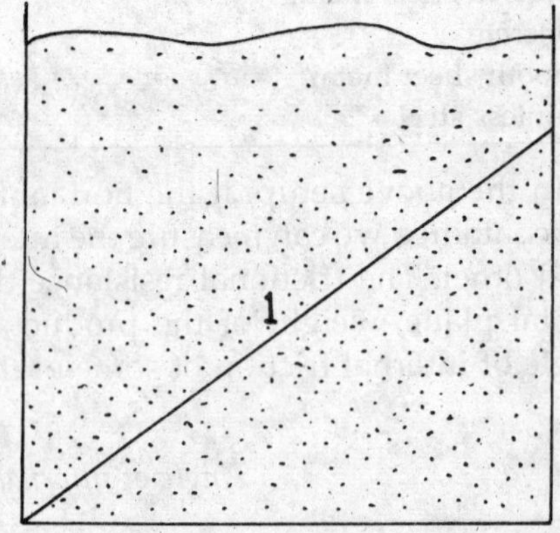

Fig. 4.10 : Deep bin
1. plane of rupture

behaves as a deep or shallow bin is decided from the concept of the plane of rupture. The plane of rupture is that surface down which a wedge of material bounded by one wall face, the free surface, and the plane of rupture would start sliding if bounding wall were to move.

A bin whose relative dimensions are such that the plane of rupture meets the grain surface before it strikes the opposite side as shown in Fig. 4.9 is called 'shallow bin'. A bin in which plane of rupture meets the opposite side before it emerges from the grain as shown in Fig. 4.10 is called 'deep bin'.

If ϕ is the angle of repose of material, then the angle of rupture may be taken as $\frac{90+\phi}{2}$ with the horizontal as per Rankine's theory. If L and h be the breadth and depth of the storage structure, then the structure can be taken to be shallow if:

$$h < L \tan\left(\frac{90+\phi}{2}\right) \qquad ...4.8$$

The Structure is deep if

$$h > L \tan\left(\frac{90+\phi}{2}\right) \qquad ...4.9$$

There are other definitions for deep and shallow bins. A grain bin referred to as a shallow bin when the depth of grain is less or equal to the equivalent diameter. In a deep bin, depth of grain is greater than the equivalent diameter. The equivalent diameter is taken as four times the hydraulic radius of the bin.

For deep bin $H_d \geq 4\,R$

For shallow bin $H_d < 4\,R$

where, H_d = depth of grain

$$R = \text{hydraulic radius} = \frac{\text{area of cross section of bin}}{\text{perimeter of bin}}$$

As per the definition given by Issacson and Boyd for deep bin

$$\frac{H_d}{D} \geq 0.75\left(\frac{1}{\mu k}\right)$$

for shallow bin $\dfrac{H_d}{D} < 0.75\left(\dfrac{1}{\mu k}\right)$

where, D = diameter of circular bin

μ = coefficient of friction

k = pressure ratio, $\dfrac{\sigma_3}{\sigma_1}\left(\dfrac{\text{lateral}}{\text{vertical}}\right)$

$$= \frac{1-\sin\,\phi_\iota}{1+\sin\,\phi_\iota}$$

ϕ_ι = angle of internal friction

Permanent storages

The modern permanent storage system should be selected for the safe keeping of stored grains and other products. The modern storage structures should be selected on the basis of first on quality and then on cost considerations. The value of the stored products over the life of the storage will far exceed the value of storage container. If the proper storage structure is not selected, the high loss rates may quickly erode any capital cost savings made in the selection of poor quality storage. A good quality storage structure may be structurally adequate, durable, hygienic and gas sealable for fumigation. Various forms of permanent storage structures like shed, silos have been designed and built. They are, however, not all equal in quality or price. In addition, some types of storage are better suited to a particular application than others.

Shed

Traditionally, horizontal sheds have been used to provide low- cost, large volume storage. Very large volume sheds have also been constructed by Central Warehousing Corporation for storing grains and other products. Sheds are usually made of steel or corrugated sheet construction with flat concrete or bitumen floors (Fig. 4.11).

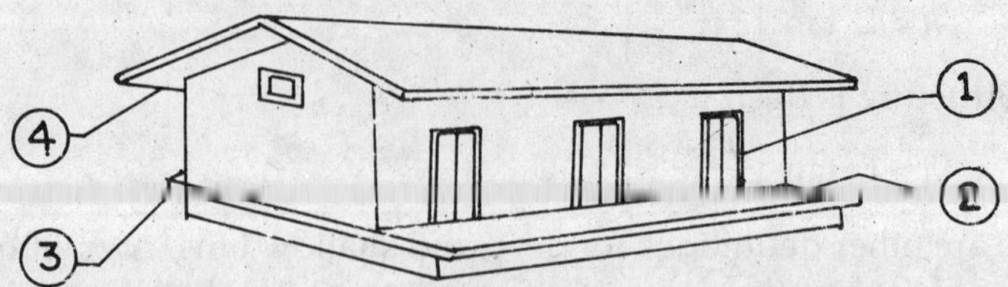

Fig. 4.11 : Diagram of modern godown (shed)
1. sealed doors 2. floor 3. rat proof slab 4. air proof roof

Due to their shape, sheds are relatively inefficient structures for grain storage. Grain loads on the walls have to be supported by girts and heavy vertical butteresses designed to resist their loads in bending. Roofs have to be supported on a network of purlins and rafters. A typical bulk ware house would be 15 m wide, 60 m long and 6 m high at the eaves. A building of this size would have a storage capacity of 2,840 tonnes. The same size building for bag storage has only 1440 tonnes storage capacity.

Belt conveyors are often used in bulk sheds for loading and unloading grains as shown in Fig 4.12. A tunnel in ground is provided for belt conveyor. This tunnel

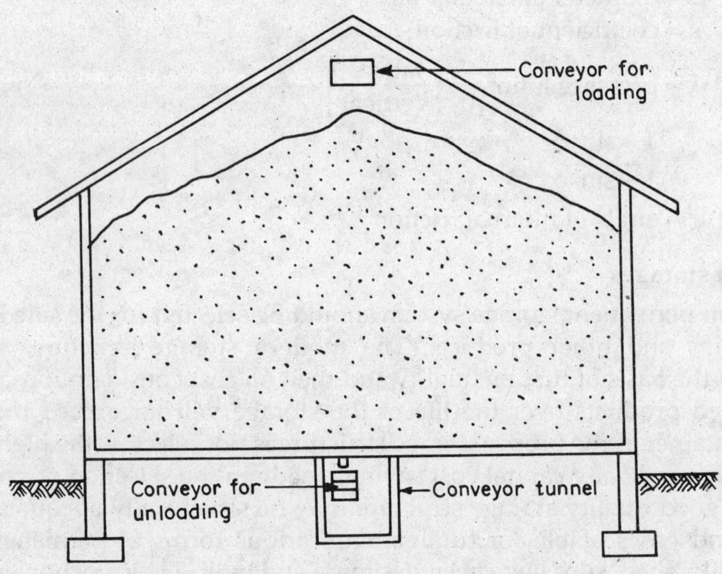

Fig. 4.12 : Cross-section of a ware house showing conveyors for loading and unloading of grains

should be wide and deep enough to allow person for service and necessary repairs. The top belt requires a tripper or plough for unloading. In still larger sheds of 50,000–60,000 tonnes capacity, suspended drag-chain conveyors are incorporated. These conveyors are suspended from motorised monorail trolleys which enable them to travel the length of the shed above the grain surface. These conveyors collect and drag the grain to the centre of the shed floor where it drops unto an unloading conveyor in a tunnel below as stated earlier. The installation of conveyors is quite expensive. Therefore, cost of sheds depends upon their size and other factors.

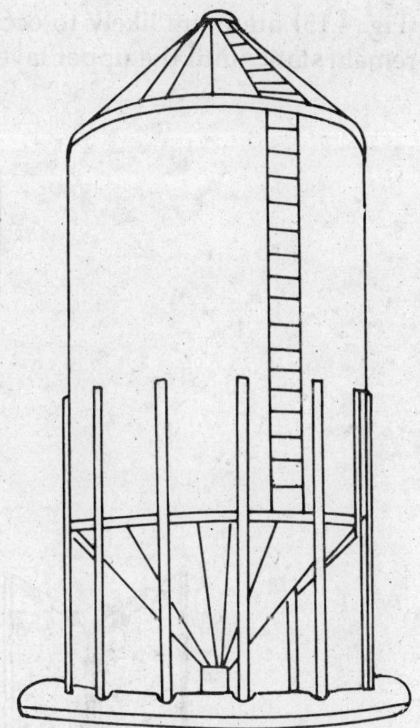

Fig. 4.13 : Hopper bottom vertical silo

Vertical Silos

Vertical silos can be circular, hexagonal, square or rectangular. They tend to be more expensive than sheds. But the extra cost is justifiable where high throughputs demand the provision of fully gravity discharge. In case of space limitations also, vertical silos are preferred because limited space precludes the use of large floor areas required for shed. Where gravity discharge is required, it is achieved by the provision of sloping floors with either single or multiple discharge points. But to reduce the risks of eccentric discharge patterns and to minimise complex construction and costs, a single central discharge is preferred.

Floor slopes should be adequate to minimise material 'hang-up'. For smooth floor surfaces, slope angles of 30 to 35° generally give satisfactory results. However, for very dusty or moist grains, slope of 40° will not guarantee complete discharge.

Hopper bottom silo (Fig. 4.13) is quite common for storing grains. Prefabricated galvanised metal silos are available in many sizes. Flat bottom vertical silos are also common. These may be fabricated by metal sheets, galvanised metal or reinforced concrete. The hopper bottom is more expensive than flat bottom bin (Fig 4.14).

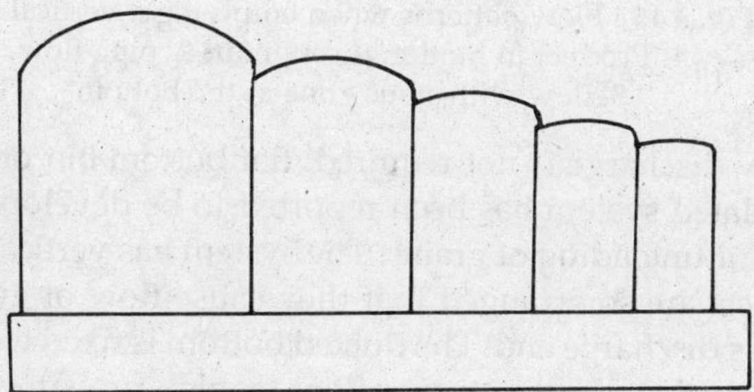

Fig. 4.14 : Flat bottom silos

In situations where silos are used as buffer storage in a continuous throughput situation, the avoidance of static product in the bottom may be critical to avoid segregation and/or deterioration. The design of silo floor to ensure discharge of product first loaded requires studies on properties such as product's unconfined compressive strength after consolidation, angle of internal friction and angle of wall friction. Hopper slope angles of 60° – 70° are necessary to achieve reliable mass flow. With normal hopper slopes of 30°–40°, pipe flow or funnel flow

(Fig. 4.15) are more likely to occur. In such flows, the bottom layers of material remain static until the upper layers of material have been discharged.

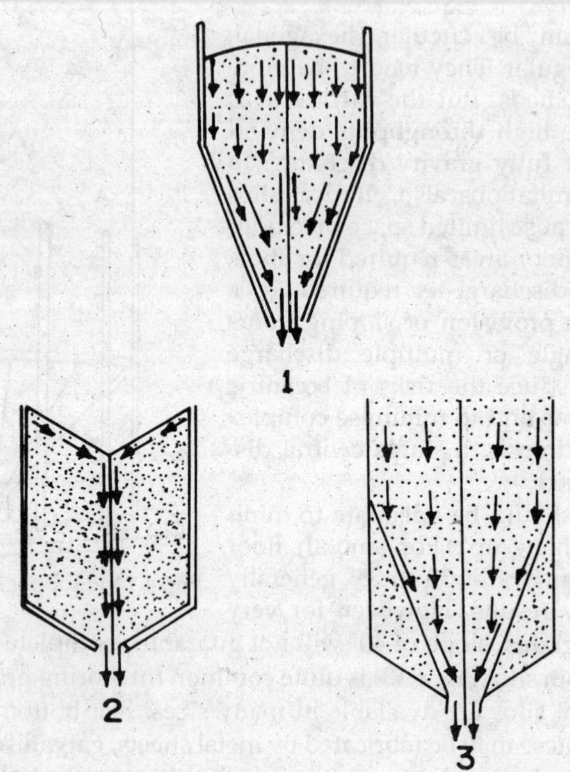

Fig. 4.15 : Flow patterns when emptying a vertical silo
1. Product in motion throughout 2. pipe flow
3. flow with static zone at the bottom

Where gravity discharge is not required, flat bottom bin offers significant cost savings. A ventilated system has been reported to be developed for small square flat bottom silo for unloading of grains. The system has vertical oval shaped metal ducts. These ducts are so arranged that they cause flow of grains in a particular direction towards discharge end. The ducted bottom is fixed to the floor of the silo. Static grain is moved across the storage floor by high pressure air inside the ducts. The system is reported to be successful but its use is primarily in square, flat bottom silo with small floor areas.

Squat Silos

A squat silo has a wall height to diameter ratio 0.5 or even less (Fig. 4.16). Squat silo can compete with sheds for low-cost quality storage. Squat silos are structurally efficient, thus can be built from lighter materials and costs are generally lower for a given storage volume. The structural efficiency is achievable by the walls carrying their loads principally in tension and by the conical roof carrying its load

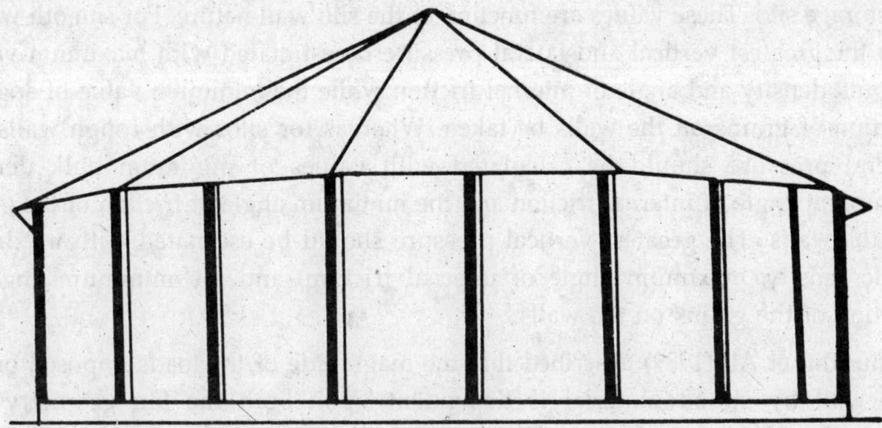

Fig. 4.16 : Squat silo

in tension, compression and shear. Squat silos as large as having 48 metres diameter, 10.5 metres high walls and 25 metres high at roof apex are being built.

Silo Design

The agricultural granular product like cereals which are mostly stored in silo, constitue a friable cohesionless mass. The equilibrium granular mass depends upon the speed of filling, the ways the grains are packed together, the height of fall of grain in silo while loading and the compressibility of the material. The above factors affect the characteristics of grain like bulk density, internal angle of friction, coefficient of friction on the walls. The equilibrium laws for grains are not well developed. It does not fit in the theory of elasticity, therefore, it is very difficult to directly determine the pressure acting on the walls.

The walls of the silos have to sustain the self weight and loads during construction. The silos walls are subjected to two main forces or thursts, (1) the horizontal thurst due to loading of material which tries to open up the silo and (2) the vertical pressure which is generated due to loading of material on the walls.

The speed of filling of granular material affects the stress development on silo. When the material is filled very rapidly without any compaction other than the compaction produced in the operation, the greatest forces are obtained for the lateral thurst while the apparent density and the angle of internal friction of the material are at minimum values. On the contrary, when the grain is filled at such a slow rate which provides every possibility of compaction, the greatest forces are obtained for the vertical pressure. In this case the bulk density and the angle of internal friction of material are at maximum values.

The characteristics values of the granular material which cause greatest stresses should be taken into account while determining the lateral and vertical pressures of storage silo. These values are function of the silo wall nature. For smooth walled silo the greatest vertical and lateral pressure be estimated with maximum values of bulk density and angle of internal friction while the minimum value of angle of friction of grains on the walls be taken. Whereas for silos with rough walls, the lateral pressure should be calculated with values of minimum bulk density, minimum angle of internal friction and the minimum angle of friction of the grains on the walls. The greatest vertical pressure should be estimated with maximum bulk density, maximum angle of internal friction, and the minimum angle of friction of the grains on the walls.

Bucklin et Al. (1989) described that the magnitude of the loads imposed on the bin wall by stored material is dependent upon both the bin geometry and properties of the stored material. These properties include bin diameter, bin wall configuration, bulk density of stored material, horizontal to vertical pressure ratio and the coefficient of friction between the material and the bin wall.

The grain pressure in bins were first calculated as being a semi liquid of same density as the grains. The lateral pressure was first calculated using the hydrostatic formula.

$$P_l = wh \qquad \text{...4.10}$$

Where, P_l = lateral pressure exerted by the grain on the bin wall

w = density of grain

h = depth of grain from the top of the bin surface

This formula has serious deficiencies because many structures buckled under the vertical load arising from the friction of the grain on the walls. The formula 4.10 was modified to incorporate a factor '*K*' known as Rankine's earth pressure coefficient. The Rinkine's formula is based on the following principle, the resistance to displacement by sliding along a given plane in a loose granular mass is equal to the normal pressure exerted between the parts of the mass on either side of the plane, multiplied by the specific constant. The formula 4.10 when multiplied by the factor '*K*' is known as Rankine's earth pressure coefficient.

$$P_l = K\,w\,h \qquad \text{...4.11}$$

Where K = Rankine's coefficient

$$= \frac{1 - \sin\phi}{1 + \sin\phi}$$

ϕ = angle of internal friction of the grain

The internal angle of friction of the grain was assumed by theoreticians and engineers, as equal to the natural angle of repose.

Therefore, $P_1 = wh\dfrac{1-\sin\phi}{1+\sin\phi}$ ...4.12

This improvement although reduced the magnitude of the lateral pressure, yet it did not consider the vertical load on the wall resulting from friction effect. The whole weight of the grain was assumed to be transferred to the bottom of the beam.

The Rankine theory with horizontal soil surface behind a vertical wall is shown in Fig. 4.17. Consider an element of soil at depth 'h'. The vertical pressure acting on this element is 'P_v' and the horizontal pressure P_l is active pressure at failure.

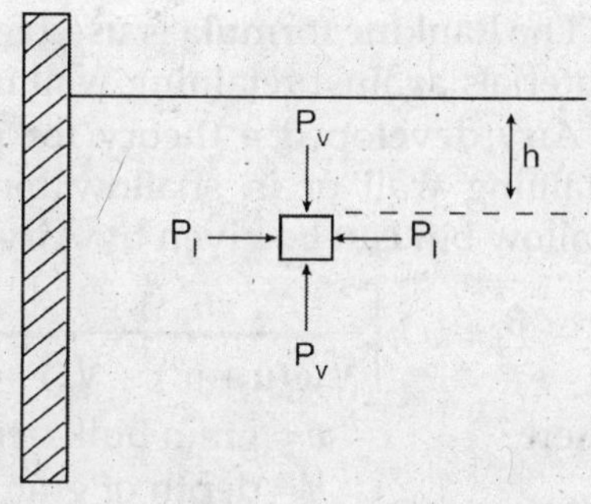

Fig. 4.17 : Pressures on an element of soil behind a vertical wall

For the Rankine theory Mohr circle may be constructed as shown in Fig. 4.18. The angle 'ϕ' is the angle of internal friction of soil. P_v and P_l are the maximum and minimum principal stresses for which the Mohr circle is constructed.

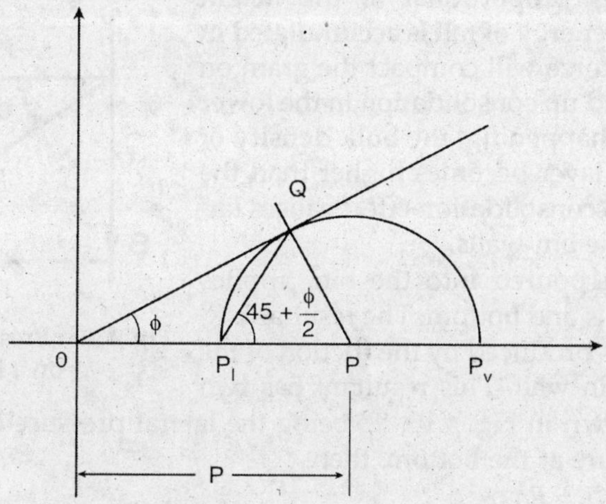

Fig. 4.18 : Mohr circle for Rankine theory

Now $$\sin\phi = \frac{QP}{OP} = \frac{\frac{1}{2}(P_v - P_l)}{\frac{1}{2}(P_v + P_l)}$$

$\therefore$ $$P_v \sin\phi + P_l \sin\phi = P_v - P_l$$

or, $$P_v(1-\sin\phi) = P_l(1+\sin\phi)$$

or, $$P_l = \frac{1-\sin\phi}{1+\sin\phi} P_v$$

or, $$P_l = K P_v$$

where, K = Rankine coefficient = Coefficient of earth pressure

$$K = \frac{1 - \sin \phi}{1 + \sin \phi} = \tan^2 \left(45 - \frac{\phi}{2}\right)$$

The Rankine formula is used for determination of pressure induced by granular materials against retaining wall in shallow bins.

Airy developed a theory for pressure induced by granular materials against retaining wall or in shallow bins. The lateral pressure exerted by grains in a shallow bin can be given by Airy equation

$$P_l = w\,h \left[\frac{1}{\sqrt{\mu(\mu + \mu')} + \sqrt{(1 + \mu^2)}}\right]^2 \quad \text{...4.13}$$

where, w = grain bulk density
h = depth of grain to point under consideration
μ = coefficient of friction of grain on grain
= $\tan \phi$, ϕ is angle of internal friction
$\mu' = \tan \phi'$, ϕ' is the angle of wall friction.

Pressure distribution in a bin

In a flowing process, a loose product like grain in bulk is not compressed. But when grain is loaded in a bin from top, it is being compressed. The increase in solidity of grain is proportional to the height because the kinetic energy of fall is accumulated in grain. This impact force will compact the grain on the bottom and build up consolidation in the lower layers. Due to this happening, the bulk density of product in bottom layer becomes higher than the upper portion. This consolidation effect causes lateral pressures on the bin walls.

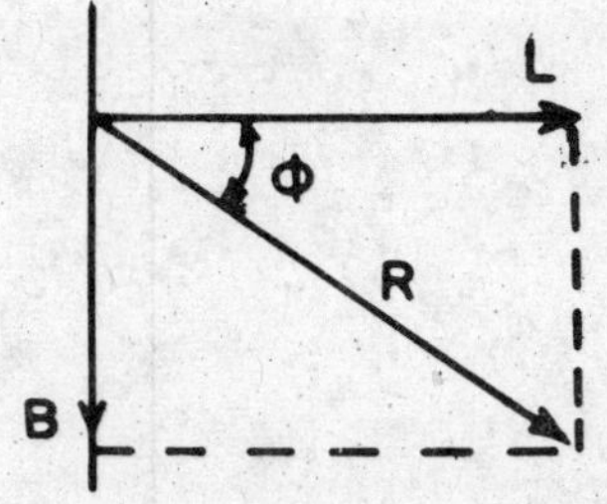

Fig. 4.19 : Pressure components on a bin wall

The grain when poured into the bin, applies pressure on the walls and bottom. The resultant 'R' of these pressures is produced by the friction of the grain against the bin wall. This resultant has two components as shown in Fig. 4.19. 'L' being the lateral pressure on the wall, and 'B' being the pressure at the bottom, then:

$$L = R \cos \phi$$

and $$B = R \sin \phi$$

Therefore, $$B = L \tan \phi$$

where, $\tan \phi$ = corresponding coefficient of friction between grain and bin wall

Thus 'B' is the equilibrium pressure by the wall friction corresponding to the lateral pressure 'L'.

At a particular depth inside the silo, the load on the bottom, or the total vertical pressure is equal to the difference between the total weight of the stored grain and the total load balanced by the friction of the material on the walls. The vertical pressure on the bin bottom increases with the level of filling upto certain value, beyond this level of filling, the vertical pressure attains a maximum value and the pressure becomes constant as shown Fig. 4.20 a.

A similar nature of curve of lateral pressure on the bin walls is achieved when the filling of grains rises (Fig. 4.20 b). This curve has a asymptote parallel to the axis of the abscissa and the ordinate represents the maximum lateral pressure.

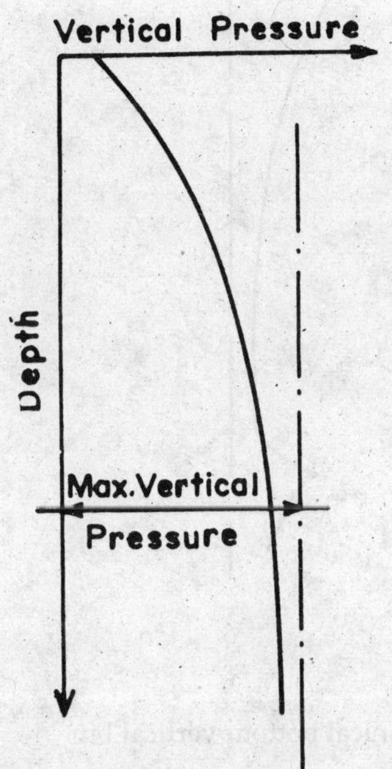

Fig. 4.20 a : Vertical pressure in a bin

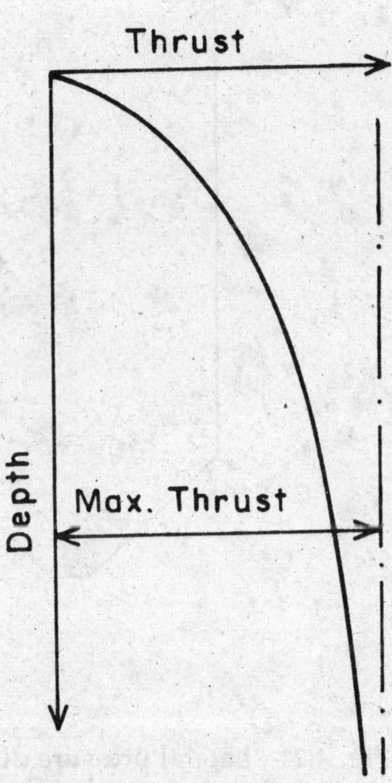

Fig. 4.20 b : Lateral pressure on a bin wall

Experimental studies have shown that when the velocity of filling is increased, the pressure on the bottom is lowered and the lateral pressures become higher. In a conical bottom vertical bin, at the transition point of the discharge hopper, the pressure reduces (Fig 4.21). The pressure again starts building up in the direction of outlet. This shows that the pressure in the discharge hopper is not dependent on the height of filling of grain in the bin. Due to such phenomenon of pressure, the flow pattern of grain from a vertical bin with gravitational discharge is influenced by the level of filling in the bin.

The function of the lateral as well as the vertical pressure is not fully known to calculate the pressures on the walls of a silo, the basic theory for calculating thrust on retaining wall has been adopted. According to this theory the ratio of vertical pressure to the lateral pressure is constant and is equal to $\frac{1+\sin\phi}{1-\sin\phi}$.

Janssen (1895) took into account the friction between the grain and bin wall and proposed the following formula for 'deep bins'.

$$P_l = \frac{wR}{\mu}\left[1 - e^{-\frac{K\mu h}{R}}\right] \qquad \ldots 4.14$$

where, P_l = lateral pressure

R = hydraulic radius

w = grain bulk density (unit weight of grain)

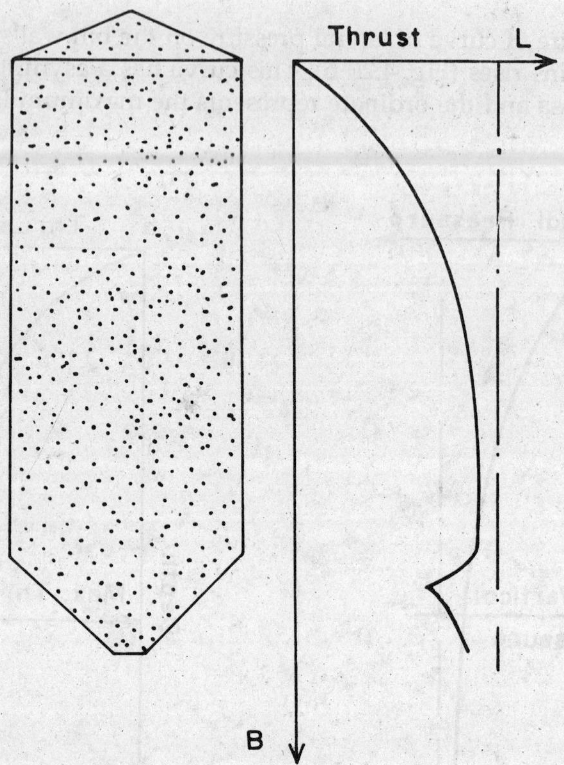

Fig. 4.21 : Lateral pressure distribution in a conical bottom vertical bin

μ = coefficient of friction of the grain on the wall

$= \tan \phi$

$$K = \frac{1 - \sin \phi}{1 + \sin \phi}$$

h = depth of grain

The Janssen formula is widely used and bin's design is safe because of a higher safety factor. Janssen assumed that 'K' was a constant throughout the grain mass.

During storage, the lateral pressure increases constantly while the bin is being filled. The lateral pressure has to be taken by the bin walls.

The vertical wall load per unit wall area indicated by the Janssen theory is given by the product of lateral pressure per unit wall area and wall friction co-efficient ($P_l \mu$). For a particular depth 'h', the total lateral wall load 'P' per unit of perimeter can be given by

$$P = \int_0^h P_l \, dh$$

The vertical wall load 'L_v' per unit of perimeter can be given by

$$L_v = \mu P = \mu \int_0^h P_l \, dh$$

Consider an elemental disc of grain in a deep bin as shown in Fig. 4.22

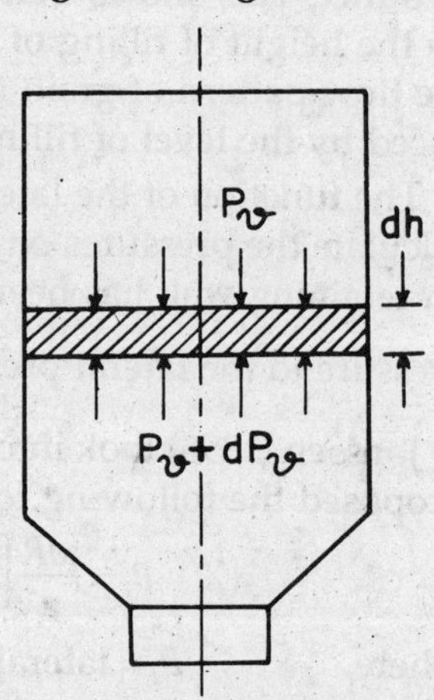

Fig. 4.22 : Vertical forces on elementary disc

If w = unit weight of grain

R = hydraulic mean radius $r/2$

$$K = \frac{P_1}{P_v}$$

$$= \frac{\text{lateral pressure}}{\text{vertical pressure}}$$

From the equilibrium of all vertical forces on the elementary disc.

The weight of elementary disc = w πr^2r dh

At equilibrium of vertical forces the elementary disc weight is equal to the sum of vertical and lateral pressure components.

i.e. $$w\pi r^2\ dh = \pi r^2\ dP_v + \mu\ P_1\ 2\pi r\ dh \quad ...4.15$$

Substituting P_1 KP_v and simplifying the Equation 4.15

$$w\pi r^2\ dh = \pi r^2\ dp_v + \mu k P_v\ 2\pi r dh$$

or, $$w\pi r^2\ dh - \mu\ KP_v\ 2\pi r\ dh = \pi r^2\ dP_v$$

or, $$dh(w\pi r^2 - 2\pi r\ \mu\ KP_v) = \pi r^2\ dP_v$$

$\therefore$ $$dh = \frac{\pi r\, x r d P_v}{\pi r(wr - 2\mu k P_v)}$$

or, $$dh = \frac{r}{wr - 2\mu K P_v} dP_v \quad ...4.16$$

Integrating the equation 4.16 from 0 to h

$$P_1 = KP_v$$

$$= \frac{wR}{\mu}\left[1 - e\frac{-\mu Kh}{R}\right] \quad ...4.17$$

Lateral and Vertical pressure

We know that the ratio of lateral pressure to vertical pressure is equal to

$$\frac{1 - \sin\phi}{1 + \sin\phi}$$

where ϕ is the angle of internal friction or equivalent to the natural angle to repose.

The functions of lateral pressure and vertical pressure are not well known, therefore, to calculate the pressures on the walls of a bin the basic theory for calculating pressure on retaining wall is adopted. As per the theory for retaining wall the ratio of lateral to vertical pressure is constant and equal to

$$\frac{P_l}{P_v} = \frac{1-\sin\phi}{1+\sin\phi} = \tan^2\left(\frac{\pi}{4}-\frac{\phi}{2}\right) \qquad ...4.18$$

In Fig. 4.18 (Mohr circle) the mean of the stresses is denoted by 'P', then radius of Mohr's circle will be $P\sin\phi$, therefore

$$P_l = P(1-\sin\phi)$$

$$P_v = P(1+\sin\phi)$$

or,

$$\frac{P_l}{P_v} = \frac{1-\sin\phi}{1+\sin\phi} = \tan^2\left(\frac{\pi}{4}-\frac{\phi}{2}\right)$$

In the limit we have:-

$$P_{v\,max} = P_{max}\frac{1}{\tan^2\left(\frac{\pi}{4}-\frac{\phi}{2}\right)}$$

$$= w \cdot R\frac{1}{\mu\tan^2\left(\frac{\pi}{4}-\frac{\phi}{2}\right)} \qquad ...4.19$$

The above theory is based on the hypothesis that the ratio of $\frac{P_v}{P_l}$ is constant. Reimbert and Reimbert (1987) conducted studies and tests on lateral and vertical pressure behaviour. They reported that the ratio of vertical to lateral pressure is not constant, instead it varies with depth and with shape of the silos. The value of ratios of $\frac{P_v}{P_l}$ and depth 'h' were plotted, and the behaviour is shown in Fig. 4.23.

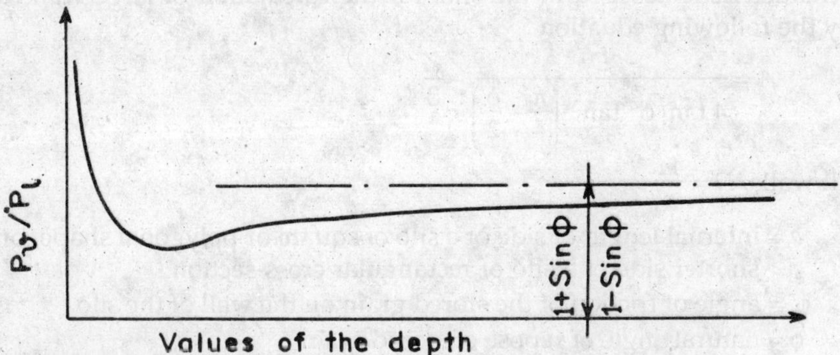

Fig. 4.23 : Plot of ratio of vertical to lateral pressures Vs depth of bin

If the basic theory is correct *i.e.* $\frac{P_v}{P_l}$ is constant, then the curve should be a straight line along x-axis, but in fact it was not achieved.

At the origin of the abscissae, the lateral pressure is zero, but the mean vertical pressure has some finite value. This value is equal to the weight of the upper cone of material divided by the cross-sectional area of the bin (Fig. 4.24). The height of the upper cone of stored material is given by;

$$Z = r \tan \phi \quad \ldots 4.20$$

where, r = radius of cylindrical bin
= radius of the circle inscribed in the polygonal outline of the cross-section.

Therefore, the value of $\frac{P_v}{P_l} \to \infty$ and the curve representing the ratio $\frac{P_v}{P_l}$ tends towards infinity having the axis of the ordinates as asymptote. The curve representing the ratio of the mean vertical pressure to the lateral pressure has a shape as shown in Fig. 4.23.

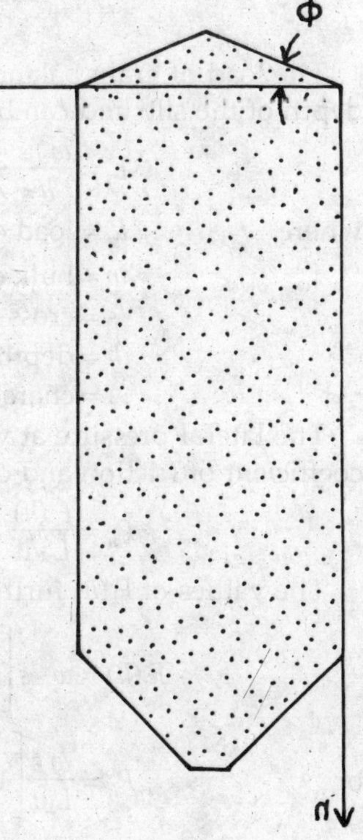

Fig. 4.24 : Cross-section of vertical bin with upper cone

The ratio of vertical to lateral pressure should be multiplied by coefficient of friction. This ratio takes into account both the friction of the material on itself and the friction between material and wall. Reimberts have also defined a characteristic abscissa as a function of the angle of internal friction ϕ, and angle of friction ϕ' of the grains on the bin wall. The value of characteristic abscissa 'A' can be given by the following expression.

$$A = \frac{R}{\mu K}$$

The characteristic abscissa in the silo, for the calculation of large wall can be given by the following equation

$$A = \frac{D}{4 \tan \phi' \tan^2 \left(\frac{\pi}{4} - \frac{\phi}{2}\right)} - \frac{Z}{3} \qquad ...4.21$$

for small walls, $D = \frac{4a}{\pi}$

where, a = internal length of side of a silo of square or polygonal shape, or the shorter side of a silo of rectangular cross-section.

ϕ' = angle of friction of the stored grain on the wall of the silo.

ϕ = natural angle of repose of stored grain

D = internal diameter of a cylindrical silo, or diameter of the inscribed circle of a polygonal silo having more than 4 sides.

The maximum lateral pressure of a bin can be given by the following expression,

$$P_{max} = \frac{w R}{\mu} \qquad ...4.22$$

The load of grain balanced by the friction on the walls of a silo is a function of depth of the silo and can be given by the following expression,

$$L_f = \frac{w \cdot s \cdot h^2}{h + A} \qquad ...4.23$$

where, L_f = load of grain balanced by the friction on the walls

w = bulk density of grain

s = cross–sectional area of silo

h = depth of silo

A = characteristic abscissa

The lateral pressure at various depth of the silo is a function of depth and the coefficient of friction and can be given by,

$$P_h = \frac{F(h)}{L \cdot \mu}$$

The values of $F(h)$ further can be given by

$$F(h) = w \cdot s \left[1 - \left(\frac{h}{A} + 1\right)^{-2}\right] \qquad ...4.24$$

or

$$P_h = \frac{w\, s}{L\, \mu}\left[1 - \left(\frac{h}{A} + 1\right)^{-2}\right] \qquad ...4.25$$

Since $\frac{w \cdot R}{\mu}$ is equal to $P_{max,}$

$$P_h = P_{max}\left[1 - \left(\frac{h}{A} + 1\right)^{-2}\right] \qquad ...4.26$$

The total load on the bottom of the silo can be expressed by,

$$L = s \cdot q_z$$

where,

$$q_z = w\left[h\left(\frac{h}{A} + 1\right)^{-1} + \frac{Z}{3}\right] \qquad ...4.27$$

Problem : A cylindrical silo of 2.5 m diameter and 20 m in height, is filled with wheat. Calculate the load on the bottom, and the lateral thrust at every 2.0 m depth on the walls. The silo is made of steel with smooth walls. The characteristics of stored wheat are as follows.

Minimum bulk density : 720 kg/m^3
Maximum bulk density : 830 kg/m^3
Minimum angle of internal friction : 25°
Maximum angle of internal friction : 30°
Minimum angle of friction on smooth sheeting : 18°
Angle of repose : 25°

Solution : Height of the upper cone of stored wheat,

$$Z = r \tan 25° = \frac{2.5}{2} \times 0.4663$$

$$= 0.5825 \text{ m}$$

Mean hydraulic radius, $R = \frac{D}{4} = \frac{2.5}{4} = 0.625$

$$\tan 25° = 0.4663$$

$$\tan 18° = \mu = 0.3249$$

$$\tan^2\left(\frac{\pi}{4} - \frac{25°}{2}\right) = 0.4058$$

Therefore the characteristic abscissa '*A*'

$$A = \frac{R}{\mu K} = \frac{0.625}{0.3249 \times 0.4058}$$

$$= 4.74 \text{ m}$$

$$P_{max} = \frac{w_{max} R}{\mu} = \frac{830 \times 0.625}{0.3249}$$

$$= 1596.6 \text{ kg/m}^2$$

Load at the bottom,
height of grain against the wall, $h = 20 - 0.5825$
$= 19.417$ m

$$L_{19.417} = 830\left[19.417\left(\frac{19.417}{4.74} + 1\right)^{-1} + \frac{0.5825}{3}\right] \times \frac{\pi \times (2.5)^2}{4}$$

$$= 16314.8 \text{ kg}$$

Values of lateral pressure at every 2.0 m depth are given in the following table.

h, m	*h/A*	$1 - (h/A + 1)^{-2}$	*Ph, Kg/m*^{2}
2.0	0.4219	0.5054	806.91
4.0	0.844	0.7060	1127.05
6.0	1.266	0.8052	1285.66
8.0	1.687	0.8614	1375.46
10.0	2.109	0.8965	1431.42
12.0	2.531	0.9197	1468.54
14.0	2.953	0.9360	1494.42
16.0	3.375	0.9477	1513.18
18.0	3.797	0.9565	1527.21
19.417	4.096	0.9614	1535.12

Problem : A square silo of side 2.5 m and height 20 m of reinforced concrete is filled with paddy. Calculate the load on the bottom of silo and the lateral thrust at every 2.0 m of depth on the walls. The characteristics of stored paddy are as follows:

Minimum bulk density : 576 kg/m^3
Maximum bulk density : 656 kg/m^3
Minimum angle of internal friction : 36°
Maximum angle of internal friction : 40°
Minimum angle of friction on concrete : 36°
Natural angle of repose of paddy : 36°

Solution : Height of the upper cone of stored paddy,

$$Z = r \tan 36° = \frac{2.5}{2} \times 0.7265$$
$$= 0.908 \text{ m}$$
$$\tan^2\left(\frac{\pi}{4} - \frac{36}{2}\right) = 0.26$$
$$\tan^2\left(\frac{\pi}{4} - \frac{40}{2}\right) = 0.217$$

Coefficient of minimum friction of the walls,

$$\tan 36° = 0.7265$$

Characteristic abscissa for the calculation of the maximum thursts,

$$A_1 = \frac{2.5}{\pi} \cdot \frac{1}{0.7265 \times 0.26} - \frac{0.908}{3}$$
$$= 3.91 \text{ m}$$

Characteristic abscissa for the calculation of the maximum vertical pressure,

$$A_2 = \frac{2.5}{\pi} \frac{1}{0.7265 \times 0.217} - \frac{0.908}{3}$$
$$= 4.745 \text{ m}$$

Mean hydraulic radius, $R = \frac{2.5 \times 2.5}{2.5 \times 4} = 0.625m$

Maximum lateral pressure

$$P_{max} = \frac{656 \times 0.625}{0.7265} = 564.35 \text{ kg/m}^2$$

Load on the bottom of the silo,
total depth of storage = 20 – 0.908 = 19.092 m

$$L_{19.092} = 656 \times 2.5 \times 2.5 \left[19.092\left(\frac{19.092}{4.745} + 1\right)^{-1} + \frac{0.908}{3}\right]$$
$$= 16822.5 \text{ kg}$$

Values of lateral pressure at every 2.0 m depth are given in the following table.

h, m	$\frac{h}{A}$	$1 - \left(\frac{h}{A} + 1\right)^{-2}$	P_h, Kg/m^2
2.0	0.5115	0.563	317.33
4.0	1.023	0.756	426.65
6.0	1.534	0.845	476.87

h, m	$\frac{h}{A}$	$1-\left(\frac{h}{A}+1\right)^{-2}$	P_{h}, Kg/m^2
8.0	2.046	0.892	503.57
10.0	2.557	0.921	519.76
12.0	3.069	0.940	530.31
14.0	3.580	0.953	537.82
16.0	4.092	0.962	542.90
18.0	4.603	0.969	546.85
19.092	4.883	0.971	548.09

Problem : Wheat weighing 900 kg/m^3 is loaded in a circular concrete silo of 3 m internal diameter and a clear height of 8 m. The angle of internal friction for wheat is 25° and that for wheat and concrete is 24°.

Applying Airy theory, calculate the maximum lateral pressure at the bottom of bin section. (ARS, 1984)

Solution : Airy formula for calculating lateral pressure against a shallow bin wall is:

$$L = w\,h\left[\frac{1}{\sqrt{\mu\,(\mu+\mu')}+\sqrt{1+\mu^2}}\right]^2$$

where, L = lateral pressure
W = bulk density of grain
h = depth of bin (point) under consideration
μ = coefficient of friction of grain on grain
= tan ϕ, ϕ is the angle of internal friction
μ' = tan ϕ', ϕ' is the angle of wall friction

$$L = 900 \times 8\left[\frac{1}{\sqrt{0.466\,(0.466+0.445)}+\sqrt{1+0.217}}\right]^2$$

$$= 900 \times 8\left[\frac{1}{0.6515+1.1031}\right]^2$$

$$= 2338.5 \text{ kg/m}^2$$

Problem : A R.C.C cylindrical grain storage bin has internal diameter of 5 m and is 20 m deep. It is completely filled with paddy weighing 600 kg/m^3. The angle of internal friction for paddy can be taken as 35° while the angle of friction between paddy and bin wall is 30°. The ratio of horizontal and vertical pressure intensity, k, is 0.4. Calculate the lateral pressure intensity at 2.0 m interval. (ARS 1990)

Solution : The lateral pressure by Janssen theory,

$$P_l = \frac{w\,R}{\mu}\left[1 - e^{-\frac{\mu K h}{R}}\right]$$

Given that, $w = 600 \text{ kg/m}^3$

$$R = \frac{D}{4} = 1.25 \text{ m}$$

$$\mu = \tan\ 30° = 0.577$$

$$K = \frac{P_l}{P_v} = 0.4$$

Substituting the values of various depths in above equation, we get

$$P_l \text{ at } 2.0 \text{ m} = \frac{600 \times 1.25}{0.577}\left[1 - e^{-\frac{0.577 \times 0.4 \times 2}{1.25}}\right]$$

$$= 401.6 \text{ kg/m}^2$$

$$P_l \text{ at } 4.0 \text{ m} = \frac{600 \times 1.25}{0.577}\left[1 - e^{-\frac{0.577 \times 0.4 \times 4}{1.25}}\right]$$

$$= 679.8 \text{ kg/m}^2$$

The values of lateral pressure at different depths are given in the following table.

depth, m	6.0	8.0	10.0	12.0	14.0	16.0	18.0	20.0
P_l, kg/m^2	870.8	1003	1094.4	1158	1202	1232.2	1253	1268.6

The properties of crop seeds related to bin loads and the coefficient of friction between grain to bin wall are given in Tables. 4.5 & 4.6. Some investigators have reported that the wide range in friction coefficient is result of grain moisture content, which has a significant influence on wall friction coefficients.

Table 4.5

Properties of some grain with respect to bin loads

Grain	*Hectolitre weight, Kg*	*Angle of repose, ϕ*	tan ϕ (μ)	*Ratio of lateral to vertical pressure, (K)*
Maize	70–76	16–27	0.29–0.51	0.60–0.65
Wheat	73–83	16–28	0.29–0.53	0.30–0.60
Rice	57	20–36	0.36–0.72	0.48
Soybean	73	16–29	0.29–0.55	0.38
Sorghum	70–76	20–33	0.36–0.65	—

Table 4.6

Coefficient of friction between grain to bin wall

Grain	*Plywood*	*Rough board*	*Steel Sheet*	*Galvanised Steel*	*Concrete*
Maize	0.25–0.50	0.30–0.35	0.20–0.60	0.2–0.4	0.50–0.60
Wheat	0.25–0.50	0.30–0.40	0.20–0.50	0.20–0.50	0.45–0.70
Rice	0.40–0.45	0.50–0.55	0.40–0.50	—	0.45–0.60
Sorghum	012–0.40	—	0.16–0.40	0.20–0.45	0.30–0.48

The layout and the design of a grain storage facility or silo building should aim towards a reliable and efficient unit, easy to survey and operate. The following two main groups of factors affect the design of silo construction.

1. *Fundamental factors :* Adequate volume of storage, proper protection of stored products and adequate methods of loading of grains, storage and discharge.

2. *Structural factors :* The silo should be strong enough to resist all the loads like dead load or the weight of structure and the attached items; live loads-forces from stored products (pressures); floor and roof live loads and wind and seismic loads.

The storage system capacity is dependent upon the quantity of grain to be stored and the number of bins to be used. This further depends on the range and characteristics of the products to be stored, the methods, rate and frequency of loading and discharge, simultaneous loading discharge etc.

Design of bin, apart from above is also dependent upon factors such as position of inlet and outlet gates like, central or eccentric outflow, the bin geometry, the hopper shape, the outlet opening etc. The required storage time affects the decision about horizontal and vertical storage structures. When high capacity gravitational outflow is needed, vertical storage facility is preferred.

Location and orientation of grain storages remarkably affect the system efficiency and suitability as a grain handling centre. If the storage structure is to be used for storing feed grains for animals, it should be located near or a convenient place to the feed processing centre, otherwise excessive handling will not only increase the capital investment on handling equipment but will also increase labour requirements. The storage facility should preferably be located such that it is accessible in all seasons. The storage installations will also have aeration and drying facilities, adequate electrical power should be made available to the site selected. While deciding for creation of storage facilities, the following factors should also be considered.

(1) Orientation of individual bins should be such that aeration or drying does not work against prevailing wind.

(2) Provision of adequate space for future expansion.

(3) Storage site should be on firm soil with good drainage.

(4) Checking for the high water tables in areas where below ground pits are to be located.

(5) The bins and floor should be at least 24-28 cm above the graded ground level. The slope of floor should be away from the bin for protecting the foundation of bins.

Overpressure during emptying

In a loaded silo, the discharge of grain occurs through a small opening. The quantity of discharged grain is insignificant as compared to stored mass. The small quantity of outflow is sufficient to produce a downward movement in the entire stocked grains. No sooner the discharge gate is opened downward movement of ensiled grain starts disturbing the equilibrium of stored mass. This results in increase of thursts on the storage structure walls. The rate of movement of grain near the walls is lesser due to friction on the walls. But at the centre of bin the downward movement of grain is faster. This phenomenon causes overpressure on storage structure walls during emptying. Dynamic loads resulting from discharge of stored grain are the most frequent reasons of structural failures of grain storage bins. The dynamic loads factor is defined as the ratio of the load during discharge to the load under static conditions.

The highest lateral pressures are measured in grain bins during unloading when grain is in motion along the bin wall. For this reason, the dynamic coefficient of friction between the grain and the bin wall is used to calculate design and the bin wall is used to calculate design lateral pressures in tall bins that unload in mass flow and in hoppers. The magnitude of the lateral pressure predicted by the Janssen's equation is strongly influenced by the value used for the coefficient of friction between the grain and the bin wall material.

When the discharge gate is opened and the solid flows out of the outlet, there is vertical expansion of solids within the flow channel. The minor pressures tend to align with the direction of expansion of solids. The channel diverges upward from the outlet, therefore, the flowing mass of solids also contracts laterally. The major pressures in flow-channel tend to align with the lateral contractions. Thus, the major pressures are lateral and the minor pressures are vertical.

Construction materials

Basically, the choice of construction materials for modern permanent storages is between steel and concrete, though brick is also used in some countries. As far material is concerned for safe and quality storage, its durability and gas tightness is of prime concern. The selection between steel and concrete can involve many considerations.

Metal bins are mainly used for smaller storage units and for provisional installation. Metal bins are preferred because; (1) their dead weight is low, (2) prefabrication of the component parts in workshop is possible, (3) standard material readily available in the market can be used, (4) they provide a smoother surface and lower coefficient of friction, (5) installation inside a building is easier, (6) complicated designs are possible and (7) when needed, they can be enlarged.

Concrete bins are used for larger permanent storage structures. The chief advantages of concrete bins are; (1) maintenance requirements are low, (2) they reduce risk of condensation, (3) corner curves are easy to construct, (4) they are water and fire proof, easy to clean. But the concrete bins have high dead weight, need stronger foundation. The interior walls of the bins are relatively rough and air bubbles are also possible. There is loss of volume by way of thickness of bin walls.

Durability

Concrete silos are not maintenance free. If their design is inadequate, they may crack, thereby allow moisture migration and loss of fumigants. Crack repair can be an expensive process. Carbon dioxide reacts readily with concrete. Carbon dioxide reacts with calcium hydroxide molecules in the cement matrix to form calcium carbonate. Calcium hydroxide is an alkaline substance which gives chemical protection against corrosion of the reinforcing steel. Carbonation removes the alkalinity and the protection.

Carbonation takes place in all concrete as a result of reaction with carbon dioxide in the atmosphere. The concentration of carbon dioxide in atmosphere is 0.03%, hence the reaction is very slow. In a chemically clean rural atmosphere, the carbonation may penetrate at less than 1 mm per year. Due to metabolic activity, grain produces carbon dioxide. Storage of grain in concrete silos thus produces an accelerated rate of carbonation. It is possible that the 200 mm thick silo may completely be carbonated in 50 years.

At controlled atmosphere for disinfestation of grain, the carbon dioxide concentration may be 60%, therefore, carbonation reaction is very vigorous. In such situation the chemical protection against corrosion of reinforcing steel may be lost in a very short time. Good compaction of concrete results in better corrosion protection because compaction controls the cracks and gives strength. The rate of carbonation can be reduced by the application of certain high build paints. But this adds both to capital investment and maintenance costs of a concrete silo.

In steel silos, the zinc galvanisation also does not provide long term corrosion protection in chemically active environments. The galvanised surfaces should be overcoated with an appropriate paint system. Painting adds cost to steel constructions and also towards its maintenance.

The welded steel silos require no special sealing provisions except at the joint between floor and wall. Since joint movement from grain pressure seldom occurs in practice, it is safer to design the walls with a rigid base. Light gauge bolted steel silos have not proven easy to make gastight due to their light flexible construction and multiple screw fixings. The joint between roof-to-wall is also difficult to seal. The cost of sealing of such silos becomes very high and thus reduces the cost competitiveness.

The concrete silo if well designed and constructed, does not require any special treatment for making gastight. To make concrete silos gastight, the circular bins should be made independent and eccentric discharge should be avoided. Independent circular concrete bins can be designed and built with minimum likelihood of wall cracks development from grain pressures. If any crack appears, it is readily accessible for sealing, which is not easily possible with inter connected bins.

Aeration and Fumigation

Aeration is the process of moving air through stored grain at low flow rates to maintain or improve its quality. Aeration can provide three major benefits in the storage of grains.

1. It cools the grain and slows down insect activity.
2. By cooling the grain, aeration prolongs the effectiveness of pesticides.
3. It can provide an appreciable drying function.

Some other benefits of aeration are—

1. It prevents storage odours. Grains stored for longer periods often develop objectionable odours. Aeration provides the grains a new, fresh smell.
2. It reduces moisture accumulation. Moisture condensation by changes in temperature and relative humidity can be reduced by aeration.
3. Application of fumigants through an aeration system is an easy and practical method for controlling insects in stored grains.

Aeration systems can be installed in both horizontal and vertical storages. The following set of basic design parameters are found suitable for cooling of dry grains by aeration.

Air flow rate : Silos 0.8 litre/second/tonne
Sheds 1.6 litre/second/tonne

Maximum duct velocity : 10 m/s

Maximum entry velocity of air into grain : 0.15 m/s

The most common air flow rates for aerating paddy range from 0.07–0.28 m^3/min./tonne of paddy. The above air flow rates for aeration are only a fraction of the air volume needed to dry grains.

From above parameters, fan output, duct size and duct surface area can be determined.

Air can be pulled or pushed through the grain in storage. The air can be moved from the top to bottom or from the bottom to the top of the bin. In the tropical climates, aeration should be done from bottom to top because any air trapped under the roof of a storage structure is usually at high temperature. This high temperature air can be forced out easily from the top. The cooler air from bottom also cools the warmer grain stored in bin.

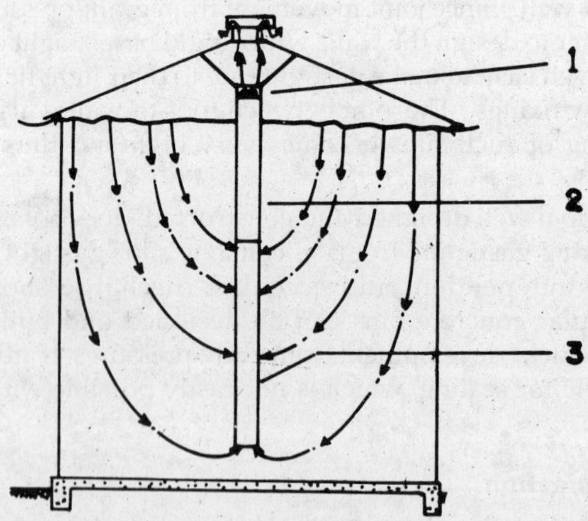

Fig. 4.25 : Central aeration system in vertical silo
1. fan 2. solid duct 3. perforated section

Axial fans are normally used for shed aeration. Non-overloading centrifugal fans should be used for vertical storages because of the higher pressure requirements. With overloading fan, it is recommended that a manually adjustable throttle valve should be used in the inlet duct to cut back the airflow while aerating partly filled silos. The principal parts of an aeration system are, (1) fans to move the air and (2) ducts for carrying air from the fans to storage system.

Various configurations of ducting can be used for air distribution inside the grain mass. Aeration systems for silos are, usually; (1) central vertical system (Fig. 4.25), (2) floor ducts (Fig. 4.26) and (3) perforated floors (Fig. 4.27).

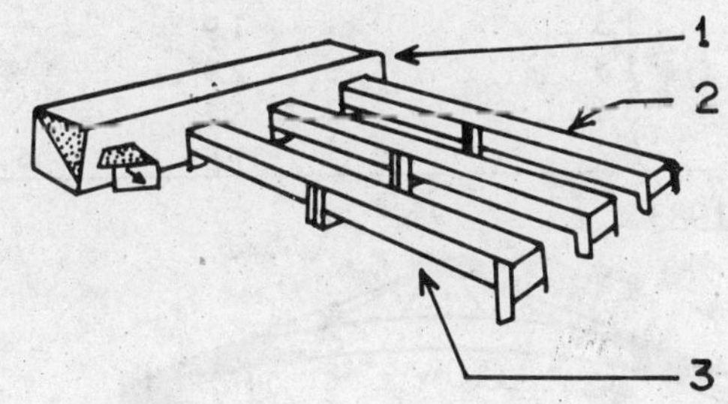

Fig. 4.26 : Floor ducts for aeration
1. main duct 2. transferable lateral ducts 3. interspace on floor

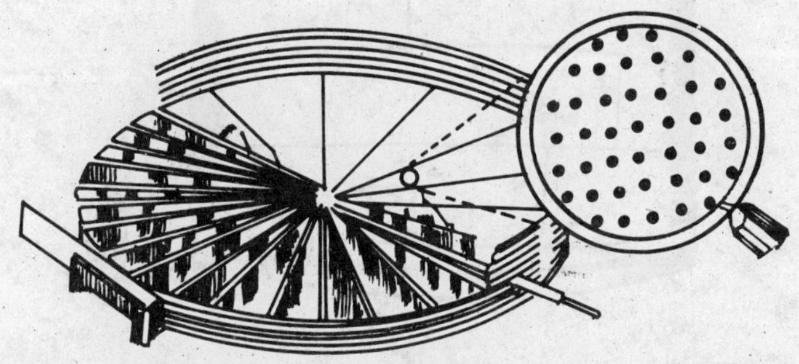

Fig. 4.27 : Perforated floor cross-section for aeration

In a vertical duct system the top part of the central duct is solid and the lower half section is perforated. A fan at the top of the duct pulls the air through the grain and perforations. The sucked air is blown out of the bin by the fan.

When on-floor ducting is used, it is often important to allow for its removal to facilitate floor sweeping. For this reason, full round ducting is usually used in sheds where grain loads are moderate. Both longitudinal and cross-floor ducting can be used in sheds. Generally ducting is made from galvanised steel sheet with 2.5 mm perforations at 6 mm centres.

Ducting in silos is normally exposed to considerably higher grain pressures. In-floor ducting requires the use of flat perforated sheet. Cost of such system is significantly higher because of requirement of a structural frame below the perforated sheet for its support. An alternative is perimeter ducting formed from quarter round corrugated ducting.

To assist in selection of fan for aerating paddy, the static pressure requirements at various depth of paddy are given in Table 4.7

Table 4.7

The static pressure requirements for various paddy depth and aeration rates

Paddy depth	*Static pressure, cm of water*		
m	0.07 $m^3/min/t$	0.14 $m^3/min/t$	0.28 $m^3/min/t$
3	0.7	0.8	0.9
6	0.9	1.2	1.8
9	1.3	1.9	3.3
12	1.8	2.9	5.7
18	3.4	6.2	13.5
24	5.7	11.5	—

Source : Wimberley, 1983

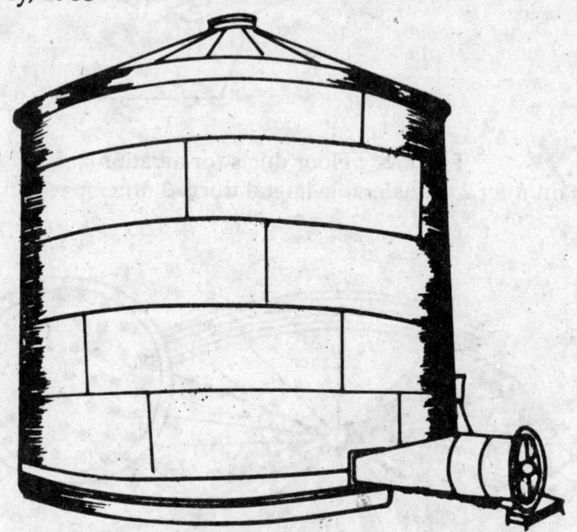

Fig. 4.28 : Fan connected with a single bin for aeration

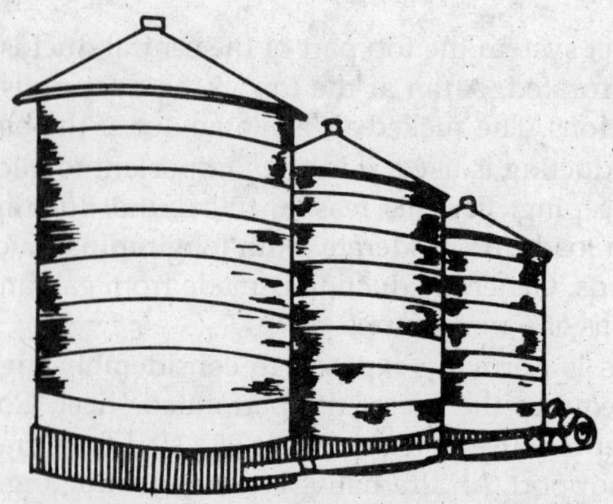

Fig. 4.29 : A large capacity fan connected with many bins for aeration

Perforated ducts may be circular, semicircular, rectangular or inverted V-shaped. The required cross-section area of a duct may be determined by the following expression

$$\text{cross-section area, } m^2 = \frac{\text{Total air volume, } m^3/\text{min}}{\text{Air velocity, } m/\text{min}}$$

The perforations should be uniformly spaced with a minimum perforated area of 10% and preferably 15%. For aerating the storage structures the aeration system may have a single fan connected with a single bin as shown in Fig. 4.28. There may be other system in which a large capacity fan is connected to a manifold system (Fig. 4.29). With such arrangement, aeration could be done in a single bin or in many bins simultaneously as per the requirement.

Often a portable fan mounted on a trolley is used. This can be moved from one duct or one manifold system to another. The initial cost of the fan and system is reduced but the operating cost is increased because of labour requirement to move and connect the fans.

Fumigation

Sealed storages can be effectively fumigated, using methyl bromide or phosphine (Fig 4.30).

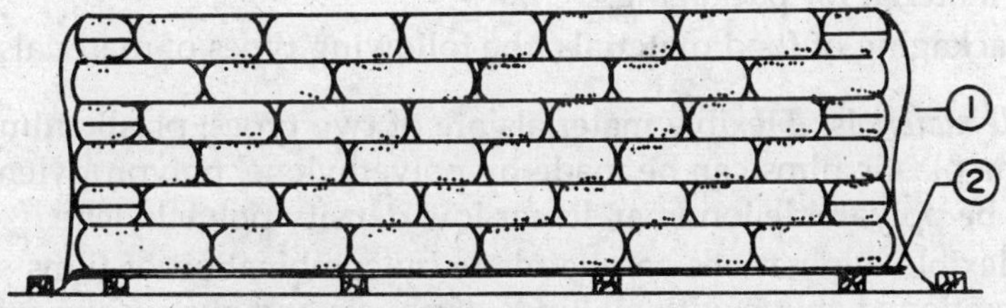

Fig. 4.30 : Grain fumigation in bags
1. plastic sheet for sealing purpose 2. jute sheet

Methyl bromide can be used to fumigate sealed vertical storages when a quick kill is required, for instance prior to unloading. Fumigation with methyl bromide requires the use of fans and ducting to recirculate the gas in the storage. Methyl bromide fumigation duct dimensions are selected using the same standards as for aeration, except the air flow rate used is lower at around 0.06 m^2/min/t.

Fumigation methods with phosphine involve the placement of pallets into the grain stream while turning the grain. Now-a-days, in modern storage system, phosphine is used in blanket form. A blanket of phosphine is placed on the grain surface and natural convection is relied upon to distribute the gas through the grain mass. Application rates between 0.5 and 1.0 g phosphine per cubic metre of storage volume are used for fumigating sealed storages. Phosphine is also used for fumigating grain in tall sealed silos, but the natural convection current is not sufficient to distribute the gas evenly throughout. Therefore, for uniform distribution of fumigant, forced recirculation of gas is necessary.

Packaging

The roles of packaging are to contain, protect and sell. A package contains a certain quantity of product for distribution through wholesaler and retailer to the consumer. Packaging provides ease in handling of the products, protects the products to maintain the original quality during storages and transport, keeping in retail and storing with the consumer. The protection generally means safe gaurd against mechanical, physical, chemical and biological activities. The sell includes attracting the buyer, identifying the product and the producer and informing the buyer about the product and its use.

The surplus foods grown in the village after harvesting are protected from spoilage and damage, packaged and transported by various means to nearby or distant markets. Unless the goods are sold with minimum spoilage with their peak flavour, appearance and nutritional value and presented in attractive way, they may not be liked. This means that food and packaging technologies become involved in the entire food supply system. This system ranges from villages, farm, plantation to the markets and consumers in cities and other places, not only in their own country but also in the distant overseas markets.

Fresh foods are biologically active materials. Many complex and varied biochemical reactions can occur causing desirable or undesirable changes e.g. in the colour of meat, the texture of fruits and vegetables, the flavour of cereals, the nutrients in dairy products. Potential biochemical changes will influence the choice of packaging. Microbial growth is a major factor in deciding the most suitable material for packaging.

For packaging of food materials, the following types of materials are available.

1. *Flexible materials* : Flexible materials are of two types; plastic films and laminations. The plastic films can be made by polyethylene, polypropylene, P.V.C. They can also be polyamide ionomer, linear low density polyethylene.

The flexible laminations can be made by combination of films selected in the range mentioned above with all kinds of papers and aluminium foil when necessary. It may include, sometimes, special coating with waxes.

Very close to flexible materials are heavy gauge sheets for thermoforming. These sheets and multilayer structures can give the possibility to thermoform small packages such as trays, tubs, cups etc. which may have very high barrier properties to moisture and gases.

2. *Rigid plastic containers* : These can be manufactured by the injection process for producing small to large containers including plastic crates. They also can be made by the blow-moulding process to produce bottle of various sizes and jerry- cans.

3. *Rigid Packages* : Rigid packages are glass bottles, metal cans, cartons (including folding boxes and set-up-boxes), corrugated cases and wooden cases. Corrugated cases can be reinforced by wooden pieces included in the corners, paperboard trays can be lined inside with a thin plastic thermoformed material when products to be packaged are wet or have high moisture content that can be absorbed by the board. Paper containers can be of plain paper bag to the corrugated case of multi wall bags.

4. Composite cans have also been developed for packaging purpose. Some processes give the possibility to use recycled waste papers sandwitched in between two liners that could be made of paper, aluminium foil or plastic. Some cans of this type may resist to vacuum, pressure and provide very good barrier to moisture.

5. Heat from the infilterated air from outside through cracks in doors, windows and through their frequent openings.
6. Miscellaneous heat sources such as
   (i) heat gain by the ducts carrying the cool air and passing through unconditioned space.
   (ii) heat transferred through interior partition of rooms in the same building which are not air-conditioned.

The sources which contribute to the latent heat are given below:

1. The latent heat load from the air entering into the air conditioned space by infilteration.
2. The latent heat load from the occupants.
3. The latent heat load from stored materials.
4. Moisture passing directly into the air-conditioned space through permeable walls at higher water-vapour pressure.

The total heat load to be removed by an air-conditioning system is the sum of the above listed heat loads and the extra loads added by the following sources.

a. The fresh air taken from outside adds the sensible and latent heat load on the air-conditioner.

b. Heat generated by the fan is added to the fresh air and it is also removed.

The steady state heat flow by conduction is given by the following equation.

$$Q = U A (T_0 - T_i), \text{ k cal/hr} \qquad \ldots 4.28$$

where, U = overall heat transfer coefficient
A = area through which heat is transferred
T_0 = Temperature of the outside air
T_i = Temperature of inside conditioned air

The overall heat transfer coefficient is given by

$$\frac{1}{U} = \frac{1}{h_0} + \frac{x_1}{K_1} + \frac{x_2}{K_2} + \ldots + \frac{1}{h_i}$$

$$= \frac{1}{h_o} + \sum_{1}^{n} \frac{x}{k} + \frac{1}{h_i}$$

where, h_0 = heat transfer coefficient on the outer surface
h_i = heat transfer coefficient on the inner surface

$\sum_{1}^{n} \frac{x}{K}$ = thermal resistance of the materials used for the construction of the wall.

The heat emitted from the bodies of people inside cold storage constitute a sizeable portion of cooling load. The heat loads added by the products to be stored are important in the design of cold storages. The heat loads in the cold storage systems are of three types which are given below

(1) Chilling load, $Q_c = \dfrac{m\, c_{pm} (T_1 - T_2)}{\text{chilling time, hr}}$ k cal/hr ...4.29

where, m = mass of the product
c_{pm} = mean specific heat of the product
$T_1 - T_2$ = initial and final temperatures difference

(2) Freezing load, $Q_f = \dfrac{m \times h_{fg}}{\text{Freezing time, hr}}$ k cal/hr ...4.30

where, h_{fg} = latent heat of freezing

(3) The cooling load below freezing temperature is given by—

$$Q_f' = \frac{m\, c_{p'm}\,(T_a - T_f)}{\text{cooling time, hr}} \text{ k cal/hr.} \quad \ldots 4.31$$

where, T_a = actual storage temperature of the product

T_f = freezing temperature of the product

$C_{p'm}$ = specific heat of the freezing product

Some food products generate heat even they are stored in cold storage due to heat of respiration by living tissues and heat of reaction of the chemical contracts of food materials. The amount of heat generated by food material is given by;

$$Q_g = m \, . \, h_g, \text{ k cal/hr} \quad \ldots 4.32$$

where, h_g = heat generated per kg of food per hour

Heat loads from equipment

Electric fans, hot plates and other lighting equipment also generate heat. Light load is one of the major load in cold storage system.

The cooling load for electrical appliances can be given by;

$$Q = kW \times 860 \times \text{use factor, k cal/hr} \quad \ldots 4.33$$

$$\text{Use factor} = \frac{\text{Actual wattage in use}}{\text{install wattage}}$$

The heat load from fans, motor and other similar equipment can be given by

$$Q = \frac{\text{B. H. P}}{\eta} \times \text{L.F.} \times 632, \text{ k cal/hr} \quad \ldots 4.34$$

where, η = efficiency of motor

L.F. = load factor = fraction of total load at which the motor is running

The outdoor air enters into the cold storage space through window cracks and through door when opened. Air infilteration factors of 5 m^3 per hour per m^2 of window and door area is taken into account.

The heat gain through the duct is calculated by the following equation.

$$Q = A_d\, U \left[T_a - \frac{T_i + T_e}{2} \right], \text{ k cal/hr} \quad \ldots 4.35$$

where, A_d = area of duct

U = overall heat transfer coefficient

T_a = temperature of ambient air

T_i = temperature of air entering the duct

T_e = temperature of air leaving the duct

The above heat load should be added into the sensible heat load of the room.

The heat load to cold storage by air infilteration can be calculated with the help of the psychrometric chart (Fig. 3.7).

'a' is the condition of outside air

'b' is the condition of air coming out of the cold storage

'c' is the conditon of air leaving the room.

The latent heat load per kg of dry air infiltered into the room = $H_a - H_c$

The sensible heat load per kg of dry air infilterated into the room = $H_c - H_b$

Based on the total heat load the tonnes of refrigeration can be calculated by:

1 tonne of refrigeration = 3000 k cal/hr
= 3489 watt

Problem : Eight tonnes of apple having specific heat of 0.80 k cal/Kg–°C is to be cooled from 25 to 14°C in 24 hours. The heat of respiration per 24 hour is 745 k cal/t. Three men will work for 4 hours and lighting load is estimated to be 100 watt. Air infilteration load is assumed as 980 k cal in 24 hours.

The cold storage measures 6 × 6 × 3 m on the inside and is constructed of bricks laid in cement mortar. Wall thickness is 40 cm and there is 10 cm thick cork insulation on the inside of the four walls. The cement plaster is 1 cm thick. The heat transfer coefficient for the ceiling is 20% more than that for the walls. The outside temperature is 30°C and the inside is maintained at 5°C. Calculate the plant capacity needed in tonnes of refrigeration.

Thermal conductivity of brick = 0.45 k cal/hr/m–°C

Thermal conductivity of cork = 0.025 k cal/hr/m–°C

Thermal conductivity of cement plaster = 0.25 k cal/hr/m–°C

Heat of respiration for men = 170 k cal/hr

There is no heat transfer through the floor

Solution :

(*i*) Heat load for cooling the apples

$= 8000 \times 0.8\ (25 - 14) = 70400$ k cal/24 hr

(*ii*) Heat of respiration from apples

$= 745 \times 8 = 5960$ k cal/24 hr

(*iii*) Heat load due to air infilteration

= 980 k cal/24 hr

(*iv*) Heat load due to lights

$$= \frac{100 \times 24 \times 3.6}{4.1868} = 2060 \text{ k cal/24 hr.}$$

(*v*) Heat load because of workers

$= 3 \times 170 \times 4 = 2040$ k cal/24 hr.

(*vi*) Heat load due to conduction of 4 walls

$$Q = U\,A\,\Delta T$$

The overall heat transfer coefficient, U, can be calculated by the following equation

$$\frac{1}{U} = \frac{1}{U_B} + \frac{1}{U_P} + \frac{1}{U_C}$$

where U_B, U_P and U_C and heat transfer coefficients for brick, plaster and cork respectively. In general—

$$U = \frac{K}{x} = \frac{\text{thermal conductivity}}{\text{thickness of layer}}$$

$$\frac{1}{U} = \frac{0.4}{0.45} + \frac{0.1}{0.025} + \frac{0.01}{0.25}$$

or, $U = 0.203$ k cal/m^2–hr–°C

Total wall surface area, $A = 4 \times 6 \times 3 = 72$ m^2

$\therefore$ Heat transfer through walls $= 0.203 \times 72 \times (30 - 5) \times 24$
$= 8769.3$ k cal/24 hr

(*vii*) Heat transfer through ceiling

$$Q = U\,A\,\Delta T$$
$$= 0.203 \times 1.2 \times 6 \times 6 \times (30 - 5) \times 24$$
$$= 5261.76 \text{ k cal/24 hr.}$$

$\therefore$ Total heat load $= 70400 + 5960 + 980 + 2060 + 2040 + 8769.3 + 5261.76$
$= 95471.06$ k cal/24 hr.

Since 1 t of refrigeration = 3000 k cal/hr

Hence, plant capacity $= \dfrac{95471.06}{3000 \times 24}$
$= 1.326$ tonnes

Problem : 110 kg or lean poultry is first cooled from 20 to 4° C, thereafter it is further cooled and frozen to – 20°C. Specific heat of poultry is 3.21 kJ/kg°C and below freezing point is 1.71 kJ/kg°C. Freezing point of poultry is – 2.8°C and the latent heat of fusion is 246.8 kJ/kg. Calculate the total heat load.

Solution : The heat loads are:-

(*i*) To cool from 20 to 4°C
$= 110 \times 3.21\ (20 - 4) = 5649.6$ kJ

(*ii*) To cool from 4°C to freezing point in freezer
$= 110 \times 3.21\ [4 - (-2.8)] = 2401.08$ kJ

(*iii*) For freezing the poultry
$= 110 \times 246.8 = 27148$ kJ

(*iv*) To cool from freezing point to storage temperature
$= 110 \times 1.71\ [(-2.8) - (-20)] = 3235.32$ kJ

$\therefore$ Total heat load $= 5649.6 + 2401.08 + 27148 + 3235.32$
$= 38434$ kJ

Problem : Find the refrigeration load expressed in tons of refrigeration which is caused by heat loss from the four side walls of a small cold room 2.4 m × 3.0m × 2.4m. The walls are made of 20 cm brick, 20 cm cork board and 1.25 cm cement. Inside wall temperature is –30 °C and outside wall temperature is 21 °C. Add a suitable safety factor for losses through joints etc. Respective thermal conductivities of brick, cork and cement plaster are 0.6 W/m °C, 0.04 W/m °C and 0.8 W/m °C. ... (ARS 1991)

Solution :

Heat load due to conduction of walls

$$Q = U\,A\,\Delta T$$

The overall heat transfer coefficient, U, can be calculated by the following equation.

$$\frac{1}{U} = \frac{1}{U_B} + \frac{1}{U_P} + \frac{1}{U_C}$$

$$U = \frac{k}{x} = \frac{\text{thermal conductivity}}{\text{thickness of layer}}$$

$$\frac{1}{U} = \frac{0.2}{0.6} + \frac{0.2}{0.04} + \frac{0.0125}{0.8}$$

$$= 0.333 + 5 + 0.0156$$

or, $U = 0.187\ W/m^2 - °C$

$$Q = 0.187[(2 \times 2.4 \times 2.4) + (2 \times 3 \times 2.4)] \times [21 - (-30)]$$

$$= 247.2\ W$$

Taking safety factor as '2' to cover the loss of heat through duct and joints.

Total heat load = 247.2 × 2

= 494.4 *W*

Since 3489 *W* = 1 ton of refrigeration

$$\therefore \quad \text{Ton of refrigeration} = \frac{494.4}{3489}$$

$$= 0.14$$

Controlled and modified atmosphere storage

The technique of modified and controlled atmosphere storage of grain involves alteration of concentration of the normal atmospheric gases present in a storage so as to give an artificial atmosphere that is insecticidal and prevents mould growth and quality deterioration of the stored product. Two classes of externally generated modified atmosphere are available: low oxygen atmosphere, generated by adding nitrogen, or the gas mixture resulting from burning hydrocarbons or high carbon dioxide atmospheres made by adding carbondioxide.

Carbondioxide generally kills insects faster than nitrogen. It can be used in situations where leakiness may be a problem. In addition to above, sorption of carbondioxide by grains or oilseeds may make it more effective against species whose immature stages feed inside the kernel.

Modified or controlled atmosphere also greatly extend the period of storage for healthy grains. These methods require gastight facilities which are more expensive and scarce in developing countries. Due to no use of chemicals for insecticidal treatments the operating costs are significantly lower than traditional storage method, but higher use of instruments is the characteristic of modified or controlled atmosphere storage.

The time required to obtain a certain level of insect mortality exposed to a given atmospheric gas composition is dependent on the temperature of the environment. In fact, from the physiological aspects at normal atmosphere development of insects can only occur within a fairly narrow range of temperatures. Humidity influences the survival of insects mainly through the effect on their water content. Dry conditions are generally unfavourable for the reproduction of most insects. However, most stored product insects can survive at very low humidities (Howe, 1965). Navarro (1974) mentioned that at 20-24% relative humidity; 3.2% oxygen or 4.3% carbondioxide in air is needed to achieve complete mortality.

Shejbal (1980) conducted storage trials with cereal grains and reported that long term preservation in nitrogen is advantage at all temperature and moisture contents (upto 30°C and 19% respectively) as compared to storage in air. Oilseed storability is higher in nitrogen atmosphers than in air, The apparent disadvantages of the nitrogen fumigation and storage technique are strictly related to the management of the storage facilities.

Storey (1980) conducted studies to assess the functional and end use properties of agricultural commodities in low oxygen storage condition. He found that the germination, milling and bread making characteristics of wheat were unaffected by exposure to a generated or combustion atmosphere composed of less than 1% oxygen, 9.0 9.5% carbondioxide and the balance principally nitrogen for periods of half to six months. Similarly, germination, milling and cooking properties of long and medium grain rice exposed as rough, brown and milled rice were unchanged after 6 months of treatment. There is no adverse effect on the quality of malt produced from barley. Almonds and raisins stored in the low oxygen atmosphere were equal or superior in flavour, quality and acceptability to samples stored in normal atmosphere for periods upto one year.

The action of low oxygen is largely dependent on temperature. At low temperatures the action of low oxygen becomes slow. This property also holds good for low oxygen atmosphere containing carbondioxide produced by burning hydrocarbons.

For increasing safe storage period of some fruits and vegetable , reducing the oxygen level in the storage air and/or increasing the carbon dioxide level as a supplement to refrigeration can provide extended storage life. For this, careful control of the concentration of oxygen and carbon dioxide level is required. If all of the oxygen is used, produce will suffocate and may develop an alcoholic off-flavour in a few days.

Danish cabbage has been reported to be kept better during four to five months at 0°C in gas mixtures with 2.5 to 5% oxygen and carbon dioxide than it has in air. Mature green tomatoes may keep green for five to six weeks at 13°C in an atmosphere at 3% oxygen with 97% nitrogen. After removal to air at 18°C, these tomatoes ripen normally with acceptable flavour.

Mushroom in refrigerated storage has been kept better for short periods in atmosphere with 5 to 10% carbon dioxide than air.

All types of controlled atmosphere storage require frequent gas analysis to determine when aeration is required to add oxygen or to remove carbon dioxide.

Airtight storage

Airtight storage are normally used to arrest the insect infestation in dry grains and to protect comparatively wet grain from mould attack. In olden period, underground pits were used to store grains in Asia, Middle East and parts of southern Europe. These underground pits were often airtight.

The chief objective of the airtight storage is to reduce the oxygen in sealed container to level which is deadly to the organism present in the grain. This restricts their activity. In general, insects are killed when oxygen in the intergranular air falls to about 2% by volume. Most of the fungi on damp grain continue to develop until the oxygen falls to about 0.2%.

In a test at slough, 60 tonnes of dry maize with sitophilus granaries were kept in a proofed concrete pit. It was found that all the insects died in the first few weeks.

The main advantage of airtight storage is that after storing the dry grain, the same can be safely stored often for years without paying much attention to it. But the essential requirement is that the silo remains airtight.

Dust hazards

Whenever grain is moved, agitated, redirected or dropped from or into a bin, transferred from one conveyor to another, dust is generated. Dust consists of small solid particles created by handling and processing. The type and composition of the dust particles depend upon the grain handled and also have positive influence on the propagation of flame in event of an explosion.

Grain dust explosions occur more often than is generally believed. The manipulation of grain produces a movement which rubs the kernels against each other and the abrasive bran coats. The rubbing actions free the dust glued to the grain surface and generate new dust from the abrasion of the bran coats. Infested grains produce greater quantities of dust upon manipulation.

The average size of dust particles is between 15 and 120 μm as determined by sieving, which is comparable to the average diameter of flour particles. The explosiveness of dust generally tends to increase as the particle size becomes smaller. Dust of 40 microns particles and less is quite dangerous and should preferably not be returned to the grain stream. The total external surface area of dust particles may be between 0.6 to 2.0 m^2 per gram. The smaller dust particles can be taken away by air velocity of nearly 1.0 m/s. The settling time of dust depends upon its concentration of particle sizes. The moisture content of grains vary between 6 to 12 percent. Chemically the grain dust consists of 30 to 70 percent starch and other carbohydrates, 6 to 20 percent proteins, 1 to 4 percent lipids, 5 to 40 percent ashes and 7 to 15 percent fibres. The bulk of the food grain dust varies from 100 to 300 kg/m^2 depending upon the type of grains and particle sizes. In 1961 the U.S. Bureau of Mines has found and reported the values of ignition temperatures of cloud of dust of wheat, maize and rice as 500°C, 400°C and 470°C respectively.The ignition temperatures of a layer of the dust of same grain was reported to be quite lower than above. The ignition temperatures of layer of dust of wheat, maize and rice were 220°C, 250°C and 220°C respectively.

Baker et al (1986) studied the grain damage and dust generation in a pressure pneumatic conveying system. They determined that fines and dust generated by conveying dry shelled maize with grain flow rates ranging from 2.5 to 25 t/hr and air velocities ranging from 15 to 30 m/s. They reported that the amount of fines and dust generated in a properly designed and operated pneumatic handling system were same as generated by other handling systems. The concentration of dust in pneumatic handling system was below the explosive limit. Brittle grains cause increased amount of fines and dust in a handling system as grain flow rate increased the amount of fines and dust generated did not change significantly; and as conveying air velocity increased, fines and dust generation increased exponentially.

Devices for suppressing dust during grain loading operations are based on the principle of arresting the flow of the grain and controlling the height of free fall below the loading spout. Spray of white mineral oil formulation is applied to grain mass to arrest the generation of dust during manipulation. To suppress dust, silos should have sealed roofs to contain dust inside.

Exercise

1. Five tonnes of apples having specific heat of 0.82 kcal/kg°C is to be cooled from 22 to 12°C in 24 hours. The heat of respiration per 24 hours is 750 kcal/t. Four men will work for 1 hours and lighting load is estimated to be 140 watt. Air infiltration load is assumed as 980 kcal in 24 hours. The cold storage measures 5 × 5 × 3 m inside and is constructed by bricks laid in cement mortar. Wall thickness is 40 cm. and there is 10 cm thick cork insulation on the inside of the flow walls. The cement plaster is 12 mm thick. The heat transfer coefficient for the ceiling is 18% more that for the walls. The outside temperature is 28°C and the inside is maintained at 5°C. Calculate the plant capacity needed in tonnes of refrigeration.
   Thermal conductivity of brick : 0.44 kcal/hr m°C
   Thermal conductivity of cork : 0.025 kcal /hr m°C
   Thermal conductivity of cement plaster : 0.25 kcal/hr m°C
   Heat of respiration for men : 160 kcal/hr
   Assume there is no heat transfer through the floor.
2. A cylindrical silo of 2 m diameter and 16 m in height, is filled with sorghum. Calculate the load on the bottom, and the lateral pressure thurst at every 2 m depth on the walls. The silo is made of galvanised sheet with smooth walls. The characteristics of stored sorghum are as follows:
   Minimum bulk density : 696 kg/m^3
   Maximum bulk density : 770 kg/m^3
   Minimum angle of internal friction : 22°
   Maximum angle of internal friction : 28°
   Minimum angle of friction on smooth sheeting : 16°
   Angle of repose : 30°
3. A RCC cylindrical grain storage bin has internal diameter of 4 m and is 14 m deep. It is completely filed with wheat weighing 740 kg/m^3. The angle of internal friction for wheat is 28°, while the angle of friction between wheat and bin wall is 25°. The ratio of horizontal and vertical pressure intensity *'K'* is 0.4. Calculate the lateral pressure intensity at 2.0 m internal.
4. A square silo of side 3.0 m and height 18m of reinforced concrete is filled with paddy. Calculate the load on the bottom of silo and the lateral thurst at every 2.0 m of depth on the walls. The characteristics of stored paddy are as follows:-
   Minimum bulk density : 560 kg/m^3
   Maximum bulk density : 670 kg/m^3
   Minimum angle of internal friction : 35°
   Maximum angle of internal friction : 40°
   Minimum angle of friction on concrete : 34°
   Natural angle of repose of paddy : 34°
5. Derive the Janssen's formula for calculation of lateral pressure induced by granular materials against wall in deep bins.

Bibliography

Baker, K.D., R.L. Stroshine, K.J. Magee, G.H. Foster and R.B. Jacko, 1986. Grain damage and dust generation in a pressure pneumatic conveying system. Trans. ASAE 29(3):840-847.

Beverloo, W.A., Leniger and J. Van de Velde, 1961. The flow of granular solids through orifices. Chem. Eng. Sci. 15:260-269.

Bisht, B.S., and B.P.N. Singh, 1989. Discharge of food grain through horizontal rectangular openings. Proceeding of SJC of ISAE, Process and Food Eng. 181-186.

Borkar, K.H, 1985. Rats and their control in houses and godowns in "Storage" (Eds.Kachru, Bisht and Patil) CIAE, Bhopal.

Bucklin, R.A., S.A. Thompson, I.J. Ross and R.H. Biggs, 1989. Apparent coefficient of friction of wheat on bin wall material. Trans. ASAE 32(5): 1769-1773.

Boumans, G, 1985. Grain handling and storage. Elsevier Science Publishers, Amsterdam.

Change, C.S., H.H. Converse and J.L. Steele, 1991. Flow rates of grain through various shapes of vertical and horizontal orifices. Trans. ASAE 34(4): 1789-1796.

Dale, AC and RN Robinson, 1954. Pressures in deep grain storage structures, Agric. Eng. 35:570-573.

Dehpour, M.B., Bachchan singh and T.C. Thakur, 1992. The gravitational flow of granular fertilisers through orifices. Indian J. Agric. Eng. 2(4):281-286.

Deming and Mehring, 1929. The gravitaional flow of fertiliser and other commuted solids. Ind. Eng. Chem. 21(7):661-665.

Domkundwar, S, 1980. A course in refrigeration and air-conditioning (Environmental Eng.) Dhanpat Rai and Sons, Delhi.

Ewalt, D.L., and F.H. Buelow, 1963. Flow of shelled corn through orifices in bin walt. Michigan Expt. Stn. Bull. 48(1):92-102.

Garg, RM and S. Gopal Krishnan, 1972. Grain silos, some structural engineering aspects. Proceedings of seminar on post harvest technology of grains and pulses, New Delhi.

Girish, GK, 1990. Factors affecting grains in storage. Regional workshop ware house management of stored foodgrains, New Delhi.

Gregory, J.M. and C.B. Fedler, 1987. Equation describing granular flow through circular orifics. Trans ASAE 30(2):529-532.

Howe, R.W., 1965. A summary of estimates of optimal and minimal condition for population increase of some stored product insects. J. Stored Prod. Res.1:177-184.

Hycle, Mary B and NJ Burrell, 1973. Some recent aspects of grain storage technology in "Grain Storage Part of a System" (Eds RN Sinha and WE Muir) AVI Publishing Co. Inc. Westport Connecticut.

Kamlesh, V.B., R.C. Maheshwari and A.P. Gupta. 1980. Dynamic stresses in deep bins during discharge. J. of Agric. Eng. XXII (3&4):73-80.

Lal Radhey, and AC Datta, 1971. Agricultural engineering (through worked examples.), Saroj Prakashan, Allahabad.

Navarro, S and M.Calderon. 1974. Exposure of Ephestia cautella (wek) pupae to carbon dioxide concentrations at different relative humidities, the effect on adult emergence and loss in weight. J.Stored Prod. Res. 10:237-241.

Newman, CJE 1987, Storage and ancillary equipment for bulk handling of grain in "Bulk handling and storage of grain in the humid tropics" (Eds Champ and Highley) ACIAR Proceedings No.22:69-100.

Pingale, SV and GK Girish, 1967. Aeration of grains. Bulletin of Grain Tech. V (3):161-166.

Pingale, SV 1976. Handling and storage of food grains, Indian Council of Agricultural Research, New Delhi.

Pradhan, S, PB Mookherjee and GC Sharma, 1965. Pusa bin for grain storage. Indian Farming, 15 No.1:14-16.

Reimbert, ML and AM Reimbert, 1987. Silos theory and practice. Vertical silos, Horizontal silos (returning Walls). Lavoisier Publishing Inc. New York.

Shejbal J, 1980. Storability of cereal grains and oilseeds in nitrogen in "Controlled atmosphere storage of grains" Ed. (J.Shejbal). Elsevier Scientific Publishing Company, Amsterdam.

Shanker, G, 1988. Traditional storage structures. Proceedings of Regional Workshop on-farm storage (facilities and design) held at Indian Grain Storage Institute, Hapur:153-161.

Stweart, BR and MG Britton, 1973. Design of farm grain storages in "Grain Storage-Part of a System" (Eds RN Sinha and WE Muir) AVI Publishing Co., Inc. Westport Connecticut.

Storey Charles L., 1980. Functional and end use properties of various commodities stored in a low oxygen atmosphere in "Controlled atmosphere storage of grains" Ed. (J. Shejbal). Elsevier Publishing Company, Amsterdam.

Vishambharan, PR, 1985. Stored grain insects and their control in "Storage" (Eds Kachru, Bisht and Patil) CIAE, Bhopal.

Wimberly, JE, 1983. Technical handbook for the Paddy Rice Post Harvest Industry in Developing Countries. IRRI. Los Banos, Philippines.

Chapter 5

Milling

Milling is a general trade name which normally means reduction of food grain into various end products like meal, flour, splitted products etc. Milling includes; cleaning, grading, separating, mixing, pearling, polishing, dehusking, size reduction etc. The meaning of the term milling varies with the crop. For example, milling of wheat refers to a grinding operation to produce flour, whereas in rice industry, milling refers to overall operations in a rice mill i.e. cleaning, dehuskng, paddy separation, bran removal and grading of milled rice. Milling also refers to extraction of juice, oil or separation of fibre etc.

Most of the agricultural products are in solid form and which is generally difficult to handle, compared to liquid and gases. In processing, solids appear in many forms as large irregular pieces or finely divided powders. These particles may be hard and abrasive, soft, brittle, dusty or sticky and plastic. According to the forms of solids, means are to be found to manipulate them into end products and possibly to improve their handling characteristics.

Food grains are characterised by their size, shape and density. Size and shape are easily specified for regular particles. For irregular particles their size and shape are not much clear and these have to be defined.

Gain shape

The shape of any individual grain particle can be expressed in terms of sphericity 'ϕ'. The factor is independent of grain size. For spherical particles of diameter D_p, the sphericity is equal to radius. The sphericity of non-spherical particles can be represented by the following equation

$$\phi = \frac{6\, v_p}{D_p\, S_p} \qquad \text{...5.1}$$

where, D_p = equivalent diameter of particle

S_p = surface area of particle

v_p = volume of particle

The equivalent diameter is defined as the diameter of a shepre having the same ratio of surface to volume as that of the actual particle.

Size

To determine the size of food grains usually three mutual perpendicular lengths of particle are measured. These may be called as longest, intermediate and smallest lengths or simply length, breadth and thickness of the particle. The equivalent diameter of granular material can be expressed as cube root of the product of three mutually perpendicular lengths.

$$D_p = (l \times b \times t)^{1/3} \quad ...5.2$$

where, l = length of particle
b = breadth of particle
t = thickness of particle

The sphericity of particle can also be estimated by the following equation

$$\phi = \frac{D_p}{l} \quad ...5.3$$

In a sample of uniform particle of diameter D_p the total volume of particle is equal to $\frac{m}{\rho_p}$

where, m = mass of particle
ρ_p = density of particle

The total number of particles, $N = \frac{m}{\rho_p v_p}$ ...5.4

The total surface area of the particles,

$$N, S_p = \frac{6m}{\phi \rho_p D_p} \quad ...5.5$$

Average size of particle in a ground product

For the measurement of the particle size in the range between 76 mm to 38 μm (micrometer), standard screens are used. These screens are known as 'testing sieves' and are made of woven wire screens. The mesh and dimensions of testing sieves are carefully standardised. The openings of these test sieves are square. In 1910 a series of test sieves called Tylar sieves were originated in the USA which were adopted by the US Bureau of Standards.

The opening size of the Sieves is based on the 200 mesh sieve. This sieve has 200 openings in one inch. The nest sieve opening is larger by $\sqrt{2}$ or 1.41 times than the previous screen. Intermediate sieves with opening ratios of 1.189 are also available. These intermediate screens are not commonly used.

For analysis of ground products a set of standard screen is arranged serially in a stack. In the bottom, a pan is placed and then on it smallest mesh size sieve is placed. On the top of the mesh the next larger one is placed and thus serially the stack is arranged.

Bureau of Indian Standards has also standardised mesh sizes for screen analysis. The size of openings of standards screens are given in Table 5.1.

Food grains are composed of cellulose, fibre, starch, fat etc. These constituents effect the size, shape and distribution of the particles in milling operation. Milling of grains usually done by abrasive or impact type of mills. Performance of a mill is characterised by the capacity, power requirement per unit of material reduced, size and shape of the product before and after reduction and the range in size and shape of the end product.

Table 5.1
Test sieves and their respective sizes

BSS	*ASTM*	*ISS*	*Width of opening in inches*	*Width of opening in mm*
	4	480	0.1870	4.750
5	6	340	0.1252	3.250
6	7	280	0.1109	2.818
7	8	240	0.0945	2.399
8	10	200	0.0800	2.032
10	12	170	0.0659	1.676
12	14	140	0.0553	1.405
14	16	120	0.0473	1.201
16	18	100	0.0394	1.000
18	20	85	0.0332	0.954
20	—	—	0.0322	0.894
22	25	70	0.0279	0.708
25	30	60	0.0233	0.592
30	35	50	0.0197	0.500
36	40	40	0.0165	0.420
40	—	—	0.0158	0.401
44	45	35	0.0133	0.351
48	—	—	0.0132	0.336
50	—	—	0.0116	0.295
52	50	30	0.0117	0.296
60	60	25	0.0099	0.251
72	70	20	0.0083	0.211
80	—	—	0.0073	0.186
85	80	18	0.0070	0.177
100	100	15	0.0060	0.157
120	120	12	0.0049	0.124
150	140	10	0.0041	0.104
170	170	9	0.0035	0.089
200	200	8	0.0030	0.075
240	220	7	0.0025	0.064
300	270	6	0.0021	0.053
325	—	5	0.0017	0.044
350	325	—	—	0.044
400	—	4	0.0015	0.038

The most common method of classification of comminuted product is screening of the ground material through set of sieves, which is also called as "screen analysis."

Screen Analysis

For determination of average particle size in ground food grains, a set of Indian Standard Screens is arranged serially in a stack. For food grain flour analysis, a set of IS Sieves No. 100, 70, 50, 40, 30, 20 and 15 with pan and cover is taken. A sample of 250 g of ground product is dried in an oven to a constant weight. The dried sample is placed in the topmost sieve and the set is placed on a sieve shaking machine and shaken for 5 minutes.

Fineness Modules

The fineness modulus indicates the uniformity of grind in resultant product. It is determined by adding the weight fractions retained above each sieve and dividing the sum by 100. An example for determination of the fineness modulus in presented in Table 5.2 A.

Table 5.2 A

Determination of Fineness Modulus

IS sieve No.	*Weight of material retained, g*	*Percent material retained*			*Fineness Modulus*	*Average particle size, mm*
100	0.0	7 × 0.0	=	0.0		$D = 0.135\ (1.366)^{F.M.}$
70	1.4	6 × 0.56	=	3.36		= 0.317 mm
50	16.7	5 × 6.71	=	33.53	$\frac{274.03}{100}$	
40	36.7	4 × 14.45	=	57.80	= 2.7403	
30	82.2	3 × 33.01	=	99.03		
20	96.0	2 × 38.55	=	77.10		
15	8.0	1 × 3.21	=	3.21		
Pan	8.7	0 × 3.50	=	0.0		
Total				274.03		

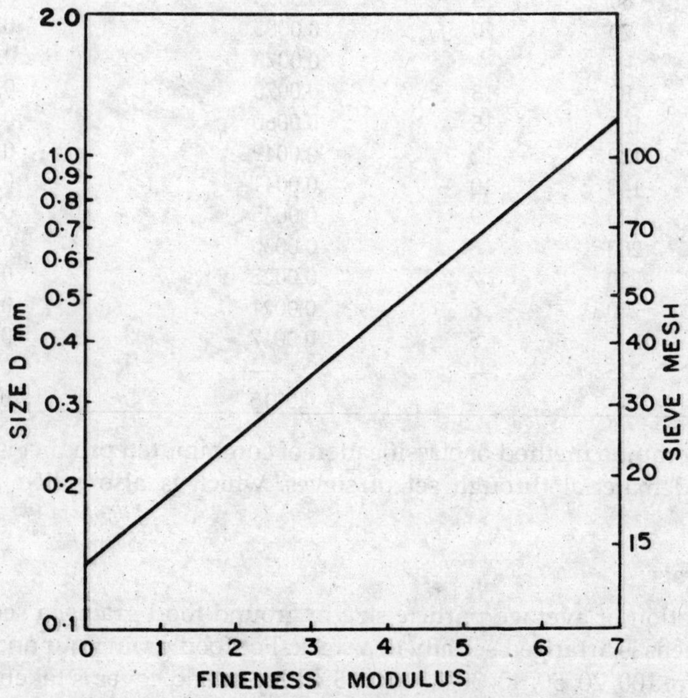

Fig. 5.1 : Relationship between the fineness modulus and average particle size

The average particle size, D_p in mm represented in terms of fineness modulus can be estimated by the following equation (Fig 5.1)

$$D_p = 0.135\ (1.366)^{\text{F.M.}} \quad ...5.6$$

To asses the particle size distribution of ground product (which is function of the proportionate amount of each size of individual particles in the powder) a relationship between the log of 'percent finer than' vs. log screen openings is drawn (Fig 5.2). This gives an idea about the size distribution of comminuted product.

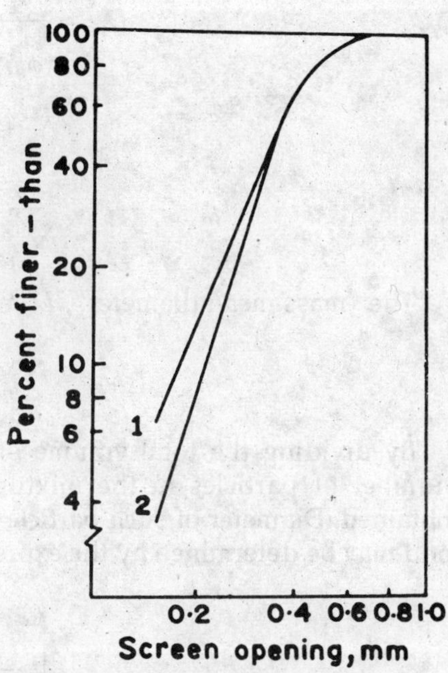

Fig. 5.2 : Log-log plot of percent finer than against screen size
1. Wheat 2. Bengal gram (chickpea)

There is a limitation in using equation 5.6 for determination of average particle size in ground product. Same set of sieves as shown in Table 5.2 should be used to obtain correct average particle size. Calculation of particle size in ground product analysed by any combination of test sieves can be done by employing following method.

As given earlier in equation 5.5 the total surface area of particles in a ground product may be given as,

$$A = NS_p = \frac{6}{\phi\, \rho_p D_p}$$

If the density of particle and its sphericity ϕ are known, the surface area of the particles in each fraction after sieve analysis may be calculated using the above equation. In this case the total surface area of particles will be the sum of particle surface ares in each fraction. Considering particle density, ρ_{p1} and its sphericity ϕ as constant, the "specific surface" of the total mixture could be represented by the following expression.

$$Ass = \frac{6m_1}{\phi\, \rho_p D_{p1}} + \frac{6m_2}{\phi\, \rho_p D_{p2}} + \ldots + \frac{6m_n}{\phi\, \rho_p D_{pn}}$$

$$= \frac{6}{\phi\, \rho_p} \sum_{i=1}^{n} \frac{m_i}{D_{pi}} \quad ...5.7$$

where, i = individual increment

n = number of increments

D_{pi} = average particle diameter in each increment

The "volume surface mean diameter", D_{vs} is one way of representing the average size of particle in a ground mixture. There is a relationship between D_{vs} and A_{ss} and can be given as below:

$$D_{vs} = \frac{6}{\phi \rho_p A_{ss}} \quad ...5.8$$

or,

$$D_{vs} = \frac{6}{\phi \rho_p \times \frac{6}{\phi \rho_p} \sum_{i=1}^{n} \frac{m_i}{D_{pi}}}$$

$$= \frac{1}{\sum_{i=1}^{n} \frac{m_i}{D_{pi}}} \quad ...5.9$$

The "mass mean diameter", D_m can be calculated from the equation as follows.

$$D_m = \sum_{i=1}^{n} m_i D_{pi} \quad ... 5.10$$

By dividing the total volume of the ground product or sample by the total number of particles in the mixture, the average volume of a particle can be obtained. Diameter of such particle is known as the "volume mean diameter", D_v and may be determined by the expression given below.

$$D_v = \left(\frac{1}{\sum_{i=1}^{n} \frac{m_i}{D_{Pi}^3}} \right)^{1/3} \quad ... 5.11$$

If the particles are of uniform size in the ground product, all the mean diameters viz D_{vs}, D_m and D_v are same, but for mixture containing dissimilar particles sizes, these diameters may vary significantly.

Number of particles in a ground material: The volume of a particle varies as cubes of its diameter

$$v_p \propto D_p^3$$

or

$$v_p = a D_p^3 \quad ... 5.12$$

Where, v_p = volume of particle

a = volume shape factor

The number of particle in a sample as given in equation 5.4 is,

$$N = \frac{m}{\rho_p v_p} = \frac{m}{\rho_p a D_p^3}$$

$$= \frac{1}{a \rho_p} \sum_{i=1}^{n} \frac{m_i}{D_{pi}^3}$$

or

$$N = \frac{1}{a \rho_p D_v^3} \left(\text{Since} \sum_{i=1}^{n} \frac{m_i}{D_{pi}^3} = \frac{1}{D_v^3} \right) \quad ... 5.13$$

A characteristic screen analysis is presented in Table 5.2 B. In the columns sieve sizes, sieve opening, average particle diameter in each increment, mass fraction over each sieve and cumulative mass fraction smaller than the sieve size, respectively, are given. Mass fraction is the amount of material retained on a particular sieve of the total mass of sample. This fraction is said as *m*, here *i* is the number of sieve starting from the bottom of the stack. Therefore, for pan, *i* is 1, and the sieve above pan is designated as $i + 1$ and so on.

The average particle size in each increment is the arithmatic mean of two successive sieve opening sizes (Table 5.2 B). The cumulative fraction smaller than D_{pi} is given in last column. As whole of the sample is placed on the top pan its value is 1.00, the amount of mass fraction on next is deducted from, 1.00 to get the cumulative fraction smaller than top sieve. Now to calculate the value of cumulative fraction smaller than second sieve from top, the value of mass fraction of third sieve from top is deducted from the value of cumulative fraction of second sieve and so on.

Table 5.2 B: Characteristic screen analysis of ground wheat

Sieve size IS	*Sieve Opening Dpi, mm*	*Average diameter in each increment Dpi mm*	*Mass fraction retained mi*	*Cumulative fraction smaller than Dpi*
200	2.032	–	0.000	1.000
170	1.676	1.854	0.015	0.985
140	1.405	1.540	0.062	0.923
120	1.201	1.303	0.104	0.819
100	1.000	1.100	0.320	0.499
70	0.708	0.854	0.208	0.291
50	0.500	0.604	0.096	0.195
40	0.420	0.460	0.078	0.117
30	0.296	0.358	0.046	0.071
20	0.211	0.254	0.031	0.040
15	0.157	0.184	0.029	0.011
Pan	–	0.078	0.011	0.000

Problem : Wheat is ground in a burr mill and screen analysis of the ground product was done. The analysis is given in Table 5.2B. Let the density of the particles be 750 kg/m^3. Calculate the followings for the ground product between 15 and 200 sieve size.

(i) Specific surface of the mixture
(ii) Number of particles per gram
(iii) Volume surface mean diameter
(iv) Mass mean diameter
(v) Volume mean diameter
(vi) Number of particles per gram between 70 and 50 sieves. The shape factors are, $a = 1.5$ and $\phi = 0.616$

Solution : Given that $\rho_p = 750\ \text{kg/m}^3$

$= 0.00075\ \text{g/mm}^3$

$$A_{ss} = \frac{6}{\phi \rho_p} \sum_{i=1}^{n} \frac{m_i}{D_{pi}} = \frac{6}{0.616 \times 0.00075} \sum_{i=1}^{n} \frac{m_i}{D_{pi}}$$

$$= 12987 \sum_{i=1}^{n} \frac{m_i}{D_{pi}}$$

and
$$N = \frac{1}{a \rho_p} \sum_{i=1}^{n} \frac{m_i}{D_{pi}^3} = \frac{1}{1.5 \times 0.00075} \sum_{i=1}^{n} \frac{m_i}{D_{pi}^3}$$

$$= 888.8 \sum_{i=1}^{n} \frac{m_i}{D_{pi}^3}$$

See the Table 5.2B, for 200/170 sieve increment the arithmatic mean diameter of two sieve openings, D_{pi} is 1.854 and for this increment the mass fraction, m_i is 0.015, therefore, $\frac{m_i}{D_{pi}}$ is equal to 0.00809 and $\frac{m_i}{D_{pi}^3}$ is 0.00235. Now we have to calculate the values of m_i/D_{pi} and m_i/D_{pi}^3 for all other nine increments except pan. The sum of all the ten increments of m_i/D_{pi} is 1.399 and that of m_i/D_{pi}^3 is 9.426. The pan fraction is not considered, thus the specific surface and number of particles per gram of particles on 15 sieve and larger than 15 sieve are calculated by dividing the value 1− m_i because for pan $i = 1$, therefore, 1 − m_i i.e. 1 − 0.011 = 0.989.

(i) Specific surface of the mixture

$$A_{ss} = 12987 \times \frac{1.399}{0.989} = 18370.89\ \text{mm}^2/\text{g}$$

(ii) Number of particles per gram

$$N = 888.8 \times \frac{9.426}{0.989} = 8471.00\ \text{particles/g}$$

(iii) Volume surface mean diameter

$$D_{vs} = \frac{1}{\sum_{i=1}^{n} m_i / D_{pi}} = \frac{1}{1.399} = 0.715\ \text{mm}$$

(iv) Mass mean diameter

$$D_m = \sum_{i=1}^{n} m_i D_{pi} = 0.912\ \text{mm}$$

(v) Volume mean diameter

$$D_v = \left(\frac{1}{\sum_{i=1}^{n} m_i \Big/ D_{pi}^{\ 3}} \right)^{1/3}$$

$$= \left(\frac{1}{9.426} \right)^{1/3} = 0.473 \text{ mm}$$

(vi) Number of particles per gram between 70/50 sieves can be given by

$$N_{70/50} = \frac{m_i(70/50)}{a\rho_p D_{p70/50}^3}$$

$$= \frac{0.096}{1.5 \times 0.00075 \times (0.604)^3}$$

$$= 387 \text{ particles/g}$$

The term 'size reduction' is associated with all the means by which solid particles are cut or broken into smaller pieces. In processing stages solids are reduced in size and shape by various methods for different purposes. Reduction in size is brought about by mechanical means without change in the chemical properties of the materials. During size reduction operation, chunks of large particles are crushed or reduced to workable size. In this process uniformity in size and shape of the individual grains or resultant product is desired, but it is very difficult to attain. Size reduction sometimes leads to increased reactivity of solids, helps separation of unwanted ingradients, also reduces the bulk of fibrous materials for easier handling.

Principles of size reduction

Crushers and grinders are the equipment mostly used for size reduction of agricultural products. An ideal size reducer should fulfill the following conditions, namely (1) large capacity, (2) should yield a predesired sized product or range of size, (3) small power input requirement per unit of product handled and (4) easy and trouble free operation. Usually the performance of any milling equipment is compared with respect to an ideal operation as standard. The characteristics of the actual equipment is compared with those of the ideal unit.

Size reduction results in the production of small particles which may be required either for larger surface area or because of their definite shape, size and number. Amount of power required to create smaller particles is one of the parameters of the efficiency of operation. Second parameter is the desired uniformity of size. The actual unit seldom yield a uniform sized product. Irrespective of uniformity of feed size the ground product consists of a mixture of various particle sizes. In some equipment there is a provision to control the magnitude of the largest particles like the hammer mill, but the fine size is beyond control. In some size reducing machine fines are minimised but they can not be eliminated altogether.

It has been found that the diameter of largest particle in comminuted product may be as large as 1000 times the diameter of smallest particle.

The major expenditure involved in crushing and grinding operation is the power requirement. Therefore, the factors controlling the power requirement are important. Work required to strain the material is temporarily stored in it in the form of mechanical energy of stress. When external force exceeds the amount of stored mechanical energy, the material is disturbed beyond its strength and finally broken into fragments. It results in generation of new surface. Solids have a certain amount of surface energy, thus for creation of new surface, work is required and it is supplied by the release of stress energy when material breaks. In size reduction process, the stress energy excess of the new surface energy created is converted into heat energy.

Crushing efficiency

The ratio of the surface energy created by crushing to the energy absorbed by the solid is referred as crushing efficiency 'η_c'

Energy requirements

The energy absorbed by a unit mass of the material is given by the following equation.

$$E_a = \frac{e\,(A_p - A_f)}{\eta_c} \qquad \text{...5.14}$$

where, e = surface energy per unit area

A_p = area per unit mass of product

A_f = area per unit mass of feed

The input energy (E) requirement for size reducing machine is greater than the energy absorbed by the solid (E_a). Some part of the input energy is used to overcome friction in the moving parts and bearings of machine, rest is used for crushing. The ratio of the energy absorbed to the input energy is known as the mechanical efficiency 'η_m'.

Then $$E = \frac{E_a}{\eta_m} = \frac{e\,(A_p - A_f)}{\eta_m\,\eta_c} \qquad \text{...5.15}$$

The power required by the machine can be calculated by the following equation.

$$P = E \times f = \frac{f\,e\,(A_p - A_f)}{\eta_m \eta_c}$$

or, $$P = \frac{6fe}{\eta_m \eta_c \rho_p}\left(\frac{1}{\phi_p\, D_p} - \frac{1}{\phi_f\, D_f}\right) \qquad \text{...5.16}$$

where, D_p and D_f = volume surface mean diameter of the product and feed, respectively

ϕ_π and ϕ_ϕ = sphericity of product and feed, respectively

ρ_p = particle density

f = feed rate

When a feed is reduced to symmetrical particles of a smaller size as shown in Fig 5.3, the energy requirements must be related to some function of the size of the feed and ground product. As per assumption both the particles are symmetrical, a common dimension is used to calculate energy requirement

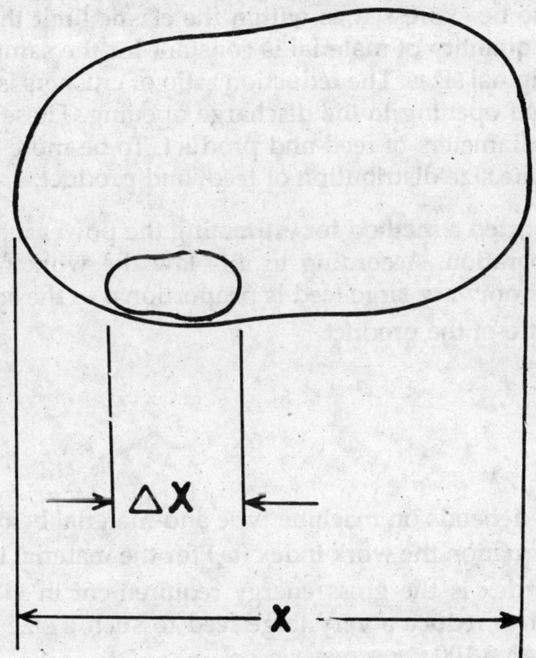

Fig. 5.3 : Reduction of feed to symmetrical smaller particle

$$\Delta E \propto \frac{\Delta X}{X}$$

or

$$\Delta E = c\,\frac{\Delta X}{X^n}$$

Therefore, the necessary energy required for size reduction is,

$$E = -c\int \frac{dx}{X^n}$$

Rittinger's and Kick's laws : A crushing law proposed by Rittinger states that the work required in crushing is proportional to the new surface created. Rittinger assumed that size reduction is essentially a shearing procedure. Therefore, energy requirement is proportional to the square of the common linear dimension and thus the values of '*n*' becomes 2. The energy requirement is given by the following equation

$$E = c\left(\frac{1}{X_p} - \frac{1}{X_f}\right) \qquad \ldots 5.17$$

where, X_p and X_f = length of product and feed, respectively.

Kick proposed another law which based on stress analysis of plastic deformation within the elastic limit. He assumed that the energy requirements for size reduction is a function of a common dimension of the material, therefore, the value of '*n*' becomes 1, and the energy requirements can be given by the following equation

$$E = c \; ln \left(\frac{X_f}{X_p} \right) \qquad \ldots 5.18$$

Kick's law can also be expressed as within the elastic limit the work required for crushing a given quantity of material is constant for the same reduction ratio irrespective of the original sizes. The reduction ratio of crushers is often expressed as the ratio of the feed opening to the discharge opening. These openings determine the maximum diameters of feed and product. To be more informative it is necessary to specify the size distribution of feed. and product.

Bond's law **: Bond reported a method for estimating** the power required for crushing and grinding operation. According to this law the work required to form particles of size 'D_p' from very large feed is proportional to the square root of the surface-to-volume ratio of the product

Since, $$\frac{S_p}{v_p} = \frac{6}{\phi D_p}$$

$$\frac{P}{f} = \frac{K}{\sqrt{D_p}} \qquad \ldots 5.19$$

where, $K =$ constant, depends on machine type and material being handled.

For use of above equation the work index (w_i) for the material being reduced is defined. The work index is the gross energy requirement in kilowatt-hour per tonne of feed needed to reduce a very large feed to such a size that 80% of the product passes through a 100 μm screen.

$$\frac{P}{f} = 0.3162 \, w_i \left[\frac{1}{\sqrt{D_p}} - \frac{1}{\sqrt{D_f}} \right] \qquad \ldots 5.20$$

where, $P =$ power in, kW

$f =$ feed rate, t/hr

$D_p =$ 80% of product passes through mesh of dia D_p, mm

$D_f =$ 80% of feed passes through mesh of dia. D_f, mm

$w_i =$ work index

The work index may be found experimentally from laboratory crushing and grinding tests.

Problem : In a wheat milling experiment it was found that to grind 4.33 mm sized grains to IS sieve 35 (0.351 mm opening), the power requirement was 8 KW. Calculate the power requirement for milling of wheat by the same mill to IS sieve 15 (0.157 mm opening) using (1) Rittinger's law and (2) Kick's law. Feed rate of milling is 200 kg/hr.

Solution : (*i*) According to Rittinger's law

$$\frac{P}{f} = c \left(\frac{1}{X_p} - \frac{1}{X_f} \right)$$

$$\frac{8}{0.2} = c \left(\frac{1}{0.351} - \frac{1}{4.33} \right)$$

$$40 = c \, (2.618)$$

or, $$c = 15.278$$

By putting the value of 'c' for second condition

$$\frac{P}{0.2} = 15.278\left(\frac{1}{0.157} - \frac{1}{4.33}\right)$$

or, $P = 18.75$ kw

(*ii*) According to Kick's law

$$E = c\ ln\left(\frac{X_f}{X_p}\right)$$

or,
$$\frac{P}{f} = c\ ln\left(\frac{X_f}{X_p}\right)$$

$$\frac{8}{0.2} = c\ ln\left(\frac{4.33}{0.351}\right)$$

$$40 = c \times 2.5125$$

or, $C = 15.92$

Putting the value of 'c' for second condition

$$\frac{P}{0.2} = 15.92\left(\frac{4.33}{0.157}\right)$$

or, $P = 10.56$ kW

Problem : How much power is required to crush 2 t/hr. of a material if 80% of the feed passes through IS sieve No. 480 (4.75 mm opening) and 80% of the product posses through IS sieve No. 50 (0.5 mm opening)? Given the work index of the material as 6.30.

Solution : According to Bond's law

$$\frac{p}{f} = 0.3162\ w_i\left(\frac{1}{\sqrt{D_p}} - \frac{1}{\sqrt{D_f}}\right)$$

Given:- $W_i = 6.30$

$f = 2$ t/hr

$D_p = 0.50$

$D_f = 4.75$

$P = ?$

$$\frac{P}{2} = 0.3162 \times 6.30\left(\frac{1}{0.5} - \frac{1}{4.75}\right)$$

or, $P = 7.129$ kW

Problem : Sorghum (3.80 mm size) was milled by a burr mill at two different gaps between the burr stones. The flour was analysed by IS sieves for particle size determination as shown in Table given below. The power required to mill sorghum at first setting was 5.0 kW.

Calculate the power requirement of the mill in second setting using (1) Rittinger's law and (2) Kick's law. The capacity of mill was 1.0 q/hr.

IS Sieve No.	*Mass fraction of flour retained over sieve, g*	
	I-setting	*II-setting*
100	—	—
70	10.1	1.5
50	16.7	13.3
40	36.0	36.1

IS Sieve No.	Mass fraction of flour retained over sieve, g	
	I-setting	II-setting
30	82.2	74.8
20	96.0	104.6
15	8.0	8.4
Pan	0.0	11.3

Solution : Determination of flour sizes of two setting by sieve analysis.

(*i*) Average particle size of product in I setting

Sieve No.	Weight of material retained, g	Percent material retained			Fineness Modulus	Average Particle size, mm
100	—	7×0	=	0.00		
70	10.1	6×4.04	=	24.24	$\frac{295.08}{100}$	$D_p = 0.135\ (1.366)^{FM}$
50	16.7	5×6.68	=	33.40		
40	36.0	4×14.40	=	57.60		$= 0.338$
30	83.2	3×33.28	=	99.84	$= 2.9508$	
20	96.0	2×38.4	=	76.80		
15	8.0	1×3.2	=	3.20		
Pan	—	0×0	=	0.00		
Total	250.0			295.08		

(*ii*) Average size of sorghum flour in second setting

Sieve No.	Weight of material retained, g	Percent material retained			Fineness Modulus	Average flour size, mm
100	0.0	7×0.0	=	0.00		
70	1.5	6×0.6	=	3.60	$\frac{264.76}{100}$	$D_p = 0.135\ (1.366)^{FM}$
50	13.3	5×5.32	=	26.60		
40	36.1	4×14.44	=	57.76	$= 2.6476$	$= 0.308$
30	74.8	3×29.92	=	89.76		
20	104.6	2×41.84	=	83.68		
15	8.4	1×3.36	=	3.36		
Pan	11.3	0×4.52	=	0.00		
Total	250.0			264.76		

Determination of Power requirements

(*i*) According to Rittinger's laws

$$\frac{P}{f} = c\left(\frac{1}{X_p} - \frac{1}{X_f}\right)$$

or,
$$\frac{5.0}{0.1} = c\left(\frac{1}{0.338} - \frac{1}{3.8}\right)$$

$$50 = c \times 2.694$$

or
$$c = 18.55$$

Putting the value of '*c*' for second setting

$$\frac{P}{0.1} = 18.55\left(\frac{1}{0.308} - \frac{1}{3.8}\right)$$

or, $$P = 0.1 \times 18.55 \times 2.984$$
$$= 5.535 \text{ kW}$$

(*ii*) According to Kick's law

$$\frac{P}{f} = c \ ln\left(\frac{X_f}{X_p}\right)$$

$$\frac{5.0}{0.1} = c \ ln\left(\frac{3.8}{0.338}\right)$$

or, $$c = 20.66$$

Putting the value of '*c*' for second setting

$$\frac{P}{0.1} = 20.66 \ ln\left(\frac{3.8}{0.308}\right)$$

or $$P = 5.2 \text{ KW}$$

Size reduction procedures

The size of agricultural products may be reduced by several ways, but mainly the following four methods are used in size reduction machines, (1) compression or crushing, (2) impact, (3) shearing, and (4) cutting.

Crushing : When an external force applied on a material excess of its strength, the material fails because of its rupture in many directions. The particles produced after crushing are irregular in shape and size. The type of material and method of force application affects the characteristics of new surfaces and particles. Food grain flour, grits and meal, ground feed for livestock are made by crushing process. Crushing is also used to extract oil from oilseeds and juice from sugarcane.

Impact : When a material is subjected to sudden blow of force in excess of its strength, it fails, like cracking of nut with the help of a hammer. Operation of hammer mill is an example of dynamic force application by impact method.

Shearing : It is a process of size reduction which combines cutting and crushing. The shearing units consist of a knife and a bar. If the edge of knife or shearing edge is thin enough and sharp, the size reduction process nears to that of cutting, whereas a thick and dull shearing edge performs like a crusher. In a good shearing unit the knife is usually thick enough to overcome the shock resulting from material hitting. In an ideal shearing unit the clearance between the bar and the knife should be as small as practicable and the knife as sharp and thin as possible.

Cutting : In this method, size reduction is accomplished by forcing a sharp and thin knife through the material. In the process minimum deformation and rupture of the material results and the new surface created is more or less undamaged. An ideal cutting device is a knife of excellent sharpness and it should be as thin as practicable. The size of vegetables and fruits are reduced by cutting.

Cereal grinding

All the cereal grains are plant seeds and contain a large centrally located starchy endosperm which is also rich in protein. Almost all the cereal grains are covered with protective outer layers such as husk and bran. The germ or embryo is located near the bottom of the seed.

For general food uses we remove hulls which are largely cellulose and indigestible to human being, the coloured bran and germ which contain most of the oil. Since the oil can be attacked by enzymes and may produce rancid condition in grain, the germ is removed. For various food purposes starchy and proteinaceous endosperm is used. Sometimes the 'B' vitamins and minerals are added to ground meal, which is known as enrichment.

Grinding of cereals may be broadly classified into two; (1) plain grinding and (2) selective grinding. In the first case grains are milled to a free flowing meal consisting of sufficiently uniform particle size, whereas in the second process the grinding operation is carried out in various stages depending upon the differences in structural and mechanical properties of components of grain. Hardness of seed effects the power requirements of grinding.

Degree of grinding : The degree of grinding is defined as the ratio of the overall surface area of the ground product to the overall surface area of the feed and is expressed as below

$$d_g = \frac{S_p}{S_f} \qquad \ldots 5.21$$

where, d_g = degree of grinding

S_p = overall surface area of product

S_f = overall surface area of feed

Hulling/Milling efficiency : The overall effectiveness of hulling is the product of coefficient of hulling "$E_{hulling}$" and coefficient of wholeness of kernel "E_{wk}" (Kupritz, 1967)

The coefficient of hulling is given by the percentage of the hulled grains obtained from the total amount of grain input.

The coefficient of wholeness of kernel is ratio of the amount of kernel, crushed grains and mealy waste obtainable by any milling system

$$E_{hulling} = 100\left(1 - \frac{n_2}{n_1}\right) \qquad \ldots 5.22$$

and

$$E_{wk} = \frac{k_2 - k_1}{(k_2 - k_1) + (d_2 - d_1) + (m_2 - m_1)} \qquad \ldots 5.23$$

where, n_1, n_2 = amount of unhulled grains before and after hulling

k_1, k_2 = amount of whole grains before and after hulling

d_1, d_2 = contents of broken kernels before and after hulling

m_1, m_2 = weight of mealy wastes before and after hulling

$k_2 - k_1$ = contents of whole kernels

$d_2 - d_1$ = contents of broken/crushed kernels

$m_2 - m_1$ = yield of mealy wastes

The overall efficiency of the milling system 'η' is given by the following expression

$$\eta = E_{hulling} \times E_{wk} \qquad \ldots 5.24$$

The above efficiency tells us about the qualitative and quantitative aspects of grain milling operation

Size reduction machinery

Size reducing devices are grouped as follows:

1. Crushers
2. Grinders
3. Fine grinders
4. Cutting machines

Crushers : These type of reducing machines squeez or press the material until it breaks. Crushers are mostly used to break large pieces of solid materials into small lumps. Crushers are used in industrial operations, like mines etc. Use of crushers in agricultural operations is limited. The crushers in use are, (1) jaw crushers, (2) gyratory crushers, and (3) crushing rolls.

Lime and other stones are first reduced by the jaw or gyratory crushers. In a jaw crusher feed is admitted between two jaws, which are open at the top like *'V'*, One of the jaws is fixed and somewhat vertical (Fig 5.4) while the other is the swinging jaw. This jaw reciprocates in a horizontal plane, and makes and angle of

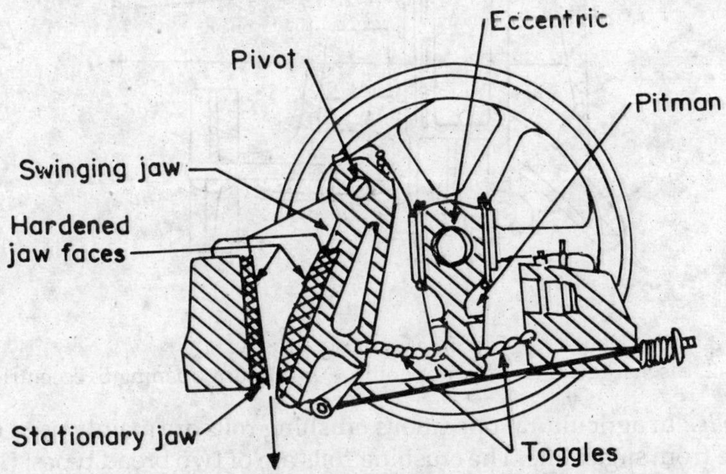

Fig. 5.4 : Jaw crusher

20 to 30° with the fixed jaw. The movable jaw is operated by an eccentric unit so as to impart great compressive force. The solids which has to be broken is caught between the two jaws. Large lumps of solid materials are caught between the upper parts of the jaws and subsequently broken and dropped into the narrower space below. The broken pieces are further reduced next time when jaws come closer. The number of strokes given to the movable jaw by eccentric unit ranges between 250 to 400 times per minute.

In a gyratory crusher the jaws between which the solid materials fed, are circular. In such crushers the material is being crushed at all times at some point. Solids are caught between *'V'* shaped space between the head and casing (Fig 5.5). The material is repeatedly broken in sufficiently small pieces to pass out from the bottom. The crushing head is rotated by an eccentric unit. The speed of crushing head ranges between 125 to 425 gyrations per minute. The discharge from the gyratory crusher is continuous, hence the driving motor is uniformly loaded. Less maintenance is required as compared to jaw crusher, also the power requirements is low.

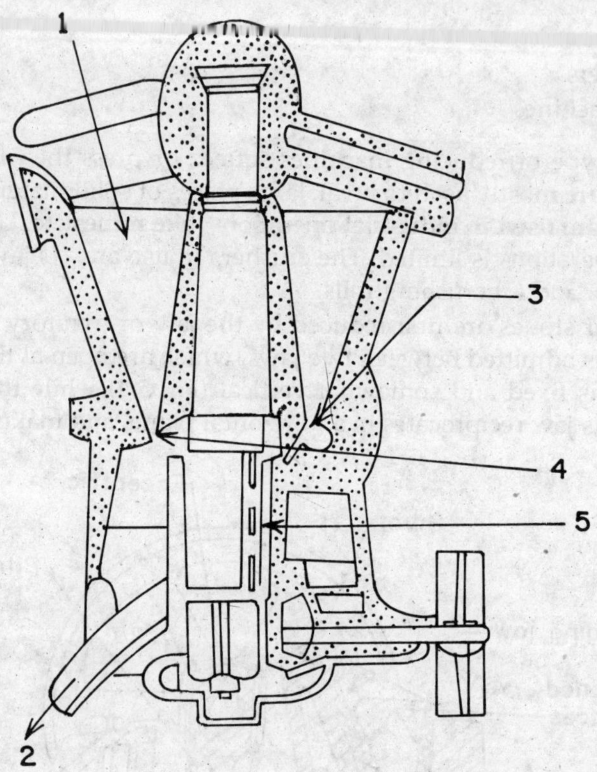

Fig. 5.5 : Gyratory crusher
1. feed 2. discharge 3. maximum opening 4. minimum opening 5. eccentric unit

Crushing rolls : In agricultural operations crushing rolls are mainly used for extraction of juice from sugarcane. The crushing rolls are of two broad types; (1) smooth-roll crusher and (2) serrated or toothed-roll crushers.

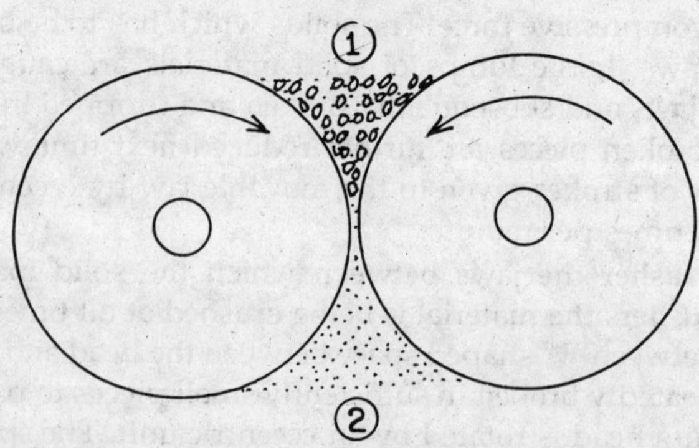

Fig. 5.6 : Schematic diagram of a smooth roll crusher

Two heavy smooth-faced metal rolls rotating towards each other at same speed on parallel horizontal axes are the working elements of the smooth-roll crusher. It is shown in Fig 5.6. The size of the particles/material that can be caught by the rolls depends upon the coefficient of friction between the material and the roll surface and can be estimated by the following equation

$$d_p = 0.04\,R + g \qquad \ldots 5.25$$

where, d_p = maximum size of particle

R = roll radius

g = half of the width of gap between the rolls

The rolls exert great force and to avoid any damage to roll surface because of some unbreakable material coming with the feed, at least one of the rolls should be spring loaded.

Apart from extraction of juice the smooth-roll crushers are used to make grits or meal from food grains. These are also extensively used for making food grains flakes.

Serrated or toothed-roll crushers : In such crushers the rolls are serrated as per need (Fig 5.7). Toothed-roll crushers are much more versatile than smooth-roll crushers. The best example of such type are the break and reduction rolls of a wheat flour milling plant. Disintegrators are the toothed-roll crushers in which the corrugated

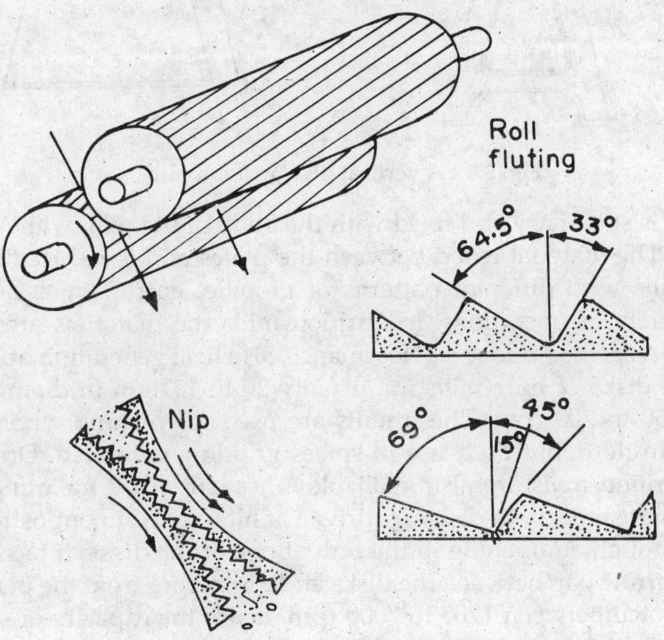

Fig. 5.7 : Serrated or toothed-roll crusher

rolls are rotating at different speeds. These machine tear the feed apart. The size reduction in serrated-roll crushers is by compression, impact and shear, and not by compression alone, as in the case of smooth roll crushers. The serrated-roll crushers can also accommodate larger particles than smooth-roll crushers.

Grinders : The grinders are used to mill the grains into powder. The grinder comprises a variety of size-reduction machines like attrition mills, hammer mills, impactors and rolling compression mills.

Attrition mills : In an attrition mill the grains are rubbed between the grooved flat faces of rotating circular disks. These mills are also known as burr or plate mills. The axis of the roughened disks may be horizontal or vertical (Fig. 5.8). In attrition

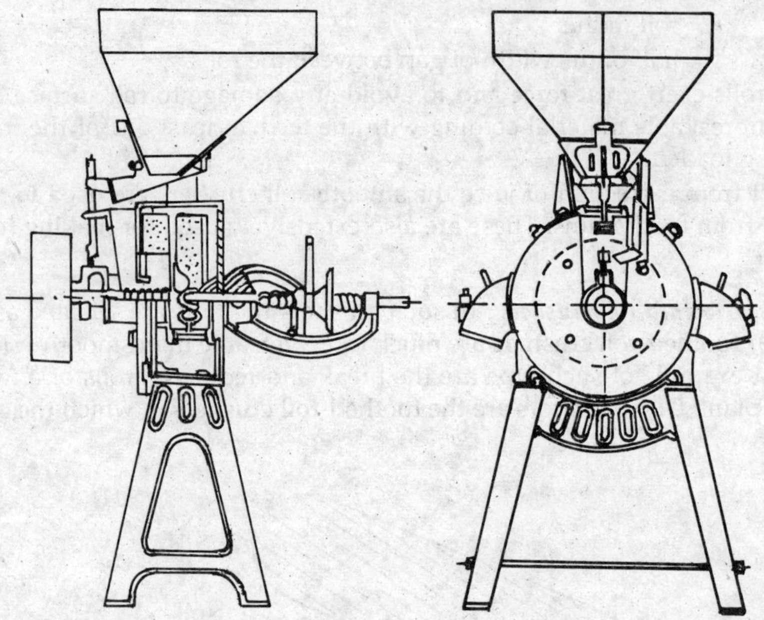

Fig. 5.8 : A vertical disk attrition mill

mill one plate is stationary and fixed with the body of the mill, while other one is rotating disk. The material is fed between the plates and is reduced by crushing and shear. Mills with different patterns of grooves, corrugations on the plates perform a variety of operations. In attrition mills the materials are slowly fed, overfeeding lowers the grinder's performance, also heat generation during milling increases. The disks of burr mills are usually 20 to 137 cm in diameter and are operated at 350 to 700 rpm. These mills are used for making whole grain and dehusked grain flour, but their use in spices grinding is limited. Double runner disks type attrition mills are also available. These are used for grinding of soft materials. In these mills both disks are driven at high speed in opposite directions. Feed enters through an opening in the hub of one of the disks, it passes outward through the narrow gap between the disks and discharges from the periphery. The disks are operated between 1200 to 7000 rpm, hence the capacity of such mills is large.

The fineness of grinding in burr mills is controlled by the type of plates and the gap between them. The spacing between the plates is adjustable and usually the arrangement is spring loaded to avoid damage to plates in case of over loading or to overcome the damage to plates by foreign material coming along with the feed. The salient features of a burr mill are its lower initial cost and lower power requirements. But foreign matter may cause damage/breakage, and operation without feed may result in burr wear.

Hammer mills : Hammer mills are used for various types of size reduction jobs. These mills contain a high-speed rotor, rotating inside a cylindrical casing. The shaft is usually kept horizontal. Materials are fed into the mill from the top of the casing and is broken by the rotating hammers and fall out through a screen at the bottom. The material or feed is broken by fixed or swinging hammers which are pinned to a rotor. The hammers are rotated between 1500 to 4000 rpm, strike and grind the material until it becomes small enough to pass through the bottom screen (Fig. 5.9). Fineness of grinding is controlled by the screen size.

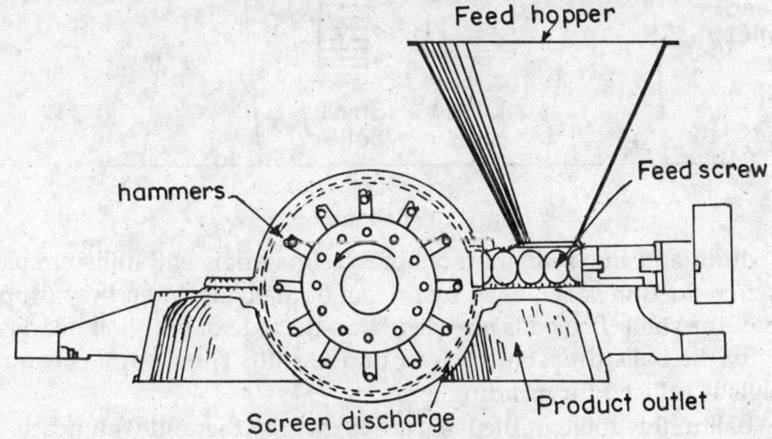

Fig. 5.9 : Hammer mill

Hammers are either rigidly fixed to the shaft or swinging. In case of swinging hammer mill there is less chances of damage of hammer if some unbreakable solid material comes to milling chamber alongwith feed. There are several designs of striking edge of the hammers. Hammer mill can grind almost anything—like tough fibrous solids, steel chips, food grains, sticky clay, hard rock etc. The hammer mill is assumed to reduce size by impact. The kinetic energy of the rotating hammers is used to disintegrate the feed. Most of the size reduction is achieved by impact of hammers, though some amount of shear also takes place between the feed and screen and other mill parts. The salient features of hammer mill are their simplicity and versatility in design and work, freedom from damage during empty operation and less chances of damage of mill due to foreign objects. The worn hammer do not significantly reduce the efficiency of mill, but the main disadvantage of the mill is its high power requirements. The capacity and power requirements of hammer will depend on the nature of feed to be ground. Commercial mills reduce solids between 60 to 240 kg/kWhr of energy consumption. Hammer mills are used for poultry feed grinding, spices grinding. It was also found suitable for grinding of wet sorghum and millets and also for potato, tapioca, banana and similar flour making.

Ball mills : The ball mill is a cylindrical or conical shell slowly rotating about a horizontal axis. Half of its volume is filled with solid grinding balls (Fig. 5.10). The shell is usually made of steel lined with high carbon steel plate, porcelain or silica

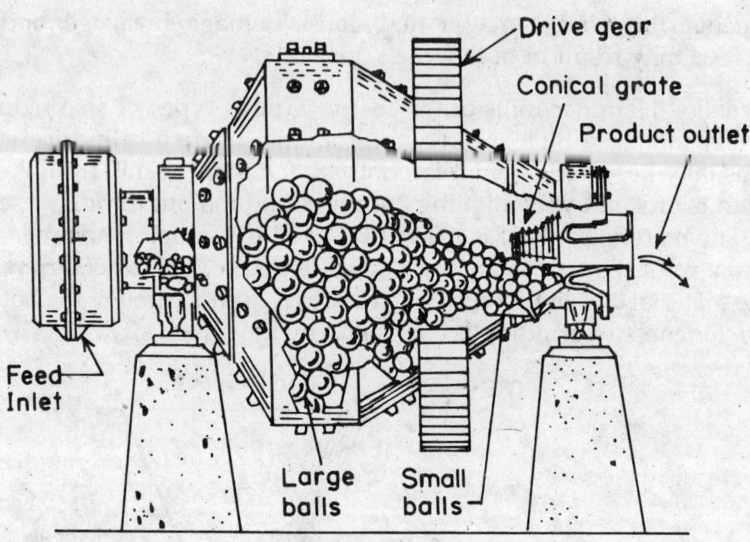

Fig. 5.10 : Ball mill

rock. For medium and fine reduction of abrasive materials ball mills are used. In a ball mill size reduction is achieved by impact of the balls when they drop from near the top of the shell. The balls are carried up the side of the shell nearly to the top. By gravity the balls drop on the feed underneath. The energy consumed in lifting the balls is utilised for grinding job.

When the ball mill is rotated, the balls are carried by the mill wall nearly to the top, where they are released by gravitational pull and drop to the bottom and picked up again. Centrifugal force keeps the ball in contact with the mill wall. Most of the grinding is done by the impact of balls. Due to centrifugal force, if the speed of rotation of mill is faster, the balls are carried to more distance. In case of too high speed, balls stick to mill wall and are not released. This is a stage of centrifuging. The rotational speed at which centrifuging occurs is known as critical speed. At this speed as the balls are released from the top, no impact occurs hence little or no grinding results. Therefore, the operating speeds must be kept less than the critical speed. The speed at which the outermost ball released from the mill wall, depends on the interaction of gravitational and centrifugal forces. The critical speed can be determined by the following equation

$$n_c = \frac{1}{2\pi}\sqrt{\frac{g}{R-r}} \qquad \ldots 5.26$$

where, n_c = critical speed, revolution/s

g = acceleration due to gravity, 9.80 m/s^2

R = radius of the mill, m

r = radius of the ball, m

The rotational speeds of the ball mills are kept at 65 to 80% of the critical speed, with the lower values for wet grinding in viscous suspension.

Problem : What would be the operating speed of rotation in revolutions per minute of a ball mill of 2,000 mm diameter charged with 100 mm balls? The ball mill is grinding solid matter.

Solution : The critical speed, $n_c = \frac{1}{2\pi}\sqrt{\frac{g}{R-r}}$

or, $n_c = \frac{1}{2 \times 3.1416}\sqrt{\frac{9.8}{2-0.1}}$

$= 0.36144$ rpm

or, $n_c = 21.58$ rpm

Therefore the operating speed for solid matter grinding

$= n_c \times 0.8$

$= 17$ rpm.

Problem : What would be the critical speed and operating speed of rotation for wet grinding in viscous suspension by a ball mill of 1600 mm diameter charged with 75 mm balls?

Solution : $n_c = \frac{1}{2\pi}\sqrt{\frac{g}{R-r}}$

or, $n_c = \frac{1}{2 \times 3.1416}\sqrt{\frac{9.8}{1.6-0.075}}$

$= 0.4034$ rpm

$= 24$ rpm

Therefore the operating speed will be 65% of 24 rpm.

or, $n_c = 24 \times 0.65$

$= 16$ rpm.

Cutting machines : Size reduction of fruits and vegetables ar mostly performed by cutting operation. To make thin slices of fruits and vegetables knife cutters are used. Few types of knife cutters have been developed for cutting slices/chips of potatoes, cassava, banana etc.

Rotary knife cutters contain a horizontal rotor rotating at 200 to 900 rpm inside a cylindrical chamber. Several knives are fixed to the rotor. Feed enters the chamber from the top and are cut by the rotating knives afterwards discharged from the bottom of the equipment.

Rietz mill or disintegrator : Rietz mill consists of a rotor inside a circular screen enclosure. The rotating shaft is usually vertical. The rotor includes a number of hammers running at a fairly close clearance. The hammers are generally rigidly fixed to the shaft, but in some cases swing hammers are also used. Feed enters the milling chamber parallel to the axis as in disk type attrition mill. The product is discharged radially out through a perforated sizing screen which surrounds the rotor (Fig. 5.11). Rietz mill's many applications are on wet materials. The advantage of this mill was found in those cases where solid content is in the range of 40 to 80 per cent. Rietz disintegrator keeps running because the close hammer clearance keeps the sizing screen open. Therefore, more fine and uniform grinding is possible. Rietz mill is able to grind materials below 15 micron size.

Rietz machines are normally supplied in rotor diameters from 10 to 60 cm (4 to 24 inches). The hammer tip speeds are in the range of 5.2 to 111 m/s and horsepower ranges from 0.5 to 200.

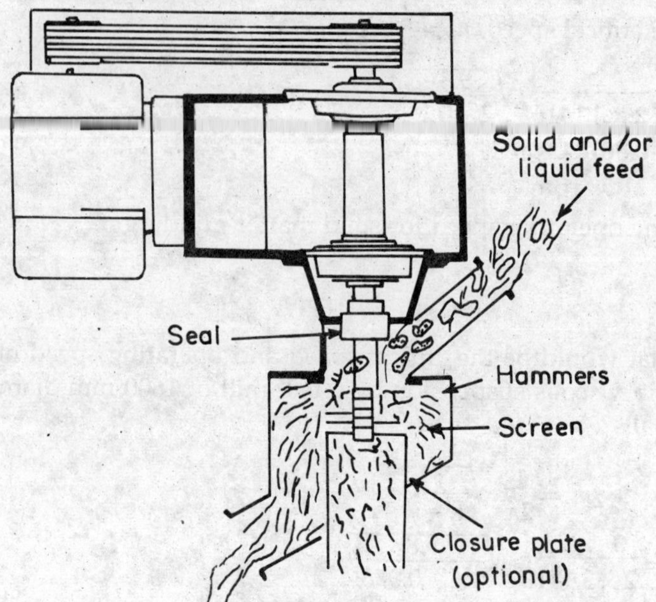

Fig. 5.11 : Rietz mill

Dispersion and colloid mills : Colloid mills are used for fine grinding. Where very little breakdown of individual particles and disruption of weakly bonded clusters are required, colloid mills are used. Purees, food pastes, pulps and other similar products are processed by this type of mill. Colloid mills fall into four main groups; the hammer or turbine, the smooth surface disk, the rough surface type and valve or orifice devices.

Concentric cylinder abrasive mills : Such type of mills are mostly used for scouring of husk or seed coverings of pulses and cereals. These work on frictional properties principle. Inside a larger drum (preferably of metal) an abrasive roller rotates. The outer cylinder can be made by a perforated pipe or from a metal sheet. The outer metal cylinder may also be fabricated as bottom half perforated, whereas the upper half portion is made from plain m.s. sheet.

Such roller machines may have a contant gap throughout its length between the outer cage and inner abrasive roller, or the gap is kept more at feeding end and continuously decreasing towards the delivery end. The material are fed from top to the rotating drum at feeding end. Milling or size reduction takes place in the annular space between perforated drum and abrasive roller and the product is discharged at the other end. Dehusking of grains is achieved by friction between the grains and abrasive surfaces and between grain to grain.

Size reduction machines operation and performance

To achieve good results from any size reducing machine the followings should be given due care, (1) feed should be of proper size and feeding rate should be uniform, (2) hard or unbreakable material is not-allowed to enter the mill, (3) after grinding the product is removed as soon as possible, and (4) there should be some arrangement to remove the heat generated during milling operation.

Determination of power requirement for a particular grinding job is difficult. The exact amount of power requirement depends on type of material, moisture content of feed, material feed rate, type and condition of mill, product particle size requirement, nature of abrasive surface etc.

Grains at higher moisture content are more difficult to grind than dry grains. The effect of moisture upon the power requirement for grinding of wheat with swinging hammer mill is shown in Fig 5.12. It was also found that fibrous materials

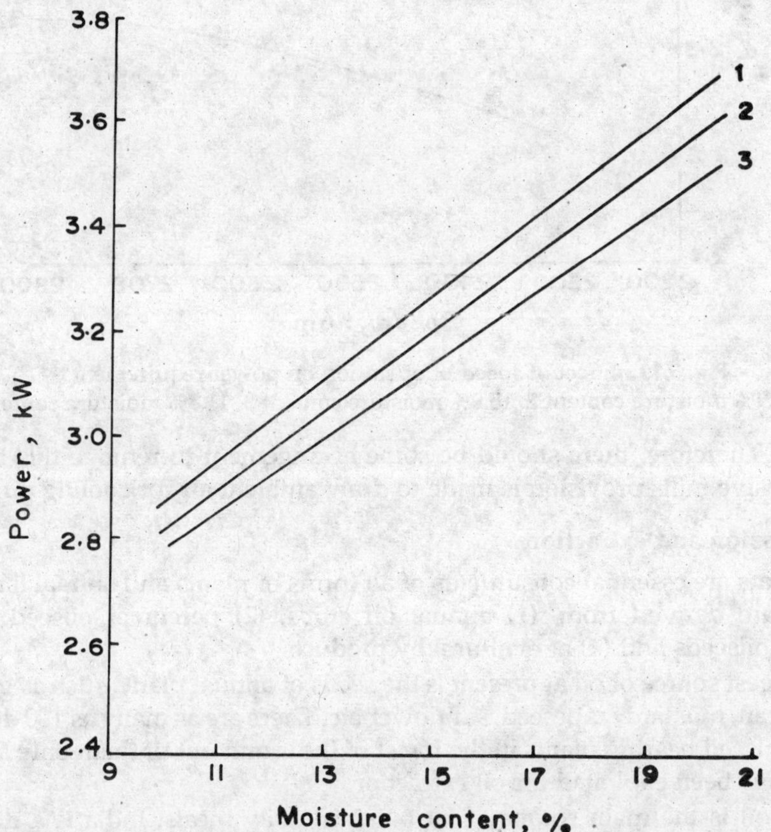

Fig. 5.12 : Effect of moisture content on power requirement
1. 2800 rpm 2. 2680 rpm 3. 2560 rpm

require more power for grinding than non-fibrous material. At increased rotational speed the power requirement also increases (Fig. 5.13).

Energy consumption : In size reduction of solids, bulk of the energy input is dissipated as heat energy. This energy raises the temperature of milled product, the mill and the air. Some of the energy is lost during vaporisation of grain moisture. Comparatively small fraction of energy input is utilised for creation of new surfaces. During grinding of grains the temperature rise is observed to be 20°C or more in emery burr mills. This temperature rise may partially decompose some

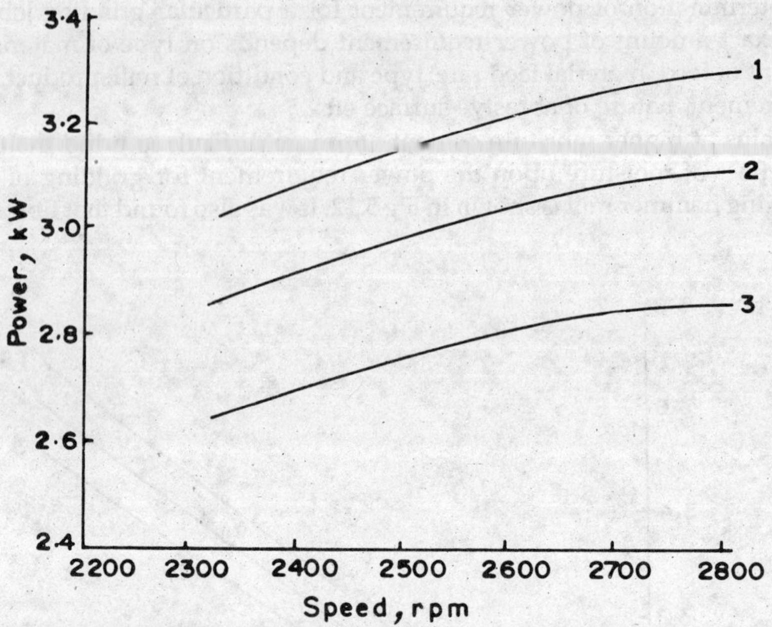

Fig. 5.13 : Effect of speed of operation on power requirement
1. 17.9% moisture content 2. 15.0% moisture content 3. 11.8% moisture content

materials. Therefore, there should be some arrangement to remove this heat. In some abrasive mills provision is made to draw ambient air for cooling purposes.

Oil expression and extraction

Oils and fats are essential constituents of all forms of plants and animal life. Most of these are derived from, (1) annual oil crops, (2) perennial oilseed plants, (3) minor oilseeds and (4) agricultural by products.

The largest source of oil at present is the seeds of annual plants such as groundnut, soybean, mustard/rapeseed, sunflower etc. There are as many as 100 different types of oilseed bearing plants in the forest of India and out of them only a dozen species have been exploited for oil extraction.

Edible oil is the main source of fat taken in daily meals. India is a deficient country as far production of oil is concerned. The need per head per day is 30 g oil to meet the minimal dietary requirement where as at present the availability is only 11 g per head per day.

Post harvest technology, out of many inputs, play an important role in increasing the edible oil production by reducing post production losses and increasing recovery of oil from the seed. For extraction of oil, mechanical expellers are the most widely used power machines. Apart from expellers solvent extraction plants play an important role. The solvent extraction plants extract oil from rice bran, soybean and expeller pressed cakes.

During the process of mechanical expression the oilseeds are compressed in various types of compression devices/equipment.

The oil is stored at different places in the oilseed in form of oil globules surrounded by tough membrane. There are two steps involved in the expulsion of oil through the expeller, (1) disintegration and (2) pressing. Under the disintegration process, oil globules are separated. The tough membrane surrounding the oil droplet is exposed and burst under pressure enabling the oil to ooze out.

Expression is the process of mechanically pressing liquid out of liquid containing solids. Screw presses, roll presses and mill, collapsible plate and frame filler presses, disk mill are some of the examples of a wide range of equipment used for expression of liquids. The advantage of mechanical expression system over chemical extraction is that it gives the liquid free of dissolved chemicals and thus a safer process.

Mechanical expression devices

(*i*) *Hydraulic press* : The hydraulic press is considered of a series of horizontal corrugated iron plates (Fig 5.14). These plates are separated by 4 to 14 premoulded oilseed cakes. Pressing is completed in two stages. In the first stage the oil samples are pressed at approximately 5 MPa for 15–20 minutes. Afterwards a pressure of 28 MPa is applied for 5–10 minutes to complete the expression process. The output

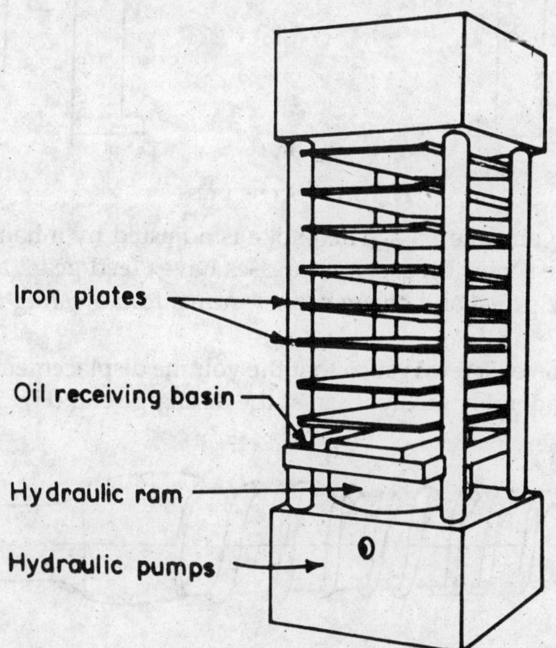

Fig. 5.14 : Hydraulic press

of the press varied depending on the sizes and seed being pressed. In recent times at commercial level processing the hydraulic press has been replaced by screw type presses.

(*ii*) *Screw press* : A screw press for oil expulsion is shown in Fig. 5.15. The press has a horizontal main shaft. The screw assembly is formed integrally with this shaft. This screw rotates within a cage or barrel. The barrel is made of case-hardened, tool steel bars or this may be made out of assembly of rings. Spacers are used between the lining bars or rings to allow drainage of the oil as the pressure on the feed material is increased. At the discharged end, a movable choke or cone controls the operating pressure. It is achieved by changing the width of annular space

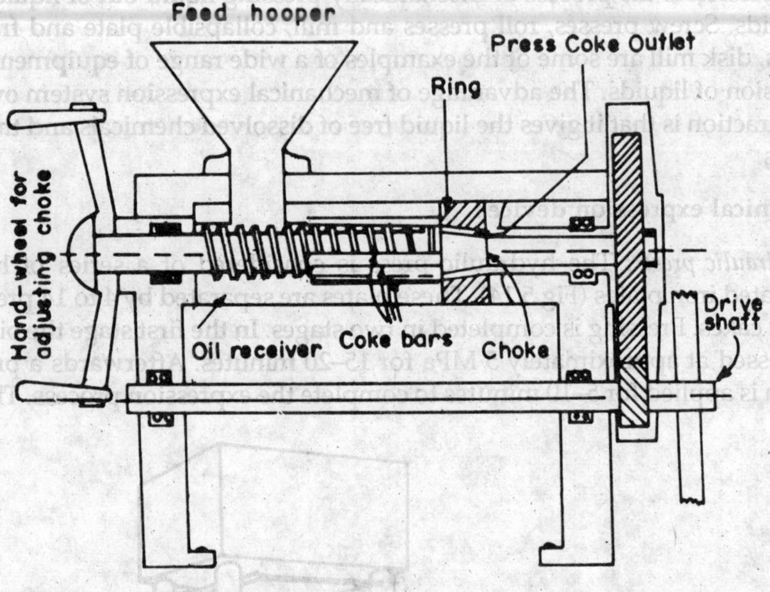

Fig. 5.15 : Screw press

through which the oil cake pass. The choke is adjusted by a hand wheel on the opposite end of the screw. Most screw presses have "feed gear" a means of ramming meal into the press, and choke gear, a means for adjusting the choke thickness.

The configuration of screw is such that the volume displacement at the feed end of the press is considerably greater than at the discharge end (Fig. 5.16). As a result

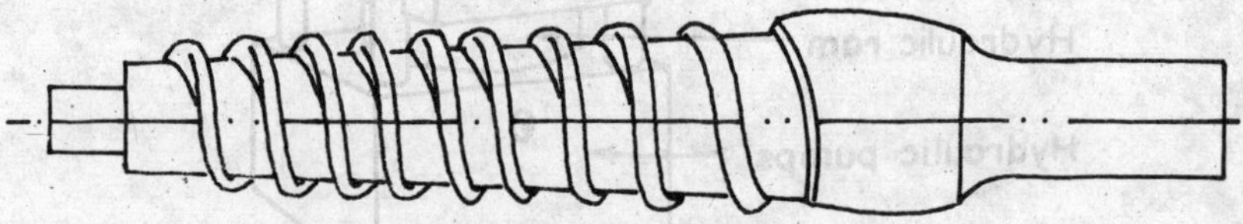

Fig. 5.16 : Configuration of a typical screw

of such configuration, as the material is conveyed from feed end to discharge end, it is subjected to increasing pressure. As pressure increases the material is compressed and oil is expelled through the spacers between the cage lining bars or rings. The "compression ratio" or a press is simply volume displaced per revolution at feed end divided by volume displaced per revolution at discharge end. The compression curve of a press is a plot of volume displacement to a base of length along the barrel. The shape of a compression curve is shown in Fig. 5.17. This curve is split into three sections, the feed section, the ram section and the plug section. By the action of screw and compressed material, radial pressure is generated along the barrel. This radial pressure can be measured by using strain gauges. The radial

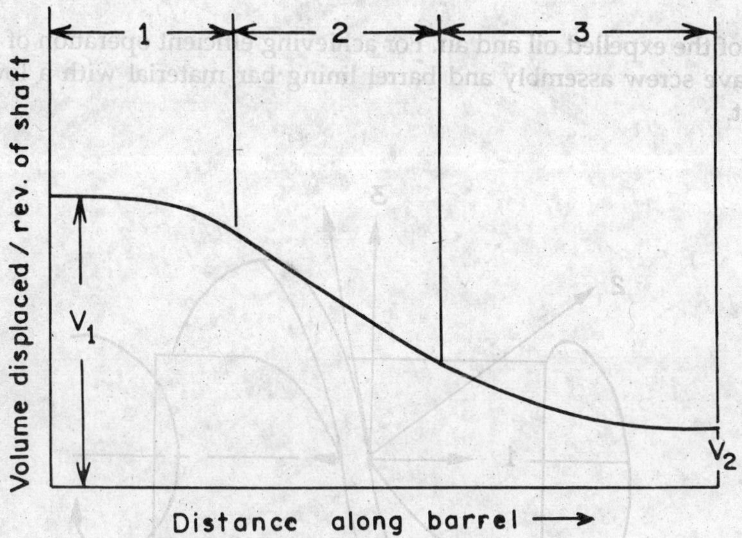

Fig. 5.17 : Compression curve of a screw press
1. feed section 2. ram section 3. plug section

$$\text{compression ratio} = \frac{v_1}{v_2}$$

pressure along a screw press barrel is shown in Fig. 5.18. The maximum radial pressure is generated at the feed end of the ram section. The axial pressure follows

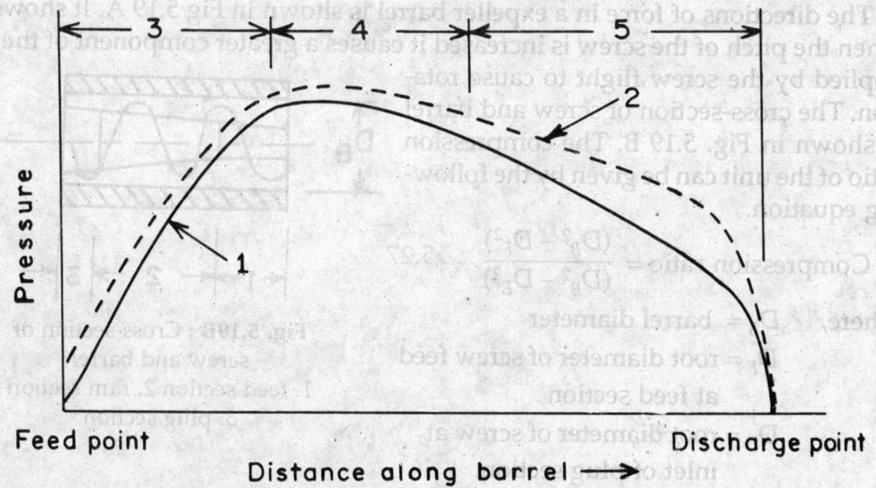

Fig. 5.18 : Radial and axial pressures in a screw press barrel
1. radial pressure 2. axial pressure 3. feed section 4. ram section 5. plug section

the radial pressure upto the beginning of the plug section. The fall off in pressure towards the discharge end is less marked. It is observed that there is a pressure gradient towards both ends of the press.

For expelling oil from a high oil content seed, the feed end of the press should be designed to dissipate the back pressure at the same time it should be able to move a sufficient volume of meal forward with minimum rotation and provide drainage of the expelled oil and air. For achieving efficient operation of a press, it should have screw assembly and barrel lining bar material with a low friction coefficient.

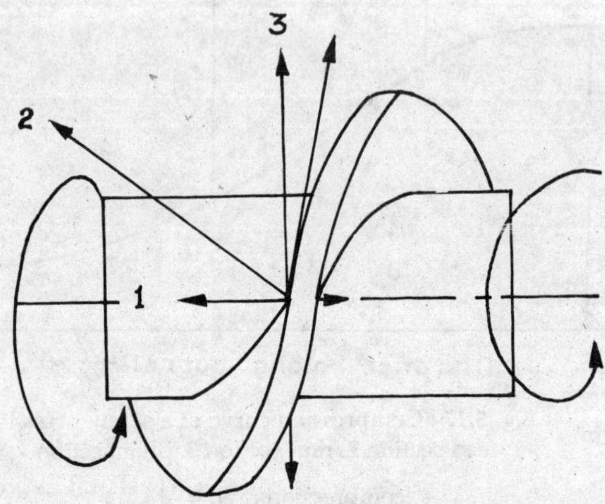

Fig. 5.19A : Directions of force in a screw press
1. force along barrel 2. force exerted by screw 3. force causing rotation

The directions of force in a expeller barrel is shown in Fig 5.19 A. It shows that when the pitch of the screw is increased it causes a greater component of the force applied by the screw flight to cause rotation. The cross-section of screw and barrel is shown in Fig. 5.19 B. The compression ratio of the unit can be given by the following equation.

$$\text{Compression ratio} = \frac{(D_B^2 - D_F^2)}{(D_B^2 - D_E^2)} \quad \ldots 5.27$$

where, D_B = barrel diameter

D_F = root diameter of screw feed at feed section

D_E = root diameter of screw at inlet of plug section.

Fig. 5.19B : Cross-section of screw and barrel
1. feed section 2. ram section 3. plug section

Ideally the flow of material along the wormshaft should be purely axial and the expelled oil should come through the spacers. But some amount of expelled oil flows reverse of the meal along the pressure gradient. The maintenance of outflow of expelled oil at related discharge points along the barrel depends on the design of slot, length of barrel, pressure drop, and viscosity of the material. Attempts were made by Koo and Baskerville to establish corelation between oil recovery from various oilseeds and parameters like operating pressure, pressing time and viscosity. Koo reported that pressing temperature, pressure time and moisture content affected the yield of vegetable oil. The pressing time has little effect on the oil yield.

Oil extraction

Extraction is the process of separating a liquid from a liquid-solid system with the use of a solvent. Extraction may also be said as a process to extract oil from oil bearing materials through the process of diffusion with the help of low boiling point solvent. This process gives a higher recovery of oil and a drier cake than expression. Extraction processing, or solvent extraction is capable of removing nearly all of the available oil from oilseed meal or flakes. Apart from giving higher yields of oil the extraction process provides meal of better preservation qualities and with higher protein quantities.

Solvent extraction technology has gained considerable importance in India. The total number of solvent extraction plants has increased from 163 in 1976 to 395 in 1990, with an annual installed capacity of about 15.6 million tonnes of oil cake.

In the solvent extraction process, the solvent is added to the well prepared materials. It is then followed by the diffusion of oil solvent mixture to the surface of solid for recovery of oil. The most commonly used solvent in the Indian Plants is the *n*-hexane. The boiling point of *n*-hexane is about 65.5°C. A screw stirrer is provided in the extractor unit to prevent accumulation of material at one place during the process. The extracted meal which comes out of the extractor bottom is removed by metering screw conveyor.

The deoiled cake coming out of the extraction unit has 25 to 40% solvent by weight. This has to be removed and recovered. This is normally done by evaporation either through the direct or indirect heating by steam or a combination of both.

Paddy Rice Milling

Paddy is the seed of certain variety of grass of Gramineae family. Botanical name of paddy is Oryza Sativa L. Oryza sativa is further divided into three sub-species viz. Javanica, Japonica and Indica. India and tropical Asian countries cultivate mostly Indica species.

Today rice has become one of the most important and major food grains of the world. Nearly 1/2 to 2/3 of total world population has partially or totally adapted rice as their main food. Asians meet half of their daily energy requirements through intake of rice. Most of the protein needs are met through rice only. The paddy grain is a living matter and represents a colloidal capillary porous body. It contains large amount of micro and macro capillaries through which water can move out or inside of the grain surface.

Paddy grain structure is shown in Fig. 5.20. It comprises of Palea and Lemma, two halves of the seed cover known as husk or hull and bottom portion known as empty glume which keeps palea and lemma in position. Rice kernel remains covered under the husk

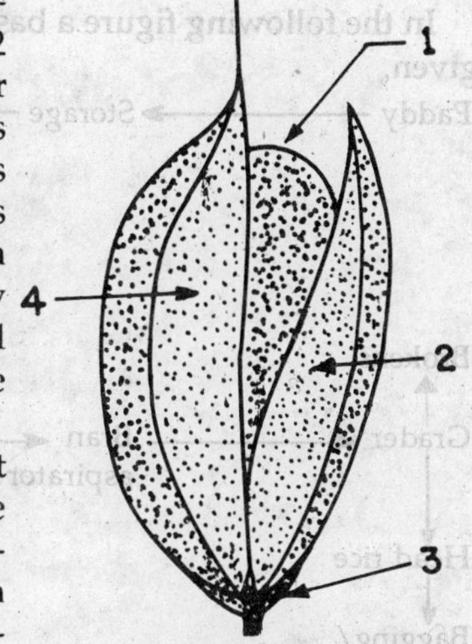

Fig. 5.20 : Paddy grain
1. rice kernel 2. palea
3. empty glume 4. lemma

Paddy undergoes certain processing treatments prior to its conversion into edible form. The various unit operations are cleaning, drying, storage, parboiling (optional) and milling. The edible portion of paddy is called rice. The kernel is enclosed by hull or husk. The weight of hull is 18–22% of the total weight of paddy grain. The endosperm weights about 70 to 72%, whereas 4–6% bran is removed during polishing. Paddy in India is traditionally milled in hullers. Engleberg huller removes husk and bran together, hence the by products are mixed. With hullers the degree of polishing can not be controlled to desired level. Use of hullers results in more breakage of grains and loss of nutrients. Government of India promulgated the 'Rice Milling Industries Regulation and Licensing Act' in 1959, according to which the use of Engleberg hullers was banned for shelling paddy.

Clean and good quality paddy can give an yield of 74% rice. But due to several reasons it has not been possible to achieve 74% yield of rice by traditional methods and machines. Use of modern machines like rubber roll sheller along with improved methods of drying, paddy separation, bran removal and graders can give higher out turn. The various unit operations and equipment used in a modern rice mill are given below.

Cleaning/cleaner	:	for removing foreign matter from paddy
Shelling/rubber-roll sheller	:	for separating husk from the paddy grain.
Husk separation/husk aspirator	:	for separation of husk from the product obtained from sheller.
Paddy separation/paddy separator	:	for separation of paddy from brown rice
Bran removal/polisher/whitener	:	for removal of bran layers from brown rice
Bran aspiration/bran aspirator	:	for removal of bran adhering to the rice kernel
Grading/grader	:	for separation of broken rice from head rice
Handling equipment	:	for conveying of paddy-rice to various processing units.

In the following figure a basic flow chart of rice processing in a modern mill is given.

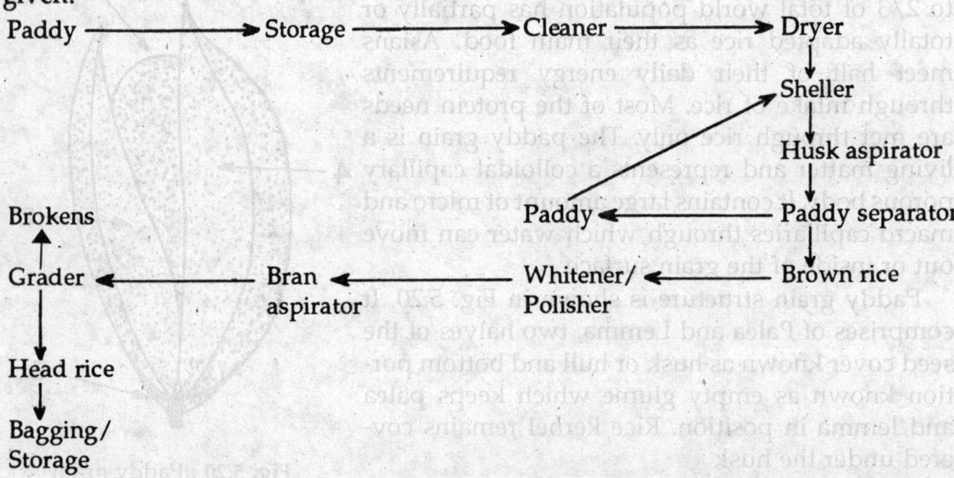

Paddy cleaning : cleaning of paddy is the first unit operation of paddy-rice milling. Cleaning adds to the quality of paddy and it is desirable for further processing operation. Cleaning refers to removal of all types of foreign matters and other impurities. These matters can cause damage to rice processing machines and their presence lower the quality of milled rice. The foreign matters can be grouped into following categories, (1) big impurities like stems, paddy earheads, stones, iron pieces etc., (2) small impurities e.g. weed seeds, dust, mud particles etc. and (3) impurities of equal size as to paddy like immature grains, stones and other matters.

For cleaning of paddy, generally the difference in physical properties of impurities and paddy are used e.g. size, weight or length of grains. Lighter foreign matters are removed or separated by air aspiration. While the bigger and smaller impurities from paddy are separated with the help of suitable screens. Impurities similar to the shape and size of paddy are broken during milling operation and then could easily be separated from the rice. Impurities equal in both weight and size as paddy are difficult to separate by mechanical means.

Parboiling

It is a premilling (optional) treatment given to paddy prior to its milling to achieve maximum recovery of head rice and to minimise breakage. Parboiling treatment was first developed in some Asian countries to reduce the milling losses. In this process paddy is soaked and the wet paddy is heated and then dried. The paddy grain is mainly composed of polygonal starch granules. The voids or intergranular spaces are filled with air and moisture. Due to these voids cracks develope and it cause breakage during milling. This breakage may be reduced by gelatinising the starch. During gelatinization process starch swells and fill the voids.

During soaking of paddy water penetrates into starch granules and results in swelling of grains. In heating the energy weakens the granule structure and more surface becomes available for water absorption and results in irreversible granule swelling. This phenomenon is called gelatinization of starch.

The temperature at which gelatinization takes place is known as the gelatinization temperature and it is specific for particular variety in the nearhood of 70°C.

The main objectives of parboiling are, (1) to increase the total and head rice yield of paddy, (2) to prevent the loss of nutrients during milling, (3) to salvage wet or damaged paddy and (4) to mill the rice according to requirements of consumers.

Parboiling of paddy is carried out in three steps, (1) soaking, (2) steaming, and (3) drying (Fig. 5.21). Theoretically soaking of paddy can be done at or below its

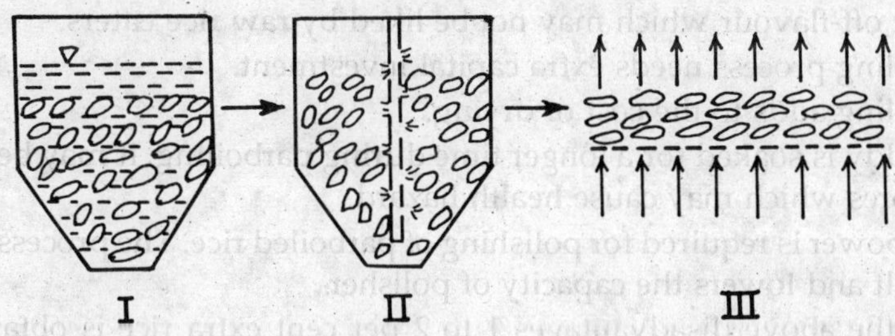

Fig. 5.21 : Various steps of parboiling of paddy
I. soaking II. steaming III. drying

gelatinization temperature. The lower the temperature used, slow is the process of soaking and vice-versa. Soaking period can be reduced by subjecting the paddy to vacuum for a few minutes before soaking and/or soaking under pressure in hot water. Heat of gelatinization of starch is supplied by saturated steam. Parboiled paddy may be dried in the shade or in the sun or with hot air. Shade drying takes longer time but gives excellent milling qualities. Rapid drying in sun or with hot air causes higher breakage during milling. The most convenient practices is to dry in two passes with a tempering period in the moisture range of 15–19% (wet basis).

Advantages of parboiling

1. The milling yield increases and the quality is improved, as there are fewer broken grains.
2. The grain structure becomes compact and vitreous, even if some kernels were entirely or partially chalky.
3. The milled rice becomes translucent and shining.
4. The shelf life of parboiled paddy and milled parboiled rice is longer than in the raw state, as germination is no longer possible and the kernel becomes hard enough to resist the attack by insects and to adsorption of atmospheric moisture.
5. The grains remain firmer during cooking and less likely to become sticky.
6. A greater amount of water is absorbed during cooking causing the rice to swell.
7. After cooking the rice absorbs less fat from added condiments, the rice keeps longer and does not become rancid easily.
8. Parboiled rice retains more proteins, vitamins and minerals.
9. Parboiled rice is more digestible and less solids are left behind in the cooking water.
10. Shelling of parboiled rice is easier.
11. Bran of parboiled rice has more oil.

Disadvantages of parboiling

1. The heat treatment during parboiling destroys some natural anti-oxidants, hence rancidity developed in parboiled rice during storage is more than that in raw rice.
2. Parboiled rice takes more time to cook than raw rice and may have characteristic off-flavour which may not be liked by raw rice eaters.
3. Parboiling process needs extra capital investment.
4. Parboiling adds to the cost of drying.
5. As paddy is soaked for a longer time during parboiling, it may be attacked by spores which may cause health hazard.
6. More power is required for polishing of parboiled rice. The process becomes difficult and lowers the capacity of polisher.

Inspite of the above disadvantages 1 to 2 per cent extra rice is obtainable by parboiling than raw rice milling. In India if 50% of the paddy production is converted into parboiled rice, it is possible to save about Rs. 1000 million by virtue of extra rice and its by products value.

Parboiling methods

The methods of parboiling may be classified as under.

A. *Traditional methods* :

(*i*) single boiling
(*ii*) double boiling

B. *Modern methods* :

(*i*) CFTRI method
(*ii*) Jadavpur university method
(*iii*) Converted process
(*iv*) Malek process
(*v*) Avorio process
(*vi*) Cristallo process
(*vii*) Fernandes process
(*viii*) Schule process
(*ix*) Rice grower's association of calfornia process
(*x*) Pressure parboiling process

Traditional methods

The traditional process consists of soaking paddy in water at room temperature for 24–48 hours or more, steaming in kettles under atmospheric pressure and drying under the sun. In a single boiling method, paddy is soaked in ordinary water for 24–72 hours and them steamed. In double boiling method, steam is first injected into raw paddy in the steaming kettle before soaking, Hot paddy raises the temperature of soaking water to 45–50°C which helps to reduce the soaking time to 24 hours. Thereafter soaked paddy is steamed. Sometimes, the soaking water is heated to about 50°C then the raw paddy is put into it and in this case first steaming is not required.

Modern methods

Some of the earlier listed modern methods of parboiling are described below.

1. *CFTRI method* : In this process, parboiling tanks are filled with clean water and heated to a temperature of about 85°C by passing steam through the coils placed inside the tank. Sometimes hot water is pumped from other sources into parboiling tanks. The resultant temperature of paddy-water mixture in the tank stays around 70°C. After soaking paddy for 3 to 3.5 hours, the water is drained out. The water discharge valve is kept open in order to remove condensed water during steaming. Soaked paddy is exposed to steam at a pressure of about 4 kg/cm^2 through the open steam coils. Soaking and steaming of paddy are done in the same tank. The parboiled paddy is taken out by opening the bottom door and dried either under sun or by a mechanical dryer.

2. *Jadavpur University method* : All the operations of this method is fully automatic and average processing time is five to six hours. Soaking of paddy is completed in high temperature water (60–70°C) within 1–3 hours, while the steaming time is limited to 3.5 minutes. After steaming and before drying the paddy is rapidly cooled. Drying takes place in a rotary steam jacketed high temperature air dryer. In this process two different systems may be applied. With the first, the soking and

steaming take place in the same tank, whereas in the second, these two operations are performed separately in a horizontal apparatus. In both the cases saturated steam is used. The steamed paddy is rapidly cooled in a draught of cold air.

3. *Pressure parboiling method* . This method of parboiling was developed at Tiruvarur in Tamil Nadu. The parboiling is achieved by penetration of moisture into the paddy in the form of water vapour under pressure. This results in gelatinization of starch of the kernel. The paddy is soaked for 40 minutes at 85–90°C. Thereafter it is steamed under pressure for 18 minutes. The water vapour which penetrates the kernel drives out entrapped air. It is reported that the whole process is completed in 1 to 1.5 hours. The rice obtained by this method has a pleasing and slightly yellowish uniform colour. Reduced soaking period of paddy is the main advantage of this method. It was also observed that such parboiled paddy has better shelling efficiency, has more fat in bran and increased storage life of rice grain.

Effect of parboiling on milling

In raw rice milling several factors are responsible for breakage of kernel. Cracking of kernel is one of the main factors for breakage. Cracks develope because of delayed harvesting, threshing or rapid drying. Immature and chalky kernels break easily. The type and design of the milling machinery influence milling outturn and quality. Rice breakage is related to milling conditions, particularly by the relative humidity, temperature and the extent of milling.

During shelling or husking operation, breakage occurs. Parboiling of paddy results in reduction of breakage. The improvement in milling quality is due to hardness imparted to kernels because of gelatinization of starch. It has been found that due to swelling of starch the cracks, incomplete grain filling, and chalkiness

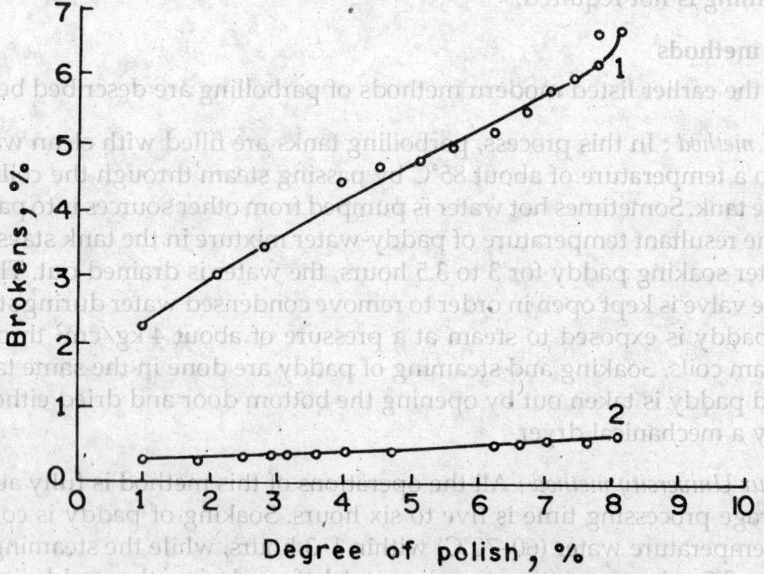

Fig. 5.22 : The broken percentage vs degree of polish
1. raw rice 2. parboiled rice

are completely healed. Such phenomenon improves the milling qualities of paddy. The most advantageous aspect of parboiling is the increase in the head yield of rice (Fig 5.22). During polishing, the polish percentage and breakage increase with time, but parboiled rice takes longer time than raw rice to attain same degree of polishing (Fig. 5.23). Parboiled rice requires three to four times as much abrasive load as raw rice for same level of polishing. As per liking for colour of rice need of polishing for parboiled rice is less as compared to raw rice. For example if consumer needs 80% bran removal, to achieve this, parboiled rice needs polishing of 3%, whereas raw rice has to be polished to 4% for same quantity of bran removal.

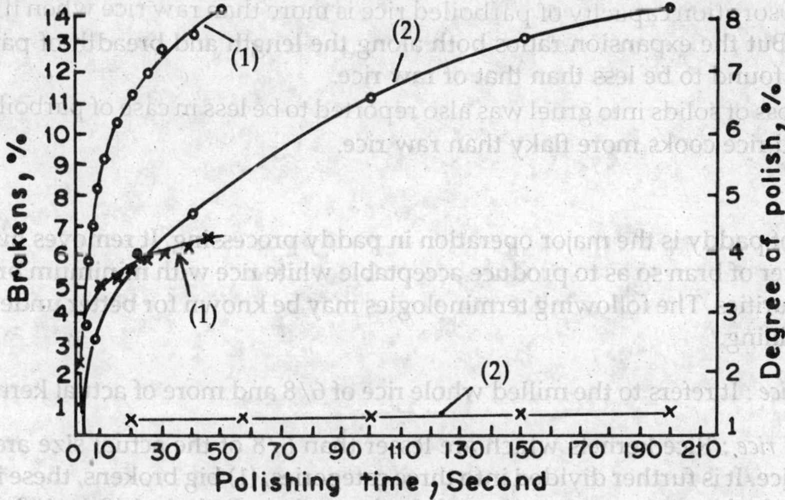

Fig. 5.23 : Effect of polishing time on degree of polish and brokens
– *o* – degree polish – *x* – *x* – broken
1. raw rice 2. parboiled rice

Effect of parboiling on nutritional qualities

The bran of parboiled rice contains less starch and more oil than raw rice bran as it comes out in the form of flakes. Therefore, separation of bran without loss of endosperm is possible. In Table 5.3 the chemical composition of raw and parboiled rice bran is shown. The nutrients of rice are found in large quantity in top layers than endosperm. Parboiled rice contains more protein, vitamins and minerals. Better nutrients availability in parboiled rice is attributed to the following factors, (1) due to hydrothermal treatment the nutrients from upper layers penetrate to endosperm, (2) parboiled rice needs lesser degree of polishing.

Table 5.3

Chemical composition of raw and parboiled rice bran

Chemical constituents, %	*Raw rice bran*	*Parboiled rice bran*
Moisture	9.8	9.1
Protein	14.2	18.5
Fat	16.6	19.8
Fibre	8.7	7.1
Ash	8.9	10.3

Parboiled rice contains less oil or fat. Due to soaking and subsequent heating or steaming certain enzymatic changes are brought about which help in release of oil from the kernel. The oil with this hydrothermal treatment move towards upper layers of kernel and thus is removed alongwith bran during milling.

Effect of parboiling on cooking qualities

Cooking qualities of rice are generally represented by time of cooking, swelling capacity, expansion ratio, gruel quality and pastiness.

Parboiled rice takes longer cooking time for required softness. Parboiled rice needs roughly double time than raw rice to attain same level of softness in cooking. Water absorption capacity of parboiled rice is more than raw rice when it is fully cooked. But the expansion ratios both along the length and breadth of parboiled rice was found to be less than that of raw rice.

The loss of solids into gruel was also reported to be less in case of parboiled rice. Parboiled rice cooks more flaky than raw rice.

Milling

Milling of paddy is the major operation in paddy processing. It removes husk and outer layer of bran so as to produce acceptable white rice with minimum breakage and impurities. The following terminologies may be known for better understanding of milling.

1. *Head rice* : It refers to the milled whole rice of 6/8 and more of actual kernel size.

2. *Broken rice* : Rice kernels which are lesser than 6/8 of the actual size are called broken rice. It is further divided into three categories, (1) big brokens, these include 4/8 to 6/8 of kernel portion, (2) small brokens, these include 1/8 to 4/8 parts of kernel and (3) points-lesser than 1/8 part of rice grain.

3. *Total rice* : It includes both head and broken rice.

Table 5.4

The advantage of improved paddy processing equipment over huller

Particulars	*Parboiled rice*		*Raw rice*	
	under-runner disk sheller	*rubber roll sheller*	*under- runner disk sheller*	*rubber roll sheller*
Additional rice obtainable over huller, %	0.8	1.6	4.1	6.6
Decrease in broken rice with respect to hullers, %	3.0	4.4	8.9	15.0
Fat in rice bran, %	20.24	23.27	15.17	18.2

The main objective of milling is to obtain maximum head rice with minimum brokens and foreign matter. Breakage of rice mainly occurs during milling operation, therefore, it is of great importance to the rice industry. Due to bad milling, breakage may be as high as half of the total rice. If has been reported that if the total efficiency of rice milling operation is increased by 2%, 1.2 million tonnes of extra rice can be made available in India. Most of the paddy in our country is being milled by rice hullers, and the losses due to hullers are estimated to about 5%.

Moreover hullers give a mixture of rice husk and bran as by product, hence it prohibits its economical use.

The advantages of improved paddy milling equipment over huller is given in Table 5.4.

Milling Equipment

In some rice growing areas rice milling is accomplished by primitive methods such as pounding the paddy in a wooden mortar and pestle followed by winnowing. It is estimated that in India $\frac{1}{3}$ of the total paddy production is processed by hand pounders and foot pounders (Fig. 5.24). Due to impact of the pounder, rice breaks and the polish assumed by the grains is non-uniform. Rice and bran get mixed

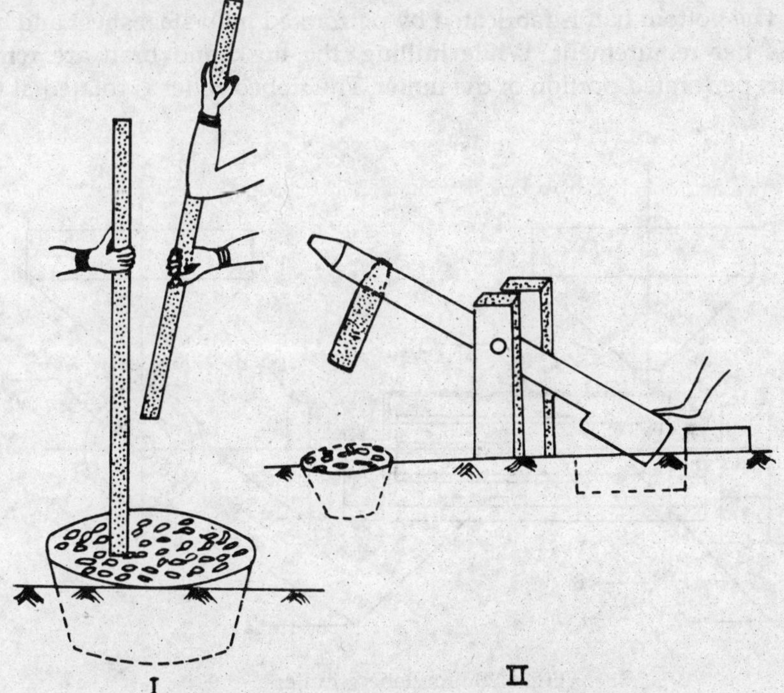

Fig. 5.24 : Hand and foot pounders for milling of rice
I. hand pounder II. foot pounder

with the husk and a portion of them is lost along with it in the process of separation thereby resulting in low milling yield. Labour requirements in above methods are high and the total rice recovery ranges between 60–65% only. But the method of milling is simple and no mechanical or electrical power is required, and pounders can be easily fabricated in villages.

Power operated milling equipment are described below.

Hulling : The purpose of hulling machine is to remove the husk from the paddy grain with minimum damage to the bran layer and as far as possible not to break the brown rice.

Due to surface characteristic of paddy it is necessary to apply friction to the grain to remove the husk. Therefore, during hulling, certain percentage of broken can not be avoided. In this respect the construction of machine, it precision, adjustment and the operation govern the optimum performance and efficiency of machine and head rice production.

The adjustment of hulling machine depends upon the variety and uniformity of grain. Its uniformity is necessary for best performance of equipment.

Huller

The most common machine used for paddy hulling in India is "Engleberg" huller. The working element of this machine is a ribbed cast iron roller. The roller rotates on its axis inside a large concentric cylinder. On the inner cast iron roller, spiral ribbed strips are mounted to 1/4 part of the length and on remaining 3/4 part 4 to 6 straight ribbed strips are mounted (Fig. 5.25). The outer cylinder is made in two parts. The bottom half is fabricated by perforated mild steel sheet and can be changed as per requirement. While hulling, the husk and bran are removed through this perforated portion of the huller. The ribbed roller is rotated at 600 to

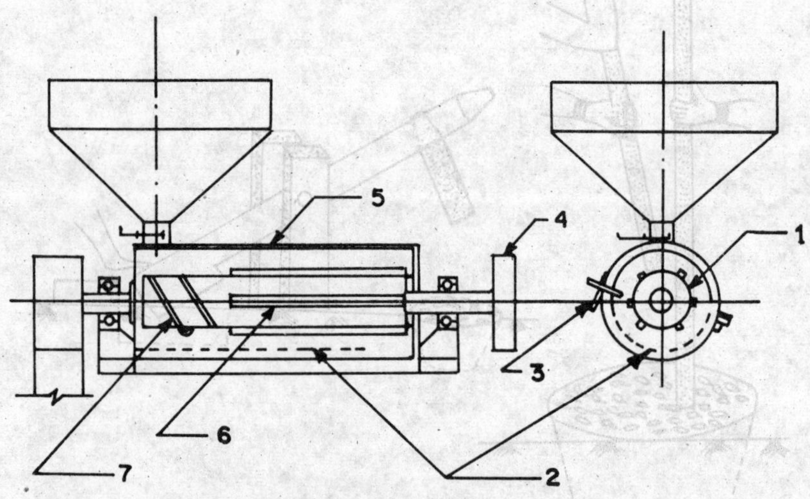

Fig. 5.25 : Engleberg huller
1. hulling roller 2. screen 3. knife 4. balancing pulley
5. casing 6. straight ribs 7. spiral ribs

900 rpm. Paddy is fed into the hopper and due to rotational direction of the flutes, it is moved around the cylinder and finally towards the outlet. Friction between the grains and the steel parts of the huller causes the husk and bran to be scrapped off. The huller does the job of husking and bran removal simultaneously, thus mixes the bran and brokens with husk. It is difficult to separate these ingredients, therefore, the mixture can not be used for oil extraction and is sold as cattle feed.

A steel blade is provided which can be moved in or out of the casing. Clearance between the blade and ribbed roller decides frictional forces generated on the grains. Friction between the grains and the steel parts of the huller causes the husk and bran to be scrapped off. The capacities of huller varies from 250 to 750 kg/hour. The average yields obtainable from huller is about 56% and 62–64% for raw rice and parboiled rice respectively. It generates around 25 to 30% brokens.

The initial investment of huller is low and requires little space and can be installed in simple structures. The machine can be locally fabricated and a semi-skilled person can operate the huller. But its power requirement per unit rice milled is higher, also it has the lowest recovery rate in total and head rice. Polishing of rice with huller can not be effectively controlled. As stated earlier because a mixture of husk and bran is obtained this by product can not be used economically.

Under-runner disk huller

The under—runner disk huller consists of two horizontal cast iron disk partly covered with an abrasive layer preferably of emery. The top disk is fixed with the body of the machines, while the bottom disk rotates. The rotating disk is vertically adjustable by which clearance between the two disks is adjusted (Fig. 5.26). As per the variety and condition of paddy, the clearance is decided. The condition of abrasive coating on the disks also affect the clearance.

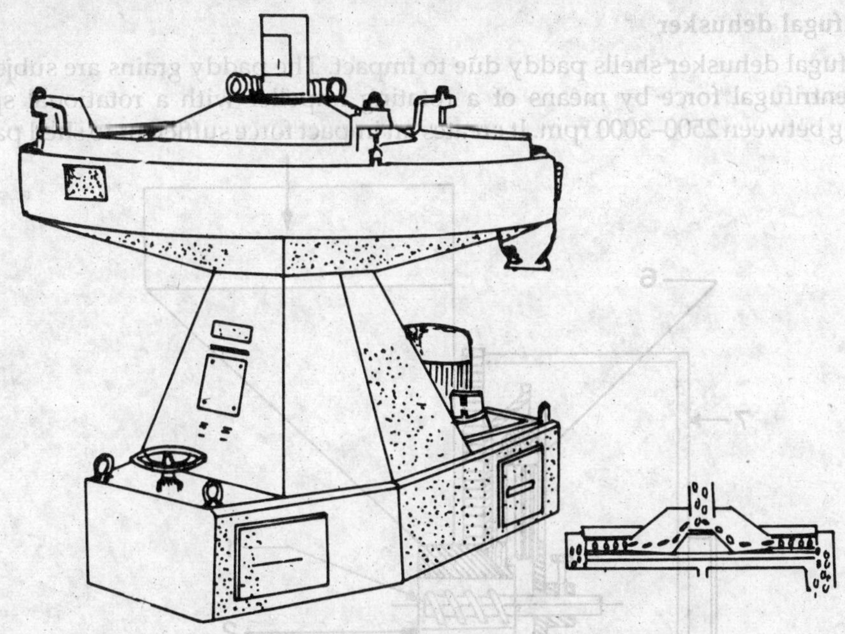

Fig. 5.26 : Under-runner disk huller

Paddy is fed into the equipment from top through a hopper. With an adjustable sleeve uniform flow of paddy is maintained, which also spreads paddy evenly to entire surface of the rotating disk. By centrifugal force the paddy is forced between the disk and dehusking takes place due to friction and pressure.

During hulling there is wear of abrasive coating and is not uniform over the entire surface of the coating. Hulling is mainly concentrated at the centre, therefore, after sometime a ridge is formed at the outer ring of the coating. The ridge causes excessive pressure on the paddy and grain breaks during hulling.

The huller shaft is normally driven by a flat or V-belt transmission. While driving the disk huller with flat belt, care must be taken to ensure that the pulling part of the belt should be perpendicular to the main transmission shaft. If such conditions are maintained, the huller will consume less energy.

Edible grade oil can be extracted from the bran obtained from disk huller. The machine has longer life and the operating cost is low. Parboiled paddy can also be processed by this machine. The total and head rice recoveries are lesser than what is obtainable by rubber-roll sheller.

The capacities and power requirements of under-runner disk huller are presented in Table 5.5.

Table 5.5

Capacities and power requirements of under-runner disk huller

Disk diameter, mm	*Capacity, kg paddy/hr*	*Horsepower requirement*
750	450–600	3.0
1000	700–1000	3.5
1250	1000–1400	4.0
1400	1600–2100	5.5

Centrifugal dehusker

Centrifugal dehusker shells paddy due to impact. The paddy grains are subjected to a centrifugal force by means of a rotating impeller with a rotational speed ranging between 2500–3000 rpm. It creates an impact force sufficient to shell paddy

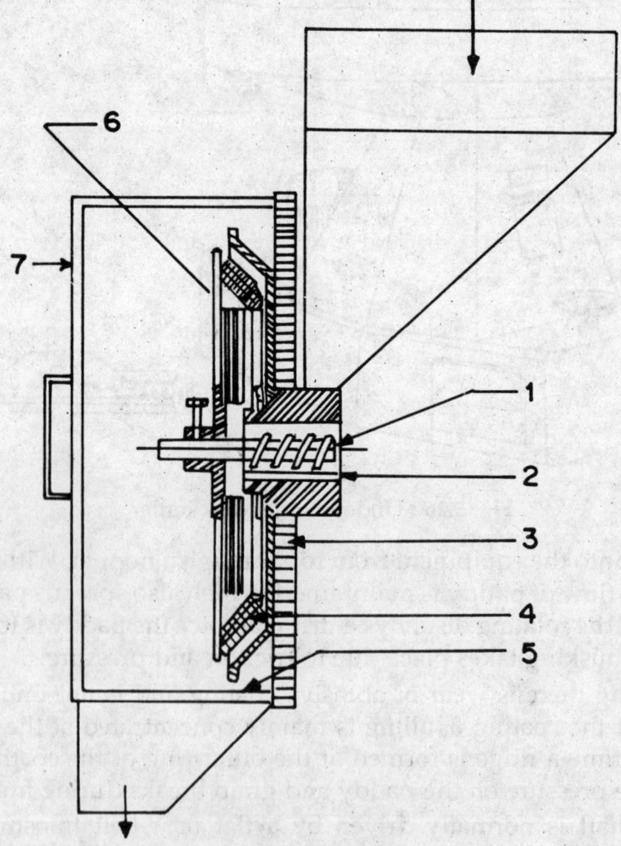

Fig. 5.27 : Centrifugal dehusker
1. feed screw 2. coller 3. crank gear 4. rubber ring
5. lining 6. impeller 7. lid

grains. Casing of the dehusker is lined with a rubber sheet (Fig. 5.27). The paddy is fed to the centre of the rotor form which is thrown towards the casing with great force and is shelled upon hitting. The salient features of the centrifugal dehusker are, its high capacity and simple construction, because there is only one moving part, the impeller. Heating of grain not takes place as there is no grinding or friction action.

The initial cost of the machine and the operational cost is low. It requires less power, 1 hp/500 kg paddy dehusking per hour. The total and head rice yield is more than the hullers. The machine needs adjustment according to paddy variety and increased shelling leads to increased breakage of grains.

Rubber-roll sheller

Rubber-roll sheller consists of two rubber rolls rotating in opposite direction at different speeds. A feeder feeds paddy uniformly to the machine (Fig. 5.28). Paddy is fed in thin layer between the rotating roll by the feeder. One of the roll is fixed

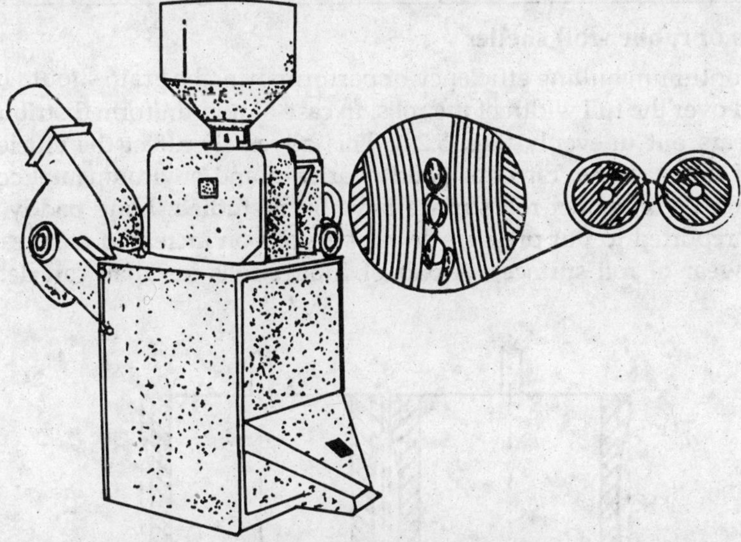

Fig. 5.28 : Rubber-roll sheller

while the other is adjustable to obtain desired clearance between them. The rolls are driven mechanically and the adjustable roll normally runs about 25% slower than the fixed one. Difference in surface speeds of the rolls develop a shearing force on grain surface resulting in the opening and breaking of husk. The clearance between the roll is kept smaller than the thickness of paddy grain. This clearance should be about half the thickness of paddy and may be adjusted subsequently by judging the shelling efficiency. If the gap between rolls is properly adjusted, this equipment can shell upto 95% of paddy fed to it. At decreased gap excess pressure results which cause more breakage of grain and can also cause colouring of shelled rice. In the modern machines, the gap between the rolls is adjustable by suction methods, therefore, as per need this gap is automatically maintained.

During shelling the faster roll should have a peripheral surface speed of 10 to 13 m/s, while the slower roll 8–10 m/s. The details of speeds, diameter, capacity and power requirement of the rolls are given in Table 5.6 and 5.7.

Table 5.6
The details of speeds of rubber rolls

Rubber-roll diameter, mm	*Width of roll, mm*	*Faster roll, rpm*	*Slower roll, rpm*
150	64	1320	900
220	76	1200	900
250	250	1000	740

Table 5.7
The capacities and power requirements of various sizes of rubber-roll

Dimension of rubber-roll, mm			*Capacity, t/hr*		*Horse power requirement, hp*
Size	*length*	*diameter*	*long-grain*	*short-grain*	
4	100	220	0.9	1.25	2.5
6	150	220	1.2	1.90	4.0
10	254	254	2.2	3.80	6.0

Constrains of rubber-roll sheller

To obtain optimum hulling efficiency or performance, the grain should be evenly distributed over the full width of the rolls. In case of non-uniform distribution, roll surface wears out unevenly (Fig. 5.29). This adversely affect the efficiency and capacity of the machine. Effect of paddy varieties and environmental conditions on performance of rubber-roll surface have been studied. Long paddy varieties have been reported to put pressure on rubber-rolls on increased area resulting in increased wear of roll surfaces. A pair of rubber rolls in Japan can dehusk 200

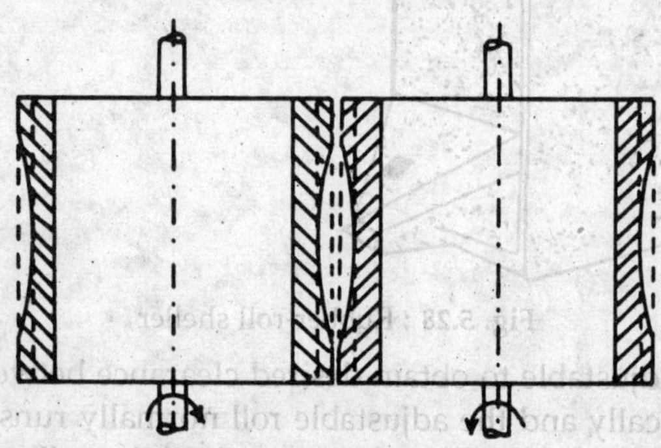

Fig. 5.29 : Unevenly damaged surface of robber rolls

tonnes of short variety of paddy, whereas the same pair can dehusk only 100 tonnes of paddy in tropical climate (high temperature and high humidity). Although the quality of rubber can be improved to increase the durability of roll in terms of tonnes of paddy hulled, but the cost incurred will be much more than the additional output. Efforts therefore be made to improve the quality of rubber-roll as well as to reduce its cost so as to made it attractive for the millers.

Husk separator

This machine is required to blow away husk from the mixture of shelled rice, husk and unshelled paddy obtained from huller/sheller. In the first stage the husk,

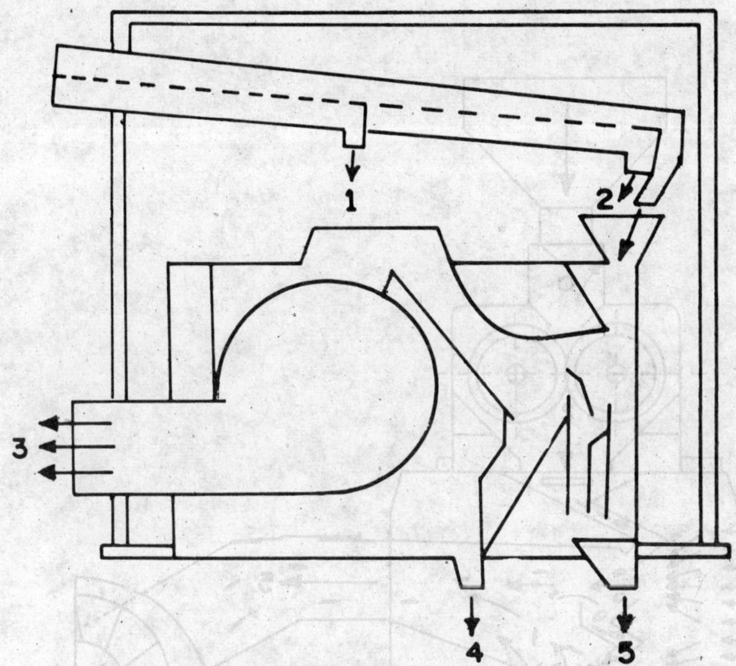

Fig. 5.30A : Husk separator with plansifter
1. bran and dust 2. brokens 3. husk 4. paddy 5. brown rice

brokens, germ and bran must be separated which is accomplished by the husk separator. It is a simple machine having a fan and an arrangement to distribute the product of sheller uniformly on an oscillating sieve with fine perforations. This is done to ensure that air flows across uniformly and blow away the husk. The broken, germs and bran are separated through perforations while the immature grains are also blown away by fan (Fig. 5.30 A).

The compact, lightweight construction of the rubber-roll husker makes is possible to combine it with husk aspirator. The rubber roll sheller does not damage the bran layer of the brown rice as well as discharge is free from bran, It produces little amount of brokens. Therefore, with rubber-roll, a plansifter is not required and only a husk aspirator is used (Fig. 5.30 B).

Paddy separator

All the paddy dehusker, like huller, disk huller, centrifugal dehusker or rubber roll sheller are not able to remove 100% husk from paddy. About 80 to 95% husking can be achieved through these machines depending on their efficiency. In addition to it the quantity of unshelled paddy with brown rice depends on the following factors:

1. Uniformity of paddy
2. Variety of paddy
3. Condition of paddy
4. Husking machine
5. Condition of husking machine
6. Operator's skill

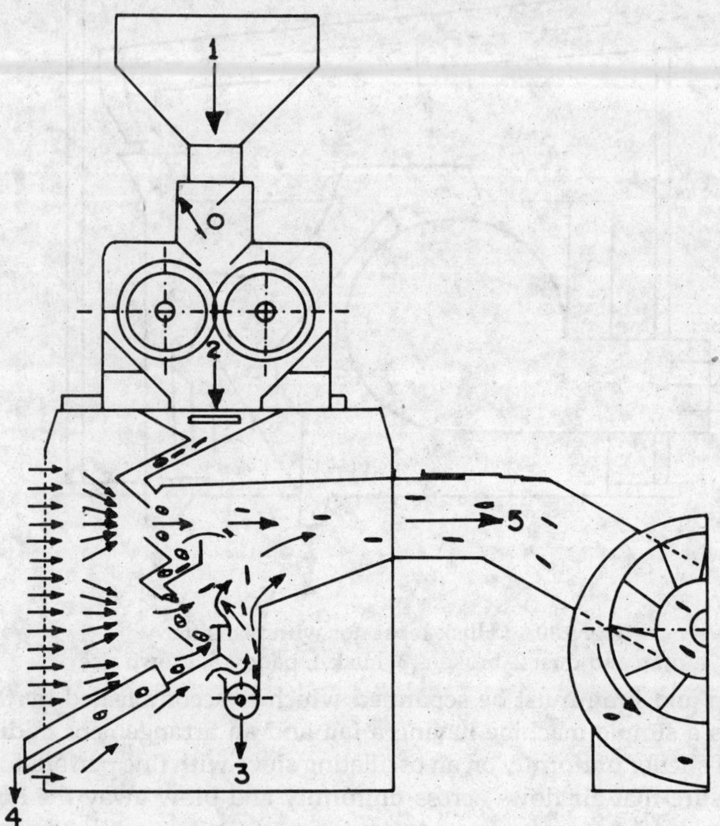

Fig. 5.30B : Rubber-roll sheller with husk separator
1. paddy 2. brown rice paddy-husk mixture
3. immature grains 4. brown rice, paddy 5. husk

The product obtained after husking is a mixture of shelled rice, unshelled paddy, husk, brokens, stone, bran and dust. With plansifter and husk aspirator dust, husk, brokens, germ and bran can be removed. For separation of brown rice and paddy, paddy separators are used. The following differences in characteristics of paddy and brown rice are used in paddy separators.

1. The average weight of paddy is lighter than brown rice.
2. The paddy grain is more buoyant than the brown rice.
3. The paddy grains are longer, wider and thicker than the brown rice.

Unshelled paddy is separated from the brown rice kernel with the following equipment.

Compartment type paddy separators

This equipment has a number of compartments arranged in a single or multiple deck (Fig 5.31), which varies according to capacity of the machine. This assembly is normally made of wood, however, machines fabricated out of steel are also

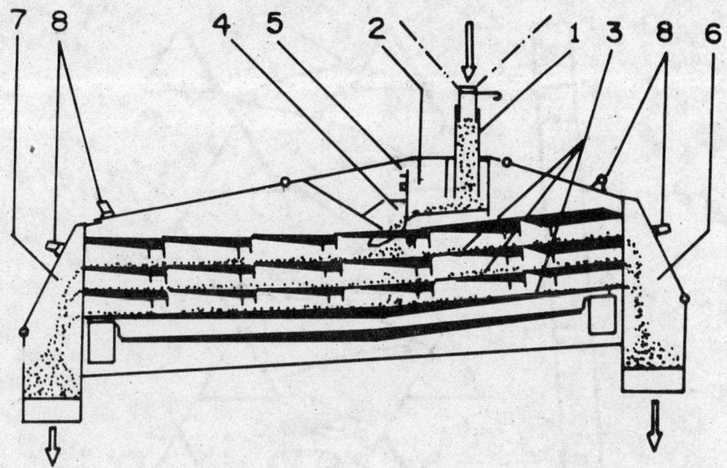

Fig. 5.31 : Cross-section of a compartment type paddy separator
1. feed duct 2. feed box 3. tiers 4. feeding gate
5. feed regulating gate 6 and 7 discharge troughs 8. inspection doors

available. The number of decks varies from 1 to 4 as per need of rice mill. The capacity of a compartment is about 40 kg. of brown rice (long grain) and 60 kg (short grain) per hour. This quantity is normally used to determine the number of compartments required for a rice mill. For separation of paddy and brown rice the machine is kept slightly inclined. The inclination depends on the percentage of paddy in the mixture.

Each compartment of the machine has a particular zig-zag shape. The sides and 'z' shaped steel flanks have an impact angle of about 30° (Fig 5.32). Paddy and rice mixture when fed to the machine, strike repeatedly the opposite walls of the compartment due to oscillating motion of the table. Paddy being lighter than rice, comes to the surface and moves to a longer distance while brown rice does not. The angular flanks force paddy grains to move upward while the brown rice moves downward along the slope of the table.

Compartment type separators are of two types, (1) fixed stroke and (2) adjustable stroke. Fixed stroke separators usually run at 100/100 rpm with table inclination of 3 to 4 degrees. Feeding of brown rice and paddy mixture is kept uniform to obtain proper separation. Speed of operation may be adjusted to achieve best performance.

The variable stroke separator has certain advantages. The capacity is increased, power requirement is low and less space is required for the installation. The following adjustments are recommended for its proper performance.

For long grain rice: 90–96 stroke at 114 rpm

For short grain rice: 75–80 stroke at 124 rpm

Compartment type separators have low power requirement. Its operating and maintenance costs are also low. The tray bottoms and compartment zig-zags can also be replaced locally after they wear out. The machine, however, is bulky, requires a strong foundation and takes considerable space in the mill.

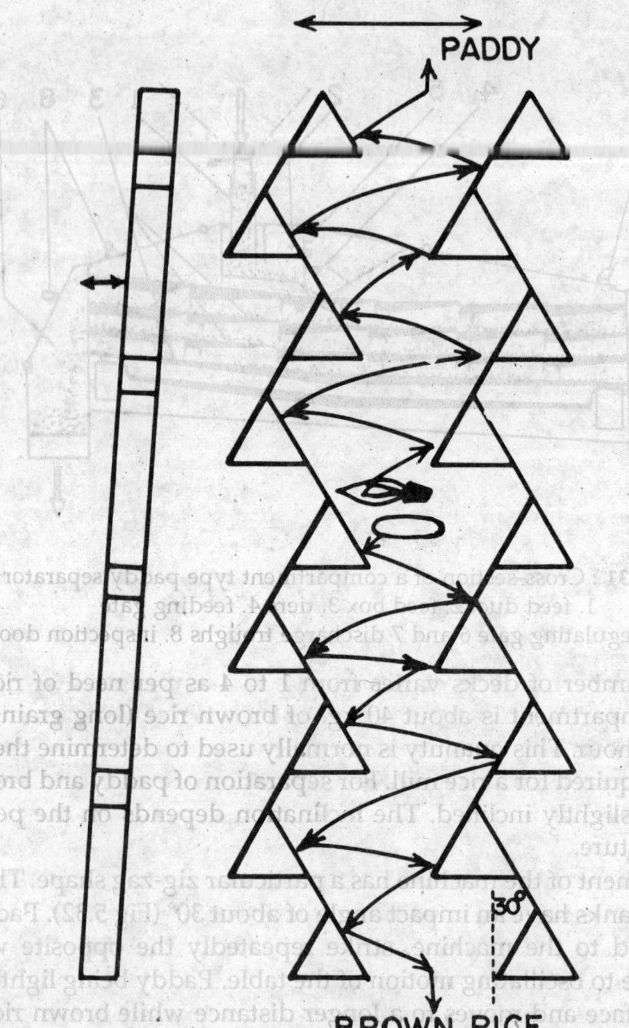

Fig. 5.32 : Internal construction of one compartment of the paddy separator

Tray separator

The tray separator consists of several indented trays mounted one above the other about 5 cm apart, and attached to an oscillating frame (Fig. 5.33). The front portion of the machine is provided with slight inclination and is not changed during operation, whereas the side inclination can be changed as per need. The tray moves up and forward, giving the assembly a low jumping movement.

Paddy moves into each tray from the inlet hopper at uniform rate. As it moves across the tray, the brown rice separated from the paddy. The brown rice has a smoother surface and a greater bulk density and moves to the top of the tray, whereas paddy moves to the lower part of the tray. Inclination of the table or tray is adjusted according to varieties and other conditions to achieve maximum separation capacity.

Grain size normally governs the capacity of machine. One typical model has 2,270 kg/hr. for long grain and 3180 kg/hr for short grain with 1 hp motor. The

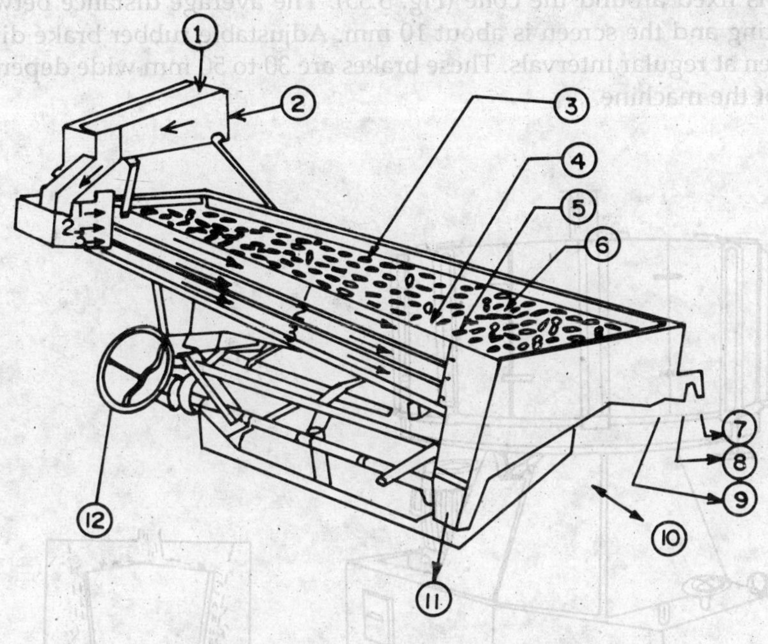

Fig. 5.33 : Tray type paddy separator
1. paddy-brown rice 2. hopper 3. paddy 4. dust 5. brown rice
6. paddy-brown rice mixture 7. paddy 8. mixture 9. brown rice
10. tray slope 11. dust 12. hand wheel to adjust the slope of tray assembly

power requirement of such machine is about half of the power required for compartment type separators.

Removal of bran

In this process, the silver skin and the bran layers of the brown rice are removed. It is also termed as "whitening".

The two processes used to remove the bran layer from the brown rice are abrasion and friction (Fig. 5.34). Abrasion process uses a rough surface which may be an abrasive stone to break and peel the bran off the grain. The friction process uses the friction between the grains themselves to break and peel off the bran.

Fig. 5.34 : Bran removal processes
I. friction II abrasion

Three kinds of whitening machines are widely used in the rice processing industries, (1) the vertical abrasive whitening cone, (2) the horizontal abrasive whitening machine and (3) the horizontal jet pearler.

The Vertical whitening cone

This machine basically consists of a cone shaped cast iron cylinder with an abrasive coating. The cone is fixed on a vertical shaft that rotates either clockwise or counter clockwise. A wire screen having mesh size as per the variety of paddy to be polished is fixed around the cone (Fig. 5.35). The average distance between the cone coating and the screen is about 10 mm. Adjustable rubber brake divide the wire screen at regular intervals. These brakes are 30 to 50 mm wide depending on the size of the machine.

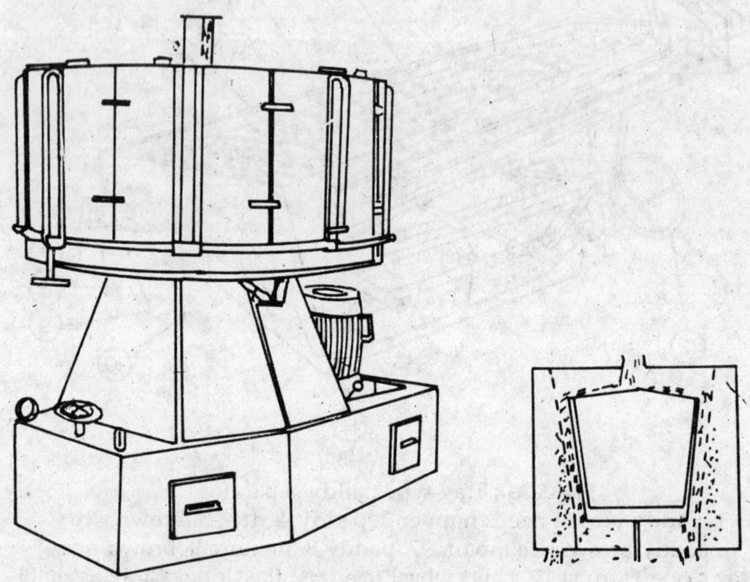

Fig. 5.35 : Vertical Whitening Cone

Brown rice is fed into the centre of the machine through a hopper. Feeding of brown rice is adjusted by a sleeve which also uniformly distributes the brown rice to entire surface of cone. The centrifugal force generated by rotation of cone feed the brown rice between the cone and wire mesh. Rubber brakes restrict the movement of rice, thus, applies pressure. As a result of pressure brown rice is pressed against the abrasive coating of the cone. This friction removes bran layer, and partly or fully whitened rice leaves the cone through a self unloading discharge spout.

Vertical adjustment of the rotating cone is done manually. The entire shaft-cone assembly is moved by moving the housing of the shaft and bearings.

The peripheral speed of the cone should be about 13 m/s, thus larger the diameter of cone, lower the speed of shaft. The abrasive surface of the cone can be replaced locally. The screens and rubber brakes can also easily be replaced. For achieving best performance of the machine it should be ensured that machine is filled with brown rice. In case the machine is run half empty, desired pressure will not be put by rubber brakes and more breakage will take place.

Vertical whitening cone removes the bran from the brown rice by friction, and the friction produces heat. For removing all the bran in one pass, the residence time of the grain is increased and the clearance between cone and screen is reduced. This results in generation of heat and causes much breakage and reduces total rice recovery. It is therefore, recommended that the bran layer be removed in three or even more passes. For example, a capacity of 1200 kg/hr can be obtained by, (1) a single pass with one 1,000 mm cone, (2) a double pass with two 800 mm cones or (3) a multi pass with three 600 mm cones. Multipass produces least amount of brokens and the largest total rice recovery. It has been found that if the rice is polished in single pass only, the head rice, large brokens and small brokens obtained are 52.5, 32.5 and 12.5 per cent respectively. Removal, if accomplished in three passes it results in 65.0, 27.5 and 7.5 per cent recoveries of head rice, large brokens and small brokens respectively.

Air aspiration through the whitener reduces breakage caused by heating and keeps dust out of the mill, Air quantities required for different cone sizes are 8 m^3/min. for 500 mm; 10 m^3/min. for 600 mm; 12 m^3/min for 800 mm; 15 m^3/min for 1000 mm and 20 m^3/min for 1250 mm cones.

When parboiled paddy is being milled, air requirement is considerably increased. In Table 5.8 the sizes, horsepower requirements for vertical cone whiteners are given.

Table 5.8

Vertical cone whitener sizes, power requirements and their capacities

Cone diameter, mm	λ_P	Capacity, kg brown rice / hr					
		Single pass		Double Pass		Triple pass	
		long grain	short grain	long grain	short grain	long grain	short grain
500	4.0	350	420	570	680	680	800
600	7.5	550	650	890	1100	1050	1250
800	10.0	750	900	1230	1450	1460	1725
1000	15.0	1000	1200	1700	2000	1900	2300
1250	20.0	1350	1600	2200	2600	2600	3000
1500	28.0	1700	2000	2700	3200	3200	3800

Note : For parboiled rice the power requirements are increased by 30%.

Horizontal abrasive whitener

The machine consists of an abrasive roll operating in a cylindrical metal perforated screen which is horizontally mounted (Fig. 5.36). The screen cylinder covers the

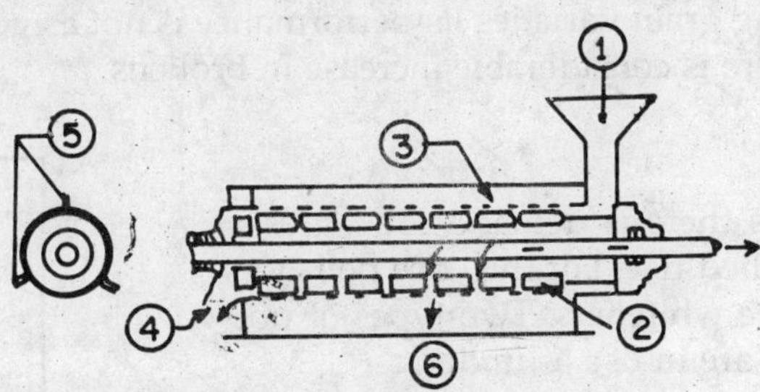

Fig. 5.36 : Horizontal abrasive whitener

1. hopper 2. abrasive roller 3. screened steel cylinder 4. white rice 5. brakes 6. bran

emery roll leaving an uniform gap where brown rice is fed through a small screw conveyor. The emery roll rotates while the screen cylinder remains stationary. Polishing is obtained due to rubbing of grains with emery roll, screen and rice grain. At the delivery end of polishing chamber one adjustable weight is placed by which degree of polishing is controlled. Adjustable brakes are also fitted to screen cylinder, to help in obtaining required polishing. Bran removed from rice escapes through holes and is aspirated out by a blower. The blower which collects and carries the bran also provides cooling to rice grains, emery roll and screen cage.

Jet pearler

The jet pearler is used to remove the final part of the bran layer and simultaneously cool the grain through an air stream of ambient temperature.

It consists mainly of a horizontal partly hollow perforated shaft on which a cast steel cylinder with friction ridges is clamped (Fig. 5.37). Just behind the two ridges

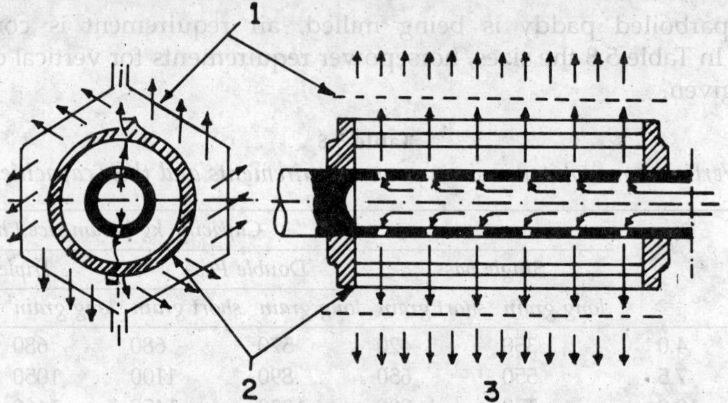

Fig. 5.37 : Cross-section of a jet pearler
1. screen 2. milling roller 3. bran

the cylinder has a long opening which allows the passage for air. This cylinder runs inside a hexagonal chamber consisting of two halves hexagonal screens with slotted perforations. A feeding screw with horizontal shaft feeds the rice into the press chamber of the machine. The clearance between the hexagonal screen and the cast steel cylinder is adjustable by a screw controlling the distance between the two halves of the screen.

The rice produced by this machine is free from bran and cool. However, for medium and long grain varieties its performance is not as good as for short grain varieties and there is considerable increase in brokens.

Rice polishers

In some markets there is demand for a glossy and highly polished rice. For this, rice polishers are used after the whiteners. Two types of polishing machines are in use as under.

Vertical polishing cone

It is a similar machine as the vertical whitening cone. The basic differences between the two are that in polishing, (1) the cone is made of steel and is covered with wood on which leather

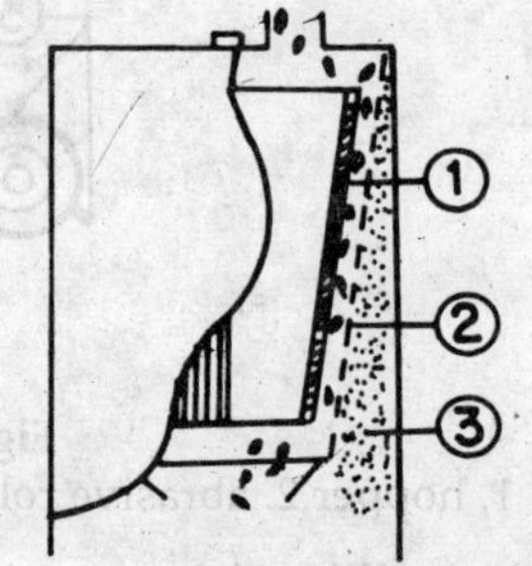

Fig. 5.38 : Vertical polishing cone
1. leather strips 2. white rice 3. bran

strips have been nailed, (2) no rubber brakes are used (Fig 5.38), and (3) the speed of rotation of the cone is about 25% lower than the whitening cone.

The operation of this machine is similar to vertical whitening cone.

The horizontal polisher

It consists of a steel cylinder on which a large number of leather strips are screwed (Fig. 5.39). The cylinder is mounted on a horizontal shaft that rotates inside a

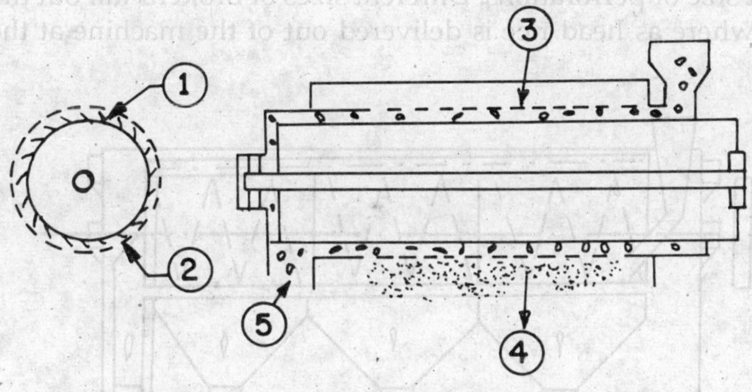

Fig. 5.39 : Cross-section of a horizontal polisher
1. leather strips 2 and 3. screen 4. bran 5. white rice

cylindrical chamber covered with slotted perforated screen. The leather strips roll the whitened rice over and over against the screen. Under slight pressure, the remaining bran is removed and the rice becomes shiny and glossier. This machine produces few brokens. Its power requirement is 30 to 40% less than that of whiteners.

Glazing

Some people/countries demand glazed rice. Such rice appears very shiny and more transparent because the surface is coated with a thin layer of talc and glucose. Normally head rice is glazed

Rice is normally glazed in batches. The glazing drum is partly loaded with rice and is slowly rotated. White shiny talc powder and glucose solution is then added to the rice. Dry talc powder about 1–1.2 % by weight is required. Glucose is added to the rice as 1:1 solution with water. About 1.0 kg glucose solution is required for 1000 kg of rice. Finally water is evaporated and the talc and glucose remain attached to rice. After glazing operation there is an increase in weight of 1000 kg rice by 60 kg. i.e. 1000 kg rice becomes 1060 kg after glazing.

Grading

Normally paddy is not graded prior to milling, but under certain circumstances the grading of paddy is desirable or even necessary. Grading results into; (1) separation of immature grains, (2) pressure adjustment of huller and (3) independent milling of graded lots.

Grading in rice milling usually refers to the separation of brokens from the head rice and the further classification of brokens.

Three machines are used for grading of milled rice, (1) adjustable reel graders, (2) indented cylinder graders or trieurs and (3) oscillating sieves.

Rotating reel graders

This consist of a rotating cylinder made up of wire screen reels or slotted metal sheet. The dimensions of these perforations are decided as per the varieties of rice. Milled rice is fed to the machine from one end through hopper. The rice slowly advances to the other end. The rotating cylinder is divided in three or more parts with different size of perforations. Different sizes of brokens fall out through these perforations where as head rice is delivered out of the machine at the other end (Fig. 5.40).

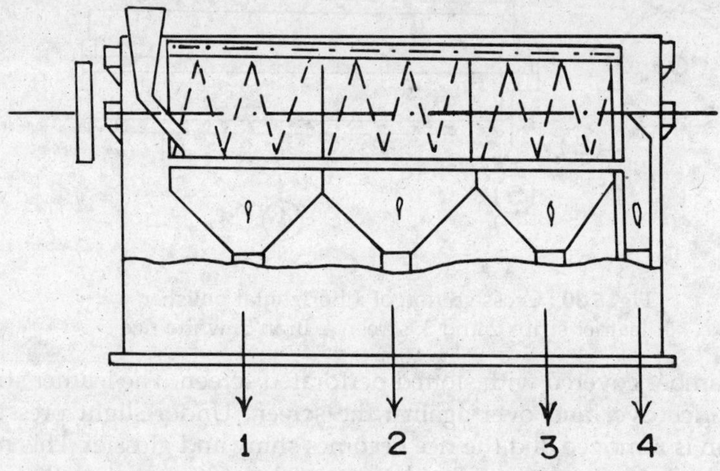

Fig. 5.40 : Rotating reel grader
1 and 2. small brokens 3. big brokens 4. head rice

Rotating indented cylinder

The machine consists of a cylinder having cup like indents on its inner wall (Fig. 5.41). This cylinder rotates at a low speed of about 30–40 rpm. It is installed on a slight inclined plane. At the centre of the cylinder a screw conveyor is fitted

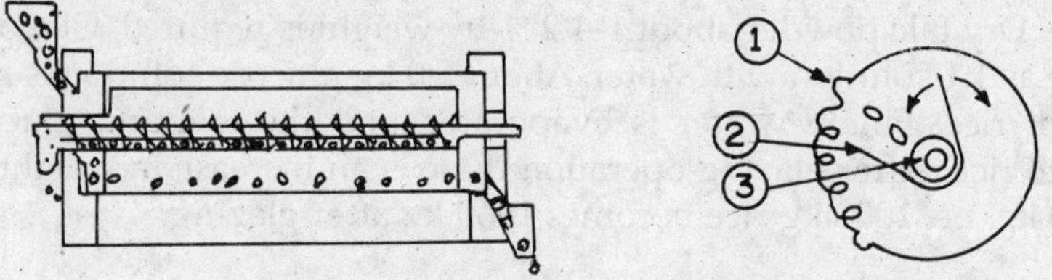

Fig. 5.41 : Indented cylinder-rice grader
1. indented cylinder 2. trough 3. screw conveyor

along with an adjustable trough. The indented cylinder grades rice according to length. As the mixture of head rice and broken rice is fed, broken rice gets accommodated inside the cups while head rice keeps on sliding downwards through the inclination. Rotary motion of the cylinder carries the brokens up inside the cups and throws them out as the cup positions become inverted. Thrown outs are collected by the adjustable trough and a screw conveyor delivers them out.

Occasional cleaning of the cups is required to remove the clogged particles. Clogging of cups leads to inefficient performance of the graders.

For best performance of indented cylinder graders two precise adjustments are necessary, (1) as per varieties of rice decision regarding appropriate cup sizes and (2) adjustment of rotational speed of the cylinder.

The indented cylinder grader must be ordered for a given capacity and specified function. For example if an indented cylinder ordered for 500 kg/hr capacity and to remove 12% large brokens, in case the brokens are more, say 20%, the machine will not be able to separate all the brokens. Again if the feeding is more than 500 kg/hr, say 800 kg/hr then also the machine will not be able to separate all the brokens even if the large brokens is limited to 12%.

Oscillating grading sieves

An oscillating sieve is a slightly inclined sieve supported by flexible wooden or steel bars. This is moved back and forth by a vertical eccentric drive from a transmission shaft. Frequency of stroke varies between 200 to 400 cycles/min.

The sieves are made from a perforated steel sheet or a wire screen of desired sizes. The brokens are separated with the help of these sieves and the head rice is retained on the sieves and delivered outside through a spout.

By knowing the following parameters, the efficiency of rice milling and quality can be calculated. Expressions usually used to estimate the milling efficiency of the rice mill and equipment are:-

1. Foreign matter, % $= \dfrac{\text{weight of foreign matter}}{\text{total weight of paddy}} \times 100$

2. Grass and weed seed, % $= \dfrac{\text{weight of grasses and weed seeds}}{\text{total weight of paddy}} \times 100$

3. Dead and immature grain, % $= \dfrac{\text{weight of dead and immature grains}}{\text{total weight of paddy}} \times 100$

4. Broken grains, % $= \dfrac{\text{weight of brokens}}{\text{weight of paddy}} \times 100$

5. Husked rice, % $= \dfrac{\text{husked grains}}{\text{total weight of paddy}} \times 100$

6. Milling recovery, % $= \dfrac{\text{weight of total rice}}{\text{total weight of paddy}} \times 100$

7. Head rice, % $= \dfrac{\text{weight of head rice}}{\text{weight of milled rice}} \times 100$

8. Broken rice, % $= \dfrac{\text{weight of broken rice}}{\text{weight of milled rice}} \times 100$

Milling efficiency gives the performance of any huller or sheller machine. To determine the milling efficiency knowledge of coefficient of hulling and coefficient of wholeness of grain is required. The product of the two coefficients is the overall milling efficiency of the machine. Coefficient of hulling gives the amount of brown rice received after hulling and the coefficient of wholeness of kernel gives the amount of whole head rice out of the total rice obtained. The performance of milling not only depends on the standard of the machine but also on condition of paddy, cracks in kernel, moisture content of grain and foreign matter present in paddy mass.

$$\text{Milling efficiency, \%} = \text{Coefficient of hulling} \times \text{coefficient of wholeness of kernel} \times 100$$

$$\text{Coefficient of hulling} = \frac{\text{weight of brown rice}}{\text{weight of paddy fed to machine}} \times 100$$

$$\text{Coefficient of wholeness} = \frac{\text{weight of brown head rice}}{\text{weight of total brown rice}} \times 100$$

Pulse Milling

Pulses are the main source of proteins for vegetarians in India, where about 15–30% of daily protein needs are supplied from edible legumes or pulses. India is the largest producer of pulses in the world, the annual yield is nearly 12 million tonnes. Pulses are consumed mostly in the form of dhal, the dehusked and split kernel. Nearly 75% of total legumes production is converted into the dhal. The availability of milled pulses has been reduced considerably because of population increase and stagnation of pulses production. In 1961 availability of dhal was 69 g per person per day which has come down nearly to 38 g per person per day in 1991. The world health organisation has recommended 80 g per person per day considering the Indian food habits. Variety, agronomic conditions of pules production, the seed size, its maturity and uniformity are few factors which influence the milling outturn. Apart from these factors, improper conditioning of pulses and machine parameters cause lower yields of dhal by 10 to 20 per cent. Dehusking is the primary process which involves removal of the fibrous seed coat. Considerable amount of breakage and powdering of the kernel occurs in dehusking and during milling operations. The efficiency of conversion of legumes to dhal by traditional milling is low. Losses of dhal can be minimised to 8–10 per cent by adopting improved technologies.

Milling of edible pulses for the production of dhal is an age old process. There are more than 10,000 pulse mills with average processing capacity of 10–20 t/day. Milling procedures vary widely from place to place. The recovery of dhal varies from 60 to 75% depending upon the type of pulses and techniques adopted by the millers. There is no standard process for milling different pulses where various parameters involved are optimised. The amount of oil mixed with pulses varies from 150 to 500 g/q of grain. Similarly addition of water also varies from region to region from 4 to 20 kg/100 kg of grain. For loosening of husk and its complete removal 3–8 passes through emery rollers are given. This action causes breakage and powdering of kernel.

Structure of pulses

There is much similarity in the structure of the seeds of various pulses. Pulses are dicot plants, however, they differ from each other in colour, shape and size. Also their seed coat thickness varies. The parts of a pulse seed is shown in Fig 5.42. The outermost layer of the seed is the testa or seed coat. The endosperm in mature seed is in the form of a layer surrounding the

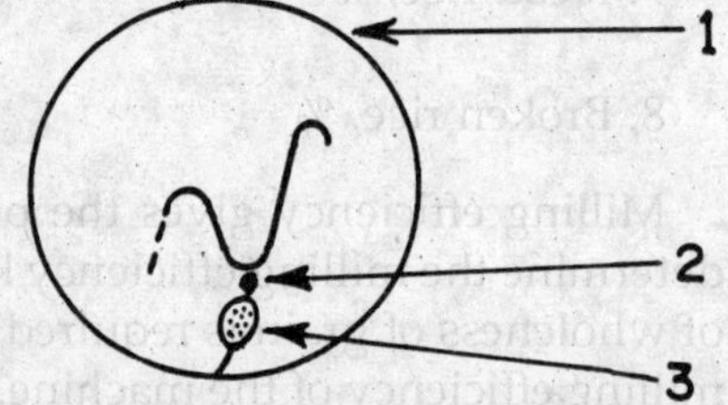

Fig. 5.42 : Parts of a pulse seed
1. seed coat 2. micropyle 3. hilum

cotyledons or embryo. There are a few external structures like hilum, micropyle and raphe. The hilum is an oval shaped scar in the middle where the seed is attached from the stalk. The micropyle is a small opening in the seed coat beside the hilum. The raphe is in a ridge from beside hilum and opposite to the micropyle.

Inside the seed coat the seed is composed of embryonic structure. The embryo consists of a radicle, a plumule or epicotyle and a hypocotyle which connects the radicle and the plumule.

Chemical composition of pulses

Pulses contain 17 to 25% of protein by weight. The protein content of pulses is about double of the cereals and more than meat, and egg. The pulses contain lipids and fats in low quantities. Pulses are good source of thiamin, niacin, calcium and iron. The food composition of major Indian pulses are presented in Table 5.9 In dry pulses very low amounts of ascorbic acid is present, but when these are soaked in water, during their germination, ascorbic acids are made available.

Pulses protein : Man and other monogastic animals get protein from pulses. The proteins of pulses are mainly glocolin, some species of pulses may also contain albumin The standard of proteins depend on the composition of amino acids. The composition of pulses protein is given in Table 5.10. The protein efficiency ratio of pulses are from 0.7 to 2.1, which is less than egg and milk protein. While cooking the nature of pulses proteins improve. Pulses contain more amount of lysine while sulfur containing amino acids are in lower quantities.

In India much of the daily protein need is met by pulses. Since the cereals have low levels of lysine, when these are mixed with pulses the PER value of diet improves. Man gets proper quality of protein through cereal and pulses diet which is easily digestible.

It is a general belief that pulses are comparatively hard to digest by monogastic people. Various studies have shown that well cooked pulses' digestibility is upto 90 per cent. Pulses cause flatulence, but if they provide only 25% of the energy to body, then only 4 to 10 times more flatus is produced by intestines. Pulses contain about 60% carbohydrate mainly starch, which in general is well-absorbed and utlised. The fat content of common dhal is between 1 to 2 per cent.

Important unit operations of pulse milling

1. *Cleaning/grading of raw grains* : In this process the raw grains received in a milling plant is first cleaned and graded. In cleaning, various inorganic and organic impurities present in the grain mass is removed with the help of air draft and screens. Destoners are also used to separate mud and stones. The cleaned grains are then graded as per their size mostly by a reel grader. The presence of impurities lower the quality of dhal and the impurities may also damage the milling machines.

2. *Conditioning* : Food legumes in India are consumed mostly in the form of dhal, the dehusked and split grains. A layer of gum between cotyledons or kernel and outer husk is present in most of the pulses. This gum layer may be thin or thick which in turn governs the degree of adherence of the seed coat to the cotyledons. The nature of gums influences the adherence of husk to kernel while the amount of gum effects the duration and severity of the conditioning process. The main

Table 5.9

The food composition of Indian pulses (per 100 g edible portion)

Grain	Protein	Fat	Mineral	Fibre	Carbohydrates	Energy	Calcium	Phosphorus	Iron	Carotene	Thianine	Riboflavin	Niacin	Vitamin
	g	*g*	*g*	*g*	*g*	*k.cal*	*mg*	*mg*	*mg*	*mg*	*mg*	*mg*	*mg*	*mg*
Bengal gram (whole)	17.1	5.3	3.0	3.9	60.9	360	202	312	10.2	189	0.30	0.15	2.9	3
Bengal gram dhal	20.8	5.6	2.7	1.2	59.8	372	56	331	9.1	129	0.48	0.18	2.4	1
Black gram dhal	24.0	1.4	3.2	0.9	59.6	347	154	385	9.1	38	0.42	0.37	2.0	0
Green gram (whole)	24.0	1.3	3.5	4.1	56.7	334	124	326	7.3	94	0.47	0.39	2.1	0
Green gram dhal	24.5	1.2	3.5	0.8	59.9	348	75	405	8.5	49	0.72	0.15	2.4	0
Lentil	25.1	0.7	2.1	0.7	59.0	343	69	293	4.8	270	0.45	0.20	2.6	0
Dry pea	19.7	1.1	2.2	4.5	56.5	315	75	298	5.1	39	0.47	0.19	3.4	0
Rajmah	22.9	1.3	3.2	—	60.6	346	260	410	5.8	—	—	—	—	—
Pigeon-pea (dhal)	22.3	1.7	3.5	1.5	57.6	335	73	304	5.8	132	0.45	0.19	2.9	0

Table 5.10

The amino acids content (g/100 g protein) of common Indian pulses

Grain	Lysine	Thrionin	Methionine	Cystine	Leucine	Isoleucine	Valine	Phenylalanine	Tryptophan
Bengalgram (chickpea)	7.04	3.52	1.28	1.28	9.28	5.12	4.96	5.76	0.80
Black gram	6.40	3.52	1.44	1.28	8.00	5.44	4.96	4.96	4.96
Green gram	7.36	3.20	1.28	0.96	8.16	5.60	4.12	5.60	0.96
Lentil	7.04	3.52	0.80	1.12	7.52	4.32	4.91	4.32	0.96
Dry pea	7.04	3.84	0.96	0.96	7.20	4.00	4.16	7.36	0.64
Pigeon peo	7.68	3.20	3.36	2.24	8.32	6.56	7.20	5.76	1.44

Source: NIN, 1971

objective of the conditioning is to loosen the husk to facilitate its separation from the kernel, thus reducing the milling losses. The conditioning process is done by various methods. There is no standard method. Conditioning of grains can be achieved by water treatment, hydrothermal treatment, use of salts and chemicals and use of heat alone. Insufficient conditioning results in incomplete loosening of husk, therefore, greater abrasive/scouring forces are necessary for complete removal of the husk. This causes higher milling losses in the form of brokens and powder with lower recovery of dhal.

3. *Dehusking and Splitting* : The conditioned grains are subjected to abrasive/scouring forces for removal of husk and for splitting of cotyledons into two equal halves. Dehusking and splitting is the most important and major unit operation of any pulse milling process. Dehusking of pulses generally carried in an abrasive roller mill. In some pulse milling plants, vertical stone chakki is also used to dehusk and split the grains. For splitting of the dehusked and moistened grains, vertical disk burr mill is used or the grains are allowed to fall on a hard or cemented surface from sufficient height. Due to impact with hard surface the dehusked grains are splitted.

4. *Polishing of dhal* : This is done to impart desirable shine and lusture to dhal. During this process a predesired quantity of edible oil and water is mixed with dhal by passing them through a screw conveyor. The presence of oil and water imparts desirable colour and shine to milled pulses.

5. *Grading of dhal* : Separation of dhal is carried mainly as per their size and soundness. The separation is usually achieved by rotating reel graders, in which separate compartments have different opening sizes of screens. Various sizes of dhal fall through the openings of these screens. Graded dhal are bagged separately as per their grades.

Pulse Milling Processes

Domestic level processes : In eastern parts of India alternate soaking and sun drying method is in use. Pulses are soaked in water for 6–8 hours, water is then drained off and the grains are spread in thin layer and sun dried for two days. The dried pulse is then milled by hand operated stone chakki. One woman can produce about 5 kg of dhal per hour. Two passes of pulse is normally required for husk removal upto 80–90%.

In the central and northern India where the dry method of milling is in use, pulses are mixed with edible oil and dried under sun for a day or two. Then moisture is added in small amounts. It is then heaped for few hours and then milled in hand operated stone chakki. One labourer can produce about 30–40 kg of dhal in eight hours.

There is a different wet milling method being practiced in southern India. The grains are steeped in water for few hours, the water is then drained off, red earth is added to it. The grains are then dried under sun in thin layers for 2 days. Such conditioned grains are milled by manual chakki. Dehusking and splitting take place simultaneously. The husk in all the above methods is removed by winnowing basket.

Dhal produced by the dry method is said to cook better while dhal produced by wet method tastes better but takes longer time to cook.

Processing by home and cottage scale methods is not efficient. These methods of processing cause more breakage and powdering of cotyledons. Further, these methods are time consuming and the dhal prepared is not properly dehusked and edges of cotyledons break during splitting.

Commercial scale methods

Some of the commercial scale methods are described below.

(*i*) *Wet milling method* : This method of processing is being adopted by comparatively smaller units. This method is mostly practiced in southern India.

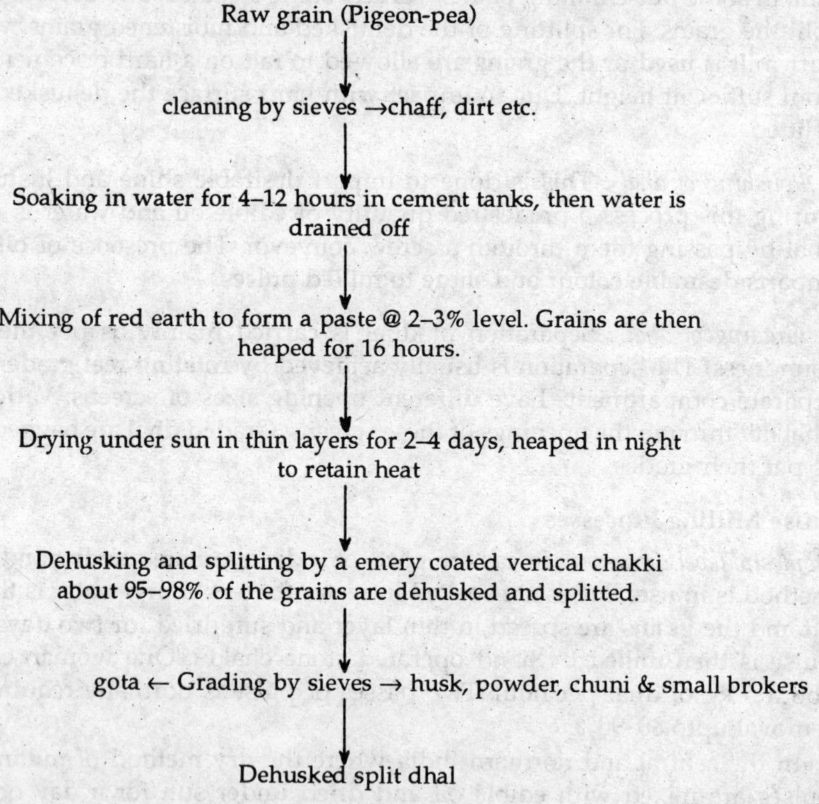

The milling quality of above method was found to be good and on an average 72 per cent dhal recovery is reported. Wet milled dhal reported to be of better taste than dry milled dhal, but as far cooking is concerned this takes longer cooking time to get some degree of dispersion. Cooking time mainly depends upon steeping time. More the steeping time better dehusking results but the resultant dhal will take longer time to cook. Treatment with red earth imparts yellow colour. It also provides mild abrasiveness and thus helps in the removal of husk from cotyledons.

Wet method of dhal milling is labour consuming and longer processing time of 4–5 days is the characteristic of such method. A batch of 6–7 tonnes is possible to process by such method.

(*ii*) *Dry milling method* : This method of milling of pulses is mostly adopted in the central and northern India.

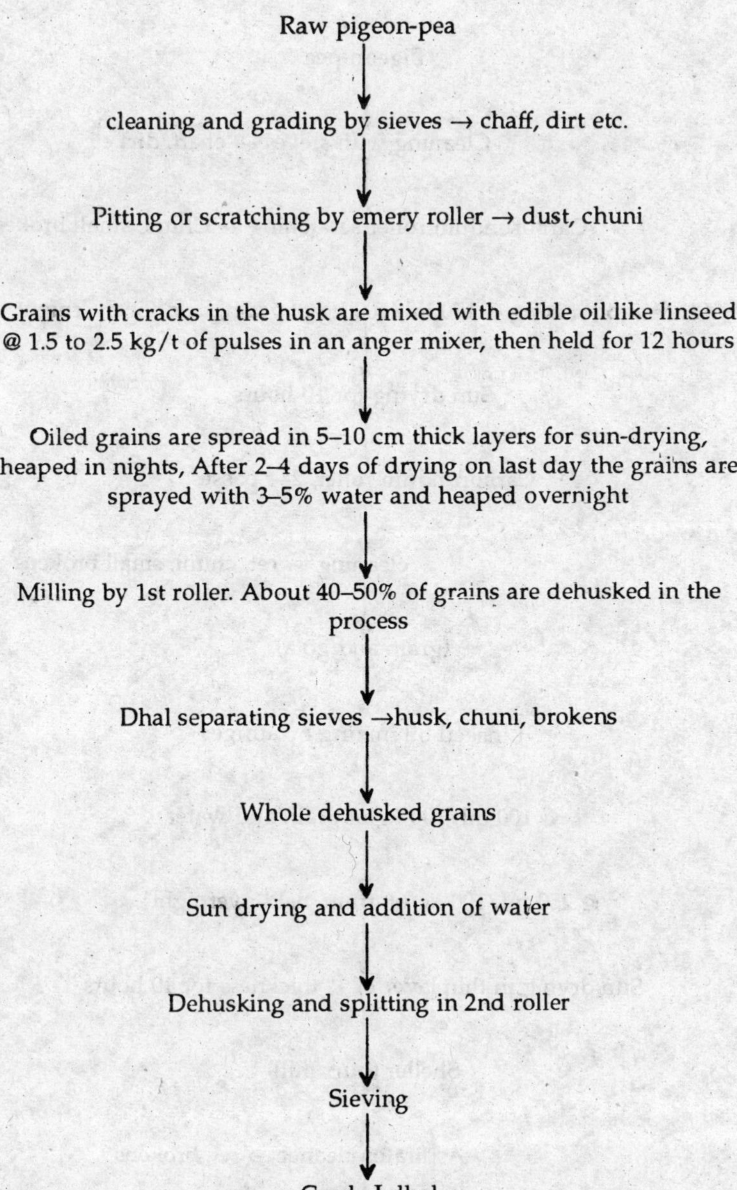

Dry method of milling vary widely from region to region. Dhal from dry milling method is said to of better cooking quality than wet milling method.

It has been observed that edible oil requirement in pulse processing varies from 150 to 500 grams per 100 kg of grain processed. The expenditure on oil alone constitute about 38% of the total processing cost. The energy requirement of dry milling method is also more. Repeated dehusking in roller mill together with repeated material handling for dehusking, grading aspirating and polishing results in consumption of more energy per unit of dhal processed.

(*iii*) Milling of pulses in M.P. (Jabalpur region)

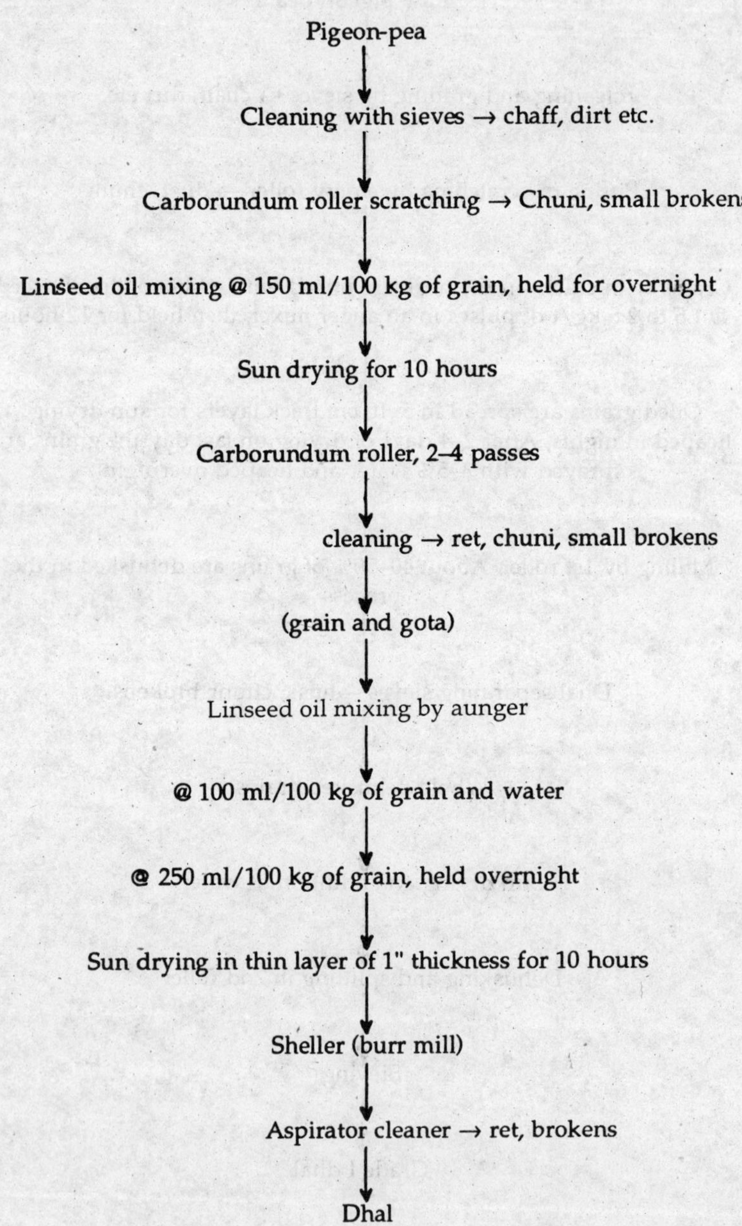

(*iv*) *CFTRI method* : This method eliminates mixing of oil and water for loosening of husk. It is claimed that it can give an overage yield of 80% dhal in lesser time and at lesser processing cost. Clean and graded grains are conditioned by dry heat treatment in two passes in LSU dryer with 120°C hot air. After each pass through dryer the grains are tempered for 6 hours in tempering bin (Kurien, 1983). The flow chart of CFTRI method is given below.

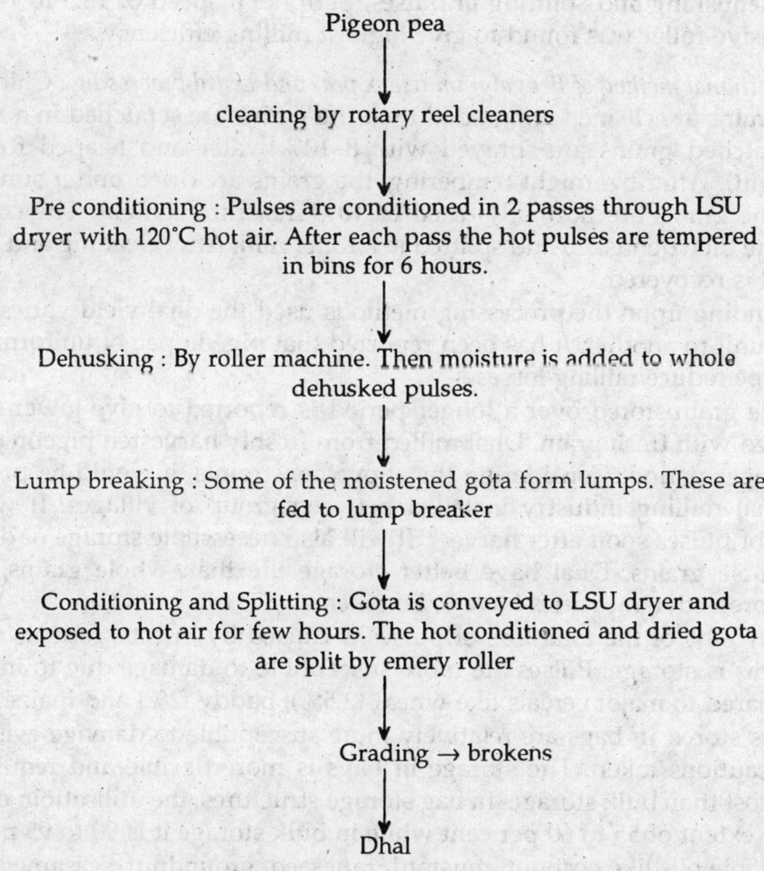

The CFTRI method requires high electrical loads for operation of dryers.

(*v*) *Pantnagar process* : In this process the pigeon-pea grains are cleaned and graded, then these are passed through an emery roller mill for scratching of grains. The pitted or scratched grains are mixed with 10% sodium bicarbonate solution. The treated grains are tempered for 4 hours in shade. Then the grains are dried under sun to 9.5% moisture content. The conditioned grains are then milled by a concentric cylinder roller mill (Saxena *et al* 1990). The milled product is cleaned and graded with the help of blower, cyclone separator and reel graders. It is reported that this method gives 80% recovery of dhal.

(*vi*) *CIAE method* : In order to eliminate the use of edible oil in pretreatment process the following process was adopted. First the pigeon pea, green gram or black gram is cleaned and graded. The grains are then fed to the roller mill developed in CIAE for scratching or pitting. During this operation scratches are developed on husk of grains. The scratched grains are cleaned. The grains are then soaked in tap water at ambient temperature for 30 minutes in case of pigeon pea and about an hour in case of black gram and green gram. Afterwards water is drained off, and the soaked grains are dried to 9–10% moisture content. Such conditioned grains are fed again to the roller mill to produce dehusked, split cotyledons.

For dehusking and splitting of pulses, peripheral speed of 13.5 to 14.0 m/s of the abrasive roller was found to give highest milling efficiency.

(vii) Traditional method of Bengalgram (chick pea) and Lentil processing : Chick pea and Lentil grains are cleaned are graded then the grains are scratched in a roller mill. The scratched grains are sprayed with 8–10% water and heaped for 12 hour (overnight). After overnight tempering, the grains are dried under sun for 2 to 3 days. The grains are normally dried to 10% moisture content. The conditioned grains are then dehusked and splitted in a roller mill. After cleaning and aspiration the dhal is recovered.

Depending upon the processing methods used the dhal yield varies from one milling unit to another. It has been reported that pigeon pea of uniform seed size and shape reduce milling losses.

Whole grain stored over a longer period is reported to give lower dhal yield compared with fresh grain. Dhal milled from freshly harvested pigeon pea stored for a longer period. Considering the above constraints, it would be profitable to have dhal milling industry installed between group of villages. It will enable milling of pulses soon after harvest. It will also necessitate storage of dhal rather than whole grains. Dhal have better storage life than whole grains, although proper protective measures have to be taken.

Nearly 80% of the total loss of pulses is caused by insects, rodents and micro organisms is storage. Pulses are more susceptible to damage due to insects (5%) as compared to major cereals like wheat (2.5%), paddy (2%) and maize (3.5%).

Pulses stored in bags are relatively more susceptible to damage even with all the precautions taken. The storage in bags is more flexible and requires lower capital cost than bulk storage. In bag storage structures, the utilisatioin of capacity is to the extent of 55 to 60 per cent while in bulk storage it is 90 to 95 per cent.

Vegetable oils like coconut, mustard, rapeseed, groundnut, seasame and castor have marvéllous protective qualities against pulse bettles. The oil is mixed at the rate of 500 g per 100 kg of pulse and it can be stored for five to six months without any damage by pulse beetle (IGSI, Hapur).

The recovery of dhal and milling losses by traditional village level chakki and commercial mills are presented in Table 5.11.

Table 5.11
Milling recoveries of pigeon-pea by traditional milling processes in M. P., Maharashtra and U.P.

State	*Milling fractions, %*	*Milling method*			
		Commercial large scale dhal mill		*Small scale mill (Village chakki)*	
		range	*mean*	*range*	*mean*
M.P.	Dhal	60–75	71.4	50–70	61.6
	rokens	2–8	3.6	5–18	10.5
	Powder	12–15	13.0	7–22	11.4
	Husk	5–22	12.0	10–25	16.5
Maharashtra	Dhal	65–72	68.2	50–70	59.0
	Brokens	4–10	5.8	8–20	12.0
	Powder	10–15	12.6	10–20	13.5
	Husk	10–20	13.4	12–20	15.5
U.P.	Dhal	60–95	70.6	50–80	62.0
	rokens	2–10	4.1	5–20	10.2
	Powder	9–18	12.7	7–20	13.4
	Husk	10–25	12.6	10–25	14.4

Source : Singh and Jambunathan, 1982.

Problems of the pulse milling industry

1. Batch process which is in use, involves excessive material handling which can be reduced by introducing continuous type processing system.
2. As sun drying is not reliable and during raining days mill has to be closed, some economical and reliable mechanical drying system is needed.
3. Improved marketing system where the millers could purchase cleaned and graded raw materials. It will reduce the cost of processing of dhal.
4. During milling lot of dust is generated inside the mill. Development and use of dust collecting and diffusing devices. Dust inside closed mill causes pollution and is hazardous to workers.
5. Separation of gota (whole dehusked kernel) from pulse grain after each pass through abrasive roller.
6. Non-standardisation of milling processes and equipment. BIS should prepare standards for milling techniques and machinery.

Factors affecting pulse milling outturn

Theoretically the endosperm of the pulse accounts for maximum 87–89% of the whole grain legumes. But in practice it is not achievable. Some of the important factors responsible for lower recovery of dhal are given below.

A. Grain parameters

(*i*) Type of grain
(*ii*) Species or strain of grain legume
(*iii*) Chemical constituents
(*iv*) Moisture content of the grain endosperm
(*v*) Shape of the grain
(*vi*) Size, bold, small
(*vii*) Seed coat surface texture

(*viii*) Thickness of the gum layer

B. Machine parameters

(*i*) Size of roller

(*ii*) Peripheral speed of the roller

(*iii*) Texture of the abrasive material

(*iv*) Hardness of the abrasive material

(*v*) Clearance between the roller and cage

(*vi*) Inclination of the roller

(*vii*) Oulet position

Pulse milling efficiency

The overall efficiency of the pulse milling system can be estimated with the help of the following equations.

1. $$E = \left(1 - \frac{M_{uh}}{M_t}\right)\left(1 - \frac{M_b}{M_t}\right) \times H_I \times 100$$

where, $H_I = \dfrac{H_a}{M_t H_t}$

E = milling efficiency

M_{uh} = mass of unhulled grains

M_t = mass of grains fed to the system

M_b = mass of brokens

H_a = actual mass of husk removed during milling

H_t = theoretical husk content of the grain

The theoretical husk content of common Indian legumes are given in Table 5.12

Table 5.12

Husk content of Indian legumes

Grain	*Husk Content,%*
pigeon pea	13.0–15.0
green gram	12.0
black gram	12.0
lentil	11.5
Bengal gram (chick pea)	11.5–13.0

2. The Kupritz formula can also be used to calculate the milling efficiency of the dhal milling system. Kupritz equation has been explained in milling section.

Problem : In pigeon pea milling experiment with concentric cylinder abrasive will the following observations were made

(*i*) Amount of unhulled grains = 2.5%

(*ii*) Recovery of whole split kernels after milling = 71.4%

(*iii*) Amount of crushed kernels = 3.6%

(*iv*) Amount of powder generated = 11%

(*v*) Amount of husk removed = 11.5%

The cotyledon to grain ratio of the grains was 86.5. Calculate the milling efficiency of the system.

Solution : (1) Using Kuprit's formula

$$\eta = E_{hulling} \times E_{wk}$$

$$= 100\left(1-\frac{2.5}{100}\right) \times \frac{71.4}{71.4+3.6+11+11.5}$$

$$= 100 \times 0.975 \times 0.7323$$

$$= 71.4\%$$

(2) Using equation no.1

$$E = \left(1-\frac{M_{uh}}{M_t}\right)\left(1-\frac{M_b}{M_t}\right) \times H_I \times 100$$

$$= \left(1-\frac{2.5}{100}\right)\left(1-\frac{3.6}{100}\right) \times \frac{11.5}{13.5} \times 100$$

$$= 0.975 \times 0.965 \times 0.851 \times 100$$

$$= 79.9\%$$

Exercise

1. In a sorghum milling experiment it was found that to grind 4.4 mm sized grains to IS sieve 30 (0.296) mm opening, the power requirement was 5 kW. Calculate the power requirement for milling of sorghum by the same grinder to IS sieve 25 (0.251 mm opening) using (1) Rittinger's law and (2) Kicks law. Feed rate of milling is 150 kg/hr.
2. Wheat of size 4.3 mm was milled in a burr mill at two different settings of gaps between the burr stones. The flour was analysed by IS sieves for particle size determination as shown in Table given below. The power required to mill wheat at first setting was 6.5 kW. Calculate the power requirement of the mill in second setting using (1) Rittinger's law and (2) Kick's law. The capacity of grinder was 90 kg / hr.

IS sirve	*Mass fraction of flour retained over sieve, g*	
	I setting	*II setting*
100	—	—
70	6.8	6.4
50	18.5	12.5
40	38.1	36.1
30	80.2	72.8
20	97.0	101.6
15	7.3	9.2
Pan	2.1	11.4

3. A ball mill of 1500 mm diameter is charged with 75 mm balls to grind solid material. Determine its speed of operation in revolution per minute.

4 Sorghum is milled in a grinder to make semolina. The screen analysis of the milled product is given in the Table below. Let the density of particles be 700 kg/m^3. Find the followings for the ground product between 15 and 240 sieve size.

(i) Specific surface of the mixture.

(ii) Number of particles per gram.

(iii) Volume surface mean diameter

(iv) Mass mean diameter

(v) Volume mean diameter and

(vi) Number of particles per gram between 85 and 70 sieve.

The shape factors given as $a = 1.6$ and $\phi = 0.78$.

Characteristic screen analysis of ground sorghum

Sieve Size	*Sieve opening D_{pi} mm*	*Average diameter in each increment D_{pi} mm*	*Mass fraction retained, m_i*	*Cumulative mass fraction smaller than, D_{pi}*
240	2.399	—	0.000	1.000
200	2.032	2.215	0.012	0.988
170	1.676	1.854	0.058	0.930
120	1.201	1.438	0.061	0.869
100	1.000	1.100	0.140	0.729
85	0.954	0.977	0.164	0.565
70	0.708	0.831	.0206	0.359
50	0.500	0.604	0.123	0.236
40	0.420	.0460	0.081	0.155
30	0.296	0.358	0.052	0.103
20	0.211	0.254	0.041	0.062
15	0.157	0.184	0.038	0.024
Pan	—	0.078	0.024	0.000

Bibliography

Ali, N and KM Sahay, 1983. Milling of rice, sorghum and pulses for more yields. Paper presented at the national seminar on contribution of agricultural machinery and power in increasing production and productivity. Dec. 28-30, CIAF Bhopal.

Autrey, HS, WW Griegorief, AM Altschul and JT Hogan, 1955. Effect of milling conditions on breakage of rice grains. J. of Agric. and Food chem. 3.593-599.

Bhattacharya KR and PV Subbarao, 1966. Processing conditions and milling yields in parboiling of rice. J. of Agric and Food Chem. 14:473-475.

Bhattacharya, KR and YMI Swamy, 1967. Conditions of drying parboiled paddy for optimum milling quality. Cereal Chemistry 44(6).592-600.

Bhattacharya, KR 1969. Breakage of rice during milling and effect of parboiling. Cereal Chemistry 46:470.

Bhole, NG and P Urde, 1980. Effect of operating parameters on performance of huller polisher. ISAE paper No. 80-0477.

Bhole, NG and BK Sarkar, 1983. Modern rice milling machinery manual. Publication No.RPEC-831, PHTS, IIT, Kharagpur.

Birewar, BR, 1982. Post harvest technology of pulses. in "Pulse production-constraints and opportunities." Oxford and IBM Publishing Co. Pvt. Ltd. New Delhi :425.

Chakravery, A and DS De, 1980. Post harvest technology of cereals and pulses. Oxford and IBH Publishing Co. New Delhi.

Desikachar, HSR, 1955 A determination of degree of polishing in rice I. Some methods for comparison of the degree of milling. Cereal Chemistry. 32:71-77.

Gariboldi, F, 1974. Rice milling equipment operation and maintenance. FAO U.N. Agric. service Bul. 22.

Govindrao, UMH., 1980. Utilisation of rice husk. A preliminary analysis. J. of Sci. and Indus. Res 89 (a) : 495-515.

Henderson, SM, and RC Hansen, 1968. Farm grain comminution hammer mill and burr mill performance analysis. Trans. ASAE 11(3) : 379-402.

Henderson, SM, and RL Perry, 1976. Agricultural process engineering. AVI Publishing Co. West port Connecticut.

Indudhar Swamy, YM and KR Bhattacharya, 1984. Breakage of rice during milling. Effect of sheller, pearler and grain types. J. of Food Sci. and Tech. 21 (1) : 8-12.

Khan, LM and MA Hanna, 1983. Expressing of oil from Oilseeds-A review. J. Agric. Eng. Res. 28 : 495-503.

Khare, RN, K Krishnamurthy and SV Pingale, 1966. Milling losses of food grains. Bul of grain tech. IV (3) : 125.

Krishnamurthy, K, 1975. Post harvest losses in food grains. But of grain tech. XVII (1) : 41.

Kupritz, YN, 1967. Technology of grain processing and povender milling. Israel program for scientific translation. Jerusalam.

Kurien, PP and HBA Parpia, 1968. Pulse milling in India I. Processing and milling of tur and arhar (Cajanus cajan). Food Sci. and Tech. 5(6) : 203.

Kurien, PP, 1983. An improved technique for milling pulses. Paper presented at SBI Officer Training Centre, Bhopal.

Luh, Bor S, 1980. Rice production and utilisation AVI Publishing Co. West port Connecticut.

McCabe, WL, JC Smith and P Harriott, 1985. Unit operation of chemical engineering, 4th ed. McGraw Hill, New York.

National Institute of Nutrition, 1971. Nutritive value of Indian foods ed. Gopalan, Shastri and Balasubramanian.

Nelson, CA and HJ Loving, 1963. Mill stream analysis. Cereal Sci. Today 8:301.

Pillaiyar, P and K. Ramachandran, 1982. Automation in rice milling—an analysis. Madras Agric Journal 69(3):203.

Ramalingam, M and N Ibrahim, 1981. Modernisation of small rice mills in India—Suggested approach. J. of Agric. Eng. XVII (3 & 4) : 119–122.

Rhind, D., 1962. The breakage of rice in milling. A review. Tropical Agric. 39(1) : 19.

Sahay, KM, 1984. A classification system for determination of particle size in grain milling. Agric. Eng. Today, Vol. 8 No. 6 :34.

Sahay, KM and BS Bisht, 1988. Development of a small abrasive cylindrical mill for milling pulses. International J. of Food Sci. & Tech. 23 : 17.

Sahay, KM, 1989. Milling technology of pulses in India. Proceedings of the National seminar of Bihar chapter of ISAE : 140–145.

Sahay, KM and BS. Bisht, 1989. Optimisation of grade of carborundum for best recovery of dhal in milling of different pulses. Agric. Eng. Today 11(6) : 10.

Sahay, KM, 1991. Milled pulses: availability, Constraints and prospects. Productivity Vol. 32 No. 2 : 258–261.

Saini, AK, 1981. Study on productivity in pulses production and dhal milling industry. NPC, Lodhi Road, New Delhi.

Saxena, RP and BPN Singh, 1990. Pantnagar system of pigeon pea milling. PHPFE/Bul. 19901. GBPUA & T, Pantnagar.

Singh, J and Y Singh, 1981. Development and evaluation of processing equipment. Tech. Bul. CIAE/PHTS/81/24 : 37–43.

Singh, Jaswant and B.D. Shukla, 1991. Post harvest technology of oilseeds. CIAE. Bhopal.

Singh, U and R Jambunathan, 1982. A survey of the methods of milling and consumer acceptance of pigeon pea in India in "Pulse production, constraints and opportunities" Oxford and IBH Publishing Co. Pvt. Ltd. New Delhi.

Siva Kumaran, K and JW Coodrum, 1987. Influence of internal pressure on performance of a small screw expeller. Agric. Eng. : 1167–1171.

Tindale, LH and SR Hill-Hass, 1976. Current equipment for mechanical oil extraction. J. Am. Oil Chemists society Vol. 53 : 265–269.

Unni Krishnan, KR, CR Viraktmath, H Krishnamurthy and KR Bhattacharya, 1982. Parboiling of paddy by simple soaking in hot water. J. of Food Tech. 17(4) : 499.

Ward, JA, 1976. Processing high oil content seeds in continuous screw press. J. Am. Oil Chemists Society Vol. 53:261–264.

Williams, PC, 1984. A study of grinders used for sample preparation in laboratory analysis of grains. Cereal Food World 29 (12) : 770–775.

Wimberly, JE, 1983. Paddy rice post harvest industry in developing countries. IRRI. Los Banos Laguna. Philippines.

Chapter 6

Material Handling and Transportation

Material handling includes a number of operations that can be executed either by hand (manual) or by mechanical means or devices to convey material and to reduce the human drudgery. Mechanical handling devices aim to lighten the work of human labour.

After harvesting, the grains are moved, transported or conveyed from place to place. In earlier periods all these operations were manual. The grains were threshed and bagged by human labour. Grains were transported several times through storage and milling plants, and the milled food products were conveyed manually to consumers. Thus, grains were handled too much involving increased costs and human drudgery. But in modern times, some mechanical devices have replaced human labour, other supplement it or in some case make possible to handle larger quantities of grains per unit human labour. The most common types of mechanical devices for grain handling are;

1. Belt conveyor
2. Bucket elevator
3. Screw conveyor
4. Pneumatic conveyor

The selection of proper conveying system is important for ease in operation and getting desired capacity for a particular product. Before selecting a conveying system, the following principles should be taken into account.

1. The conveying device has to be selected according to the characteristics of the products being conveyed
2. The stability of the conveyor must be ensured under all normal working and climatic conditions.
3. The capacity of conveying and speed rating should be maintained at specified limits.
4. The deadload of the conveyor should be low in relation to the weight of transported product.
5. In a conveying system possibility of use of gravity should be taken into consideration.

6. The capacity of handling/conveying equipment should match with the capacity of processing unit or units.
7. Spillage of conveyed products should be avoided. Pollution of the environment due to noise or dust by the conveying system should also be avoided.

Belt conveyors

A belt conveyor is an endless belt operating between two pulleys with its load supported on idlers. The belt may be flat for transporting bagged material or V-shaped or some other enclosed shape for moving bulk grains. The belt conveyor consists of a belt, drive mechanism and end pulleys, idlers and loading and discharge devices (Fig. 6.1)

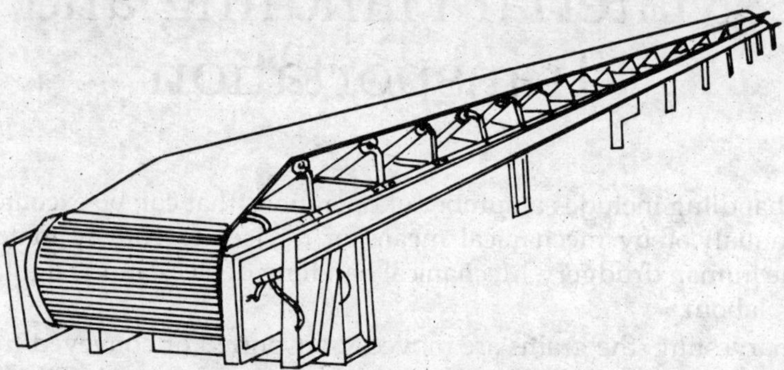

Fig. 6.1 : Diagram of a belt conveyor

Belt conveyors have antifriction bearing, therefore, these have a high mechanical efficiency. Material carried by belt conveyor lie still on the surface of belt or there is no relative motion between the product and belt. This results in generally no damage to material. Belt can be run at higher speeds, so, large carrying capacities are possible. Horizontally the material can be transported to longer distance but there is a limit to carry the material on elevation. A properly designed and maintained belt conveyor has long service life and low operating costs. Compared to other types of horizontal conveying system, the initial cost of belt conveyor is high for short distances. But for longer distances, the initial cost of belt conveying system is competitive or low. For these reasons belt conveyors are used to carry grains in processing plants.

The design of belt conveyor system is based on available space, horizontal conveying length and conveying lift, characteristics of the material to be conveyed and capacity requirement. On the basis of overall requirement and information, the following will be determined to design a belt conveyor, belt width, belt speed, required horsepower, maximum belt tension and breaking strength of the belt, diameter of the pulleys and idlers and quality of belt (thickness).

The first step in the design of a belt conveyor with a specified conveying capacity is to determine the speed and width of the belt.

The belt speed should be selected to minimise product spillage or removal of fines due to velocity of the belt. For transportation of grains, the belt speed should not increase 3.5 m/s. Generally, for grain conveying, belt speed of 2.5 to 2.8 m/s is recommended. The selection of belt width will depend upon the capacity requirement, speed of operation, angle of inclination of belt conveyor, trough angle and depth. The capacity of belt conveyor can be given by;

Capacity, m³/hr = (area of cross-section, m²) x (belt-speed, m/min) x 60...6.1

The load cross sections of troughed belt is shown in Fig. 6.2.

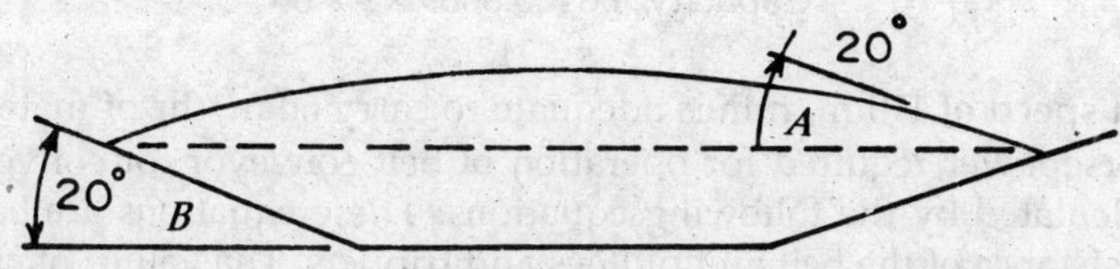

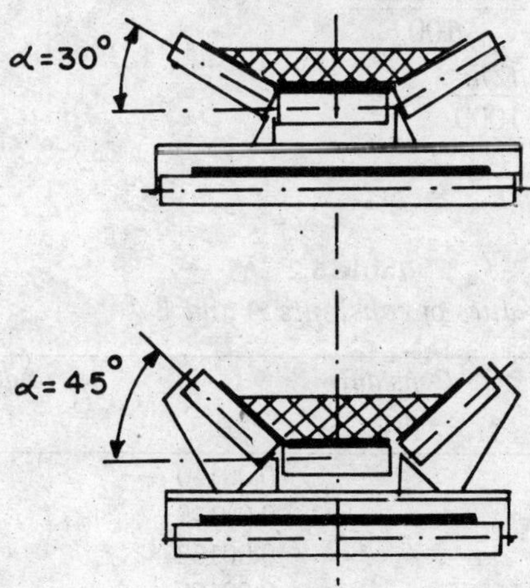

Fig. 6.2 : Various troughing configurations

A trough angle (*B*) of 20° is best suited for paddy and most other grains. Other common trough angles are 30° and 45°. Paddy forms a surcharge angle (*A*) of 20°. Other common surcharge angle ranges between 5° and 30°. The cross-sectional area of loaded belt and belt speeds are given in Table 6.1.

Table 6.1

Cross-sectional area of loaded belt and operational belt speeds

Belt width, cm	*clear margin, cm*	*Cross-sectional area, m^2 for 20° surcharge angle*	*Operational speed, m/min*	
			Normal	*Maximum*
30.5	4.1	0.0072	61	122
35.6	4.3	0.0089	61	122
40.6	4.6	0.0122	61	137
45.7	4.8	0.0161	76	137
50.8	5.1	0.0204	76	152
61.0	5.6	0.0308	91	183

Problem : The capacity of a toughened belt conveyor is 60 m^3/hr. With the help of Table 6.1, calculate the belt width and belt speed.

Solution : Let the belt width be 35.6 cm. The cross- sectional area for 20° surcharge angle of 35.6 cm wide belt is 0.0089 m^2

$$\text{Capacity, } 60 = 0.0089 \times s \times 60$$

or

$$s = 112.35$$

or the belt speed of 113 m/min is adequate to carry 60 m^3/hr of material.

The horsepower required for operation of belt conveyor for conveying grains can be calculated by the following equations. These equations are based on lift, friction resistance of the belt and pulleys and trippers. The values of constants are given in Table 6.2.

$$HP_1 = \frac{\text{Belt speed, m/min}}{0.3048} \times \frac{(A+B)\ (3.281\ L)}{100} \qquad \text{...6.2}$$

$$HP_2 = \text{Capacity, } t/hr \times \frac{0.48 + 0.01\ L}{100} \qquad \text{...6.3}$$

$$HP_3 = \frac{\text{Lift, } m}{0.3048} \times 1.015 \times \frac{t/hr}{1000} \qquad \text{...6.4}$$

where L = length of belt, m

A and B = constants

Table 6.2

Values of constants A and B

Belt width, cm	*Constants*		*Additional horsepower for tripper*
	A	*B*	
36	0.20	0.00140	0.70
41	0.25	0.00140	0.85
46	0.30	0.00162	1.00
50	0.30	0.00187	1.40
60	0.36	0.00224	1.70
76	0.48	0.00298	2.50

The majority of belt conveyors for transporting bulk material use some type of rubberised conveyor belt made up of carcass. The pull of load is taken by the longitudinal strength of belt while the transverse strength supports the load. The belt is protected from damage by a rubber cover. The thickness of top rubber cover varies with thickness and wear resistance requirements.

Belt conveyor idlers : The efficiency of belt conveyor is largely dependent on idlers. For higher efficiency of belt conveying systems, the idlers must be accurately made and provide a rigid framework. This will maintain a permanent, well balanced smooth running alignment.

There are, in general, three kinds of belt carrying idlers which are used in handling of bulk materials. The type of idlers affects the cross-sectional load on the belt.

1. The flat belt idlers (Fig 6.3) are used for granular materials having an angle of repose of not less than 35°. Flat belt idlers are preferred for low capacity requirements where inexpensive or low-cost conveying is desired.

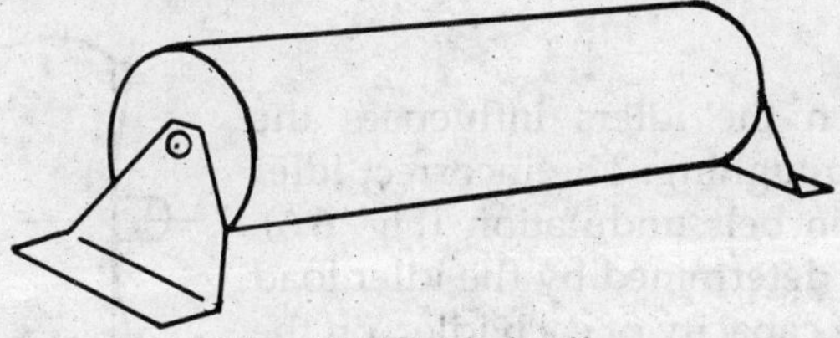

Fig. 6.3 : Flat belt idler

2. Troughing idlers with 20° trough (Fig. 6.4). It is used for conveying all kinds of bulk materials.

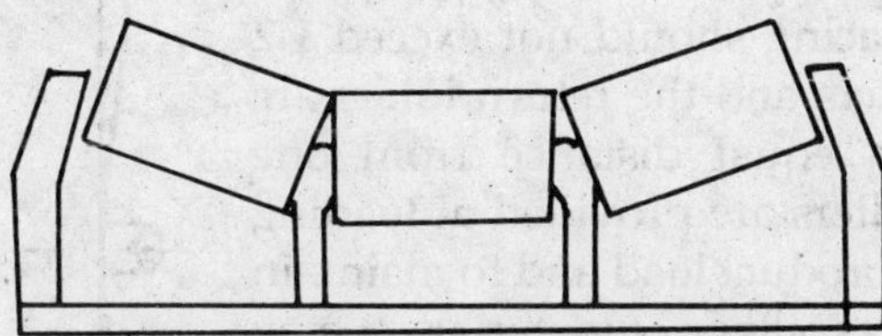

Fig. 6.4 : Idlers for 20° troughed belt

3. Troughing idlers with 35° and 45° trough (Fig. 6.5). It is mainly used for transportation of small particle light weight materials like grain, cotton seed etc. It is also used for carrying heavier, medium size lumps like crushed stones.

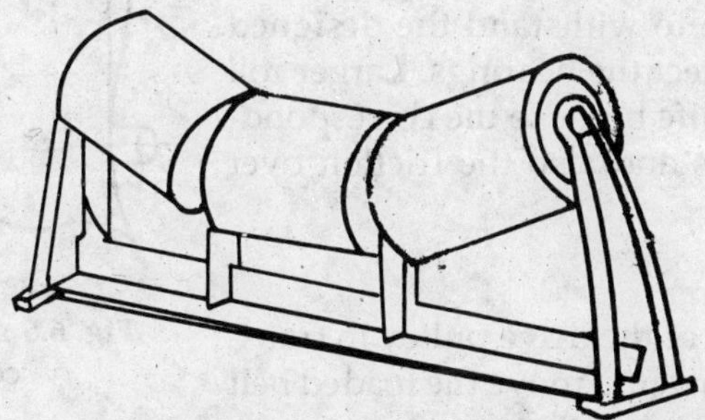

Fig. 6.5 : Idlers for 35° troughed belt

To serve specific purpose, the idlers can be self-aligning idlers, impact idlers and transition idlers. The transition idlers are used for highly stressed belts. These are troughing idlers which regulates the angle of inclined rolls for providing a gradual transition from a fully troughed section of the belt to a flat faced pulley. The relationship between idler diameter and belt speed is given in Table 6.3.

Table 6.3

Idler diameter, belt speed and belt width for different purposes

Idler diameter, cm	*Belt speed, m/min*	*Belt width, cm*	*Purpose*
10.0	91.0	50.8	All types of material
		61.0	upto material weight of 800 kg/m^3
		76.0	upto material weight of 1600 kg/m^3
12.5	152.0	61.0	
		91.0	upto material weight of 1600 kg/m^3
15.0	244.0	91.0	Handling all types of material

Idler Spacing

The spacing between the idlers influences the retension of correct troughing. The incorrect idler spacing may result in belt undulation (Fig. 6.6). The pitch of idlers is determined by the idler load rating or the carrying capacity of each idler, on the sag of the belt between the idlers, belt tension and belt speed. As a token, the space between the successive idlers should be approximately equal to the width of belt. The spacing should not exceed 1·2 metres. The upper idlers and the return idlers are usually placed at an equal distance from one another. Additional idlers are provided at loading points to support the product load and to maintain the trough. This avoids spillage of materials. It may also be necessary to install shock absorbing idlers at the loading points. Since, in a belt conveyor, the number of idlers involved is large, the efficiency of belt conveyor depends largely on the correct design and selection of the idler rollers. The main functions of rollers are to withstand the designed work loads and to protect the bearings. Larger roll diameters give longer life because the corresponding 'rpm' becomes less and also the friction over idlers is reduced.

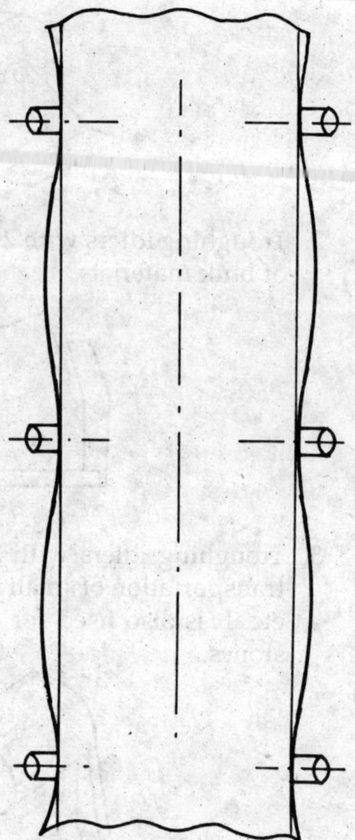

Fig. 6.6 : Sagging of belt conveyor

Belt tension

The tension developed at the drive pulley in transmitting the required power to move the loaded belt is known as effective tension. The effective tension is the sum of tension to move the empty belt, the tension to move the load horizontally and the tension to lift the material. The effective tension is related with the power required to move the belt and belt speed in the following manner.

$$\text{Effective tension, } T_e = \frac{kW}{S} \qquad \ldots 6.5$$

where, kW = power in kilowatt

S = belt speed, m/s

The idler spacing and the minimum operating belt tensions are given in Table 6.4 for 20° troughing idlers.

Table 6.4

Idlers spacing and minimum belt tensions for 20° troughing idlers

Belt width, cm	*Minimum belt tensions at the loading end for 20° sag limitation, kg*							
	800 kg/m³ material			*load on belt, kg/m*	*1600 kg/m³ material*			*Load on belt, kg/m*
	150	*120*	*90*		*135*	*120*	*90*	
40	182	159	136	10.4	227	204	182	20.8
45	204	182	159	13.4	295	250	204	26.8

Belt width, cm	Minimum belt tensions at the loading end for 20° sag limitation, kg							
	800 kg/m³ material			load on belt, kg/m	1600 kg/m³ material			Load on belt, kg/m
	150	120	90		135	120	90	
50	227	204	182	16.4	363	318	227	32.8
60	318	273	227	25.3	545	454	386	50.6
75	500	409	318	40.2	864	727	545	80.5
90	727	591	454	58.1	1136	1000	772	116.3

The maximum permissible belt tensions while using vertical convex curve are given in Table 6.5. In case of inclined belt operation, the spacing of troughing idlers is about one-half of the normal spacing.

Table 6.5

Maximum permissible belt tensions for vertical convex curve

Belt width, cm.	Maximum permissible tension, kg	Permissible belt tension at entry to curve, kg				
		Radius of curve, cm				
		609	914	1524	2286	3048
60	1636	818	1090	1272	1409	1454
	2181	1090	1454	1727	1863	1954
75	2454	909	1409	1818	2045	2136
	3272	1227	1909	2454	2727	2863
90	3909	1000	1954	2727	3090	3318
	5227	1272	2590	3636	4181	4409

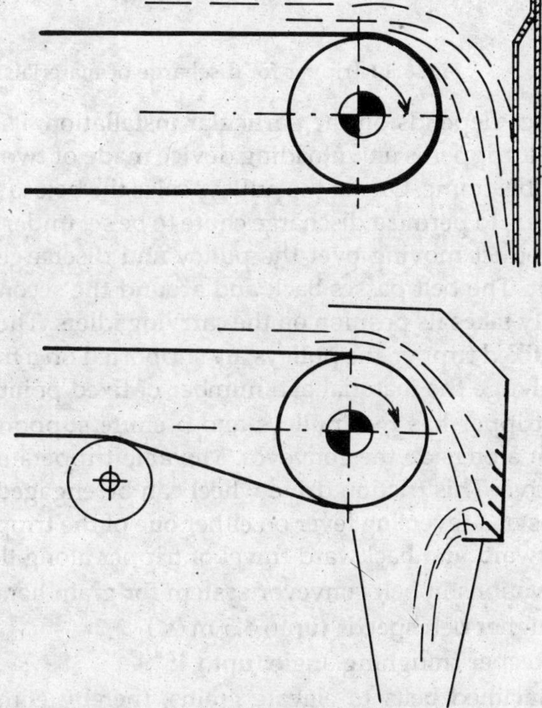

Fig. 6.7 : Belt conveyor unloading arrangements

Grains are mostly discharged from the belt conveyor over the end pulley or at any other point along the conveyor by a scraper plough or a throw-off carriage known as a tripper.

While leaving the belt over the end pulley, product flow will describe the path of a parabola. This parabolic form of unloading influence the design of discharge chutes (Figs. 6.7). The shape of parabola is dependent on belt speed and pulley diameter. When discharge is desired at more than one point along the length of belt conveyor, discharge ploughs are used. Discharge plough diverts the flow of products over the side of the belt. These are generally used in flat and horizontal conveyors. Scrapers or ploughs are diagonally mounted usually at 30° across the direction of travel. The scraper or plough may damage the belt if speed is higher than 60 m/min.

Belt conveyors can be discharged at various locations by means of a movable tripper (Fig. 6.8). Trippers are available as hand propelled, self propelled or auto-

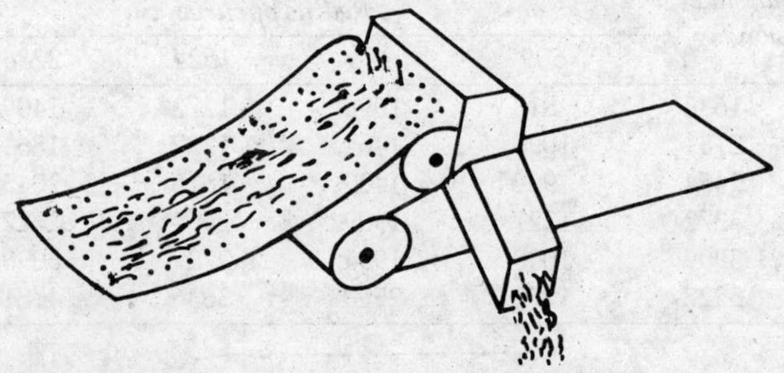

Fig. 6.8 : Tripper for discharge of materials

matic. The choice depends on the particular installation, its capacity and system operation. A belt tripper is an unloading device made of two pulleys supported in a fixed or movable frame. One of the pulleys raise the belt to sufficient height from the carrying idlers to permit a discharge chute to be set under the pulley. This chute receives the product moving over the pulley and discharges it to either or both sides of the belt. The belt passes back and around the second pulley beneath the chute and finally takes its position on the carrying idlers. The tripper may be fixed or movable. In fixed tripper, the pulleys are supported on a fixed frame. This could be used to discharge the material at a number of fixed points.

A movable tripper has two pulleys and a chute supported on a frame, with wheels that run alongside the conveyor. Movable trippers may be self propelled by a friction drive. This friction drive wheel can be engaged to a driving friction wheel by means of a reversing lever on either one of the tripper pulley shafts. This enable both forward and backward travel of tripper along the belt.

Recent innovations in belt conveyor system for grain handling are:

1. Use of higher belt speeds (upto 4.5 m/s)
2. Use of steeper troughing angle (upto 45°)
3. Use of inclined belts to elevate grains, thereby eliminating the need for bucket elevators.

High capacity belt conveyors have a conveying capacity of 2000 tonnes grain per hour. The belt width is 1500 mm and have a 45° troughing angle. Belt speed is around 3.5 m/s and the longest of the elevators is approximately 200 metres between pulleys. Inclined conveyors have a maximum feed on angle of 10° and conveying angle of 12°.

Bucket Elevator

A bucket elevator consists of buckets attached to a chain or belt that revolves around two pulleys one at top and the other at bottom. The vertical lift of the elevator may range between few metres to more than 50 m. Capacities of bucket elevators may vary from 2 to 1000 t/hr. Bucket elevators are broadly classified into two general types, (1) spaced bucket elevators and (2) continuous bucket elevators. The above two types are further subdivided into various classes.

The spaced bucket elevators are further classified as, (1) centrifugal discharge elevators, (2) positive-discharge elevators, (3) marine leg elevators and (4) high-speed elevators. The continuous bucket elevators are classified as (1) super capacity bucket elevators and (2) internal-discharge bucket elevators.

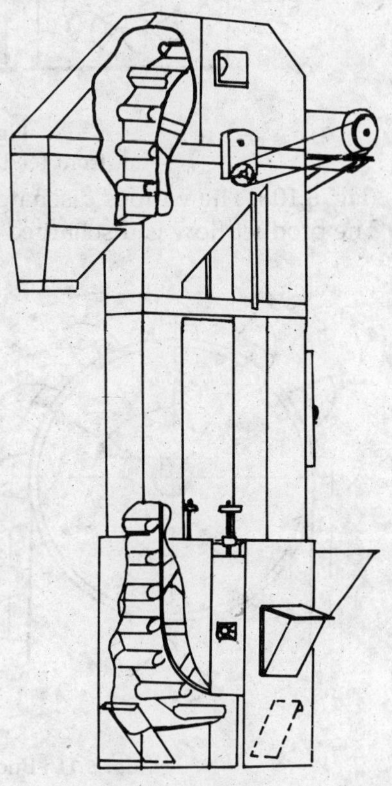

Fig. 6.9 : Bucket elevator

The spaced-bucket centrifugal discharge type is most commonly used for elevating the grains. A centrifugal discharge bucket elevator is shown is Fig. 6.9. The bucket elevator is a very efficient device for the vertical conveyance of bulk grains. Bucket elevators with belts are employed in food industries for vertical conveyance of grains, derivatives and flours. Bucket elevators are usually mounted at a fixed location, but they can also be mounted in a mobile frame. Bucket elevators have high capacities and it is a fairly cheap means of vertical conveyance.

It requires limited horizontal space and the operation of conveying is enclosed in housing, thus it is dust free and fairly quite. The bucket elevator has limited wear problem since the product is enclosed in buckets.

In a bucket elevator, the conveyor belt with buckets runs over pulleys at the upper and lower ends. The top pulley is driven pulley while the lower pulley is return and tension pulley. Buckets are usually made of steel or plastic and are bolted onto the belt. The buckets may be enclosed in a single housing called leg, or two legs may be used. The return leg may be located at some distance from the elevator leg. The housing or legs are also made of steel, are welded or bolted together, and are dust tight. The curved hood is designed for proper centrifugal discharge of the grains. The boot can be loaded from the front or back or both

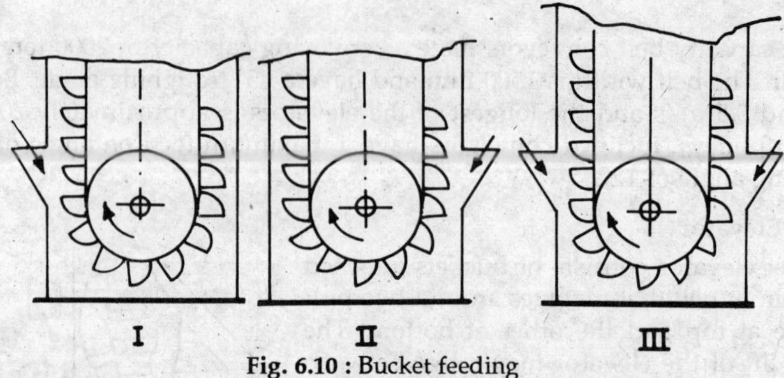

Fig. 6.10 : Bucket feeding
I. front feed II. back feed III. combined feed

(Fig. 6.10). The various discharge types of bucket elevators are shown in Fig. 6.11. The product flow is discharged either by means of gravity or centrifugal force.

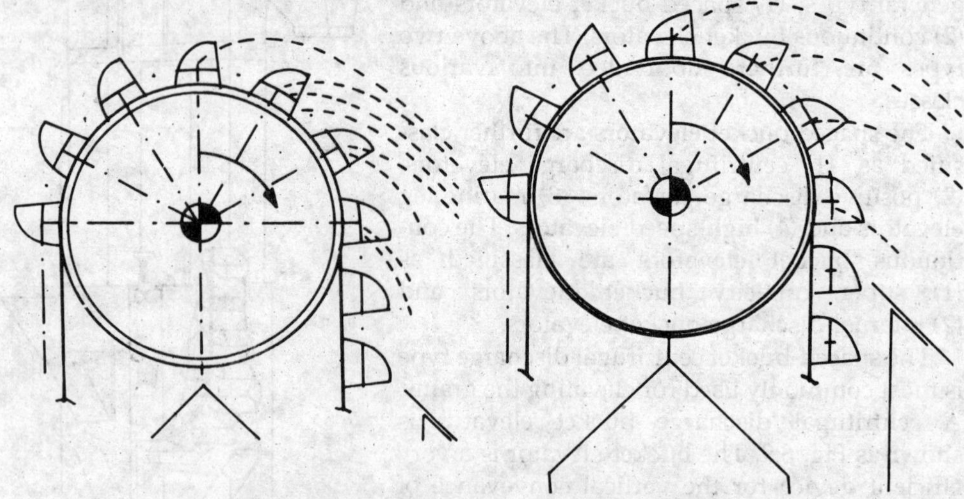

Fig. 6.11 : Bucket elevator's discharge methods
1. low speed gravitational discharge 2. high speed centrifugal discharge

The bucket elevator's capacity mainly depends on bucket size, conveying speed, bucket design and spacing, the way of loading and unloading, the bucket and the characteristic of bulk material. Belt speed is the first critical factor to consider. Bucket elevators with a belt carrier can be used at fairly high speeds of 2.5 to 4 m/s. The speed of the belt depends on the head pulley speed. If the belt speed is too low, the discharge of the grains becomes more difficult, with too high speed the buckets are not fed well. A properly designed bucket elevator driven at the correct speed will make a clean discharge.

The gravitational discharge occurs with non-adhesive bulk material elevated at low speed and by means of buckets mounted closely together. In such discharge, the contents of a bucket flow over the rear side of the previous bucket. With purely centrifugal discharge, complete contents of a bucket are projected towards the discharge chute. Such type of discharge is obtainable with high belt speeds and smaller diameter drive pulleys. In elevating of grains the discharge from bucket elevators is a combination of centrifugal and gravitational discharge. Part of the bucket contents is projected by the centrifugal force, the rest flows out by gravity.

When a product mass turns around a pulley, it is influenced by two forces, (1) gravitational force, which is oriented downwards and (2) centrifugal force.

The magnitude of the centrifugal force which is oriented outward can be given by

$$c_f = \frac{WV^2}{g.r} \quad \text{...6.6}$$

where, W = weight of grain
V = velocity of product mass
g = acceleration due to gravity
r = radius of the wheel plus one–half of the projection of the bucket

For optimum centrifugal discharge, and calculation of the speed of head pulley, the resultant force is zero, it means that the centrifugal force is equal to the force of gravity, or, $c_f = W$

Hence $W = \dfrac{WV^2}{gr}$ or, $V = \sqrt{gr}$

Since Velocity, $V = \dfrac{2\pi n r}{60}$ (n = rpm) ...6.7

$$\frac{2\pi n r}{60} = \sqrt{gr}$$

or, $$n = \frac{60\sqrt{g}\sqrt{r}}{2\pi\sqrt{r}\sqrt{r}} = \frac{29.9}{\sqrt{r}} \quad \text{...6.8}$$

The bucket elevator's capacity may be calculated by the following equation.

Elevator capacity, m^3/hr = bucket capacity, m^3 × number of bucket per meter of belt × belt speed, m/min. × 60 ...6.9

$$\text{Capacity, t/hr} = \frac{\text{capacity, m}^3\text{/hr} \times \text{material density, kg/m}^3}{1000} \quad \text{...6.10}$$

The main parts of a bucket elevator are, (1) elevator head section, (2) elevator boot section, (3) elevator legs, (4) belts for bucket elevator and (5) buckets.

Head Section

The shape of the head section is dependent upon the bucket design, pulley diameter and belt speed. The head section should be of the proper shape and size with smooth contours. A high speed conventional bucket elevator's head section is shown in Fig. 6.12. The discharge side of the head should be shaped so that material thrown from the buckets may not deflect into the down leg. When the product is not thrown well clear of the buckets into the discharge chute, it will fall in the down leg. This is called as "back logging". The back logged material has to be reelevated, thus it reduces the capacity of the elevator. To avoid back-logging, an adjustable cut off plate is provided close to the lip of bucket.

Lagging on the elevator head pulley is needed in pulling heavy loads. Proper lagging increases the coefficient of friction between the pulley and belt. With the drive pulley of the elevator, the angle of wrap and the friction between the belt and pulley is important. To increase the friction coefficient, lagging of head pulley with rubber is required.

The shaft of the drive pulley is always fixed to the pulley and bearing outside the casing construction. The head shaft must be heavy enough to resist bending

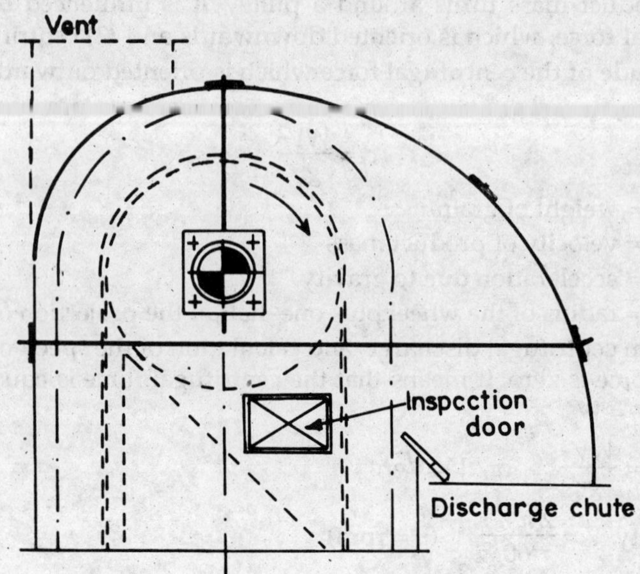

Fig. 6.12 : Diagram of bucket elevator head section

and to provide required torque carrying capacity. The shaft must be level. Anti-friction and properly lubricated bearings should be used.

Boot Section

Bucket elevator boots should be of bolted assembly to allow for proper maintenance and replacement of pulley, shaft and other accessories. In the boot section of most bucket elevators, there is provision for tightening of belt as required. It is also required to train the belt that it runs true and does not rub on either side of

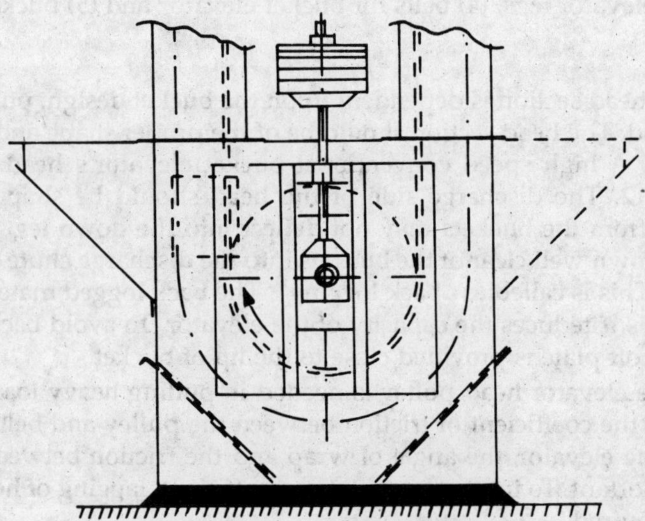

Fig. 6.13 : Bucket elevator boot section with automatic gravity take-up

the boot. On small bucket elevators where gravity take-up is not practical, a manual screw type take-up is most often used. The screw take-up consists of two spindles which can be tightened by a hand wheel. A manual take-up does not ensure proper automatic tensioning. On tall heavy capacity legs, an automatic take-up boot pulley is used (Fig. 6.13). This provides correct belt tension at all times. In the boot section, the loading chute should be located at such point that the pick-up of the product by the buckets takes place above the centre line of the return pulley.

Grain entry may be on either side of the boot. Additional power is required to overcome the "dredging effect" while pulling the buckets through the grain in the boot. In the boot section clean out gates should be provided on both sides of the boot. This is necessary for cleaning of boot.

Elevator Legs

The up and down moving string of buckets in bucket elevators are enclosed in elevator legs. The elevator legs limit the emission of dust. These legs are constructed as all welded, bolted or riveted units. The strings of up and down moving buckets can either run in a common leg or in separate legs. Balanced pressure inside the housing of bucket elevator is attributed to single leg or trunk. With double legs, a balanced pressure can be obtained by ducts connecting on different levels of the upgoing and downgoing trunk. Cross- section of different types of elevator legs is shown in Fig 6.14. Bucket elevators require adequate maintenance,

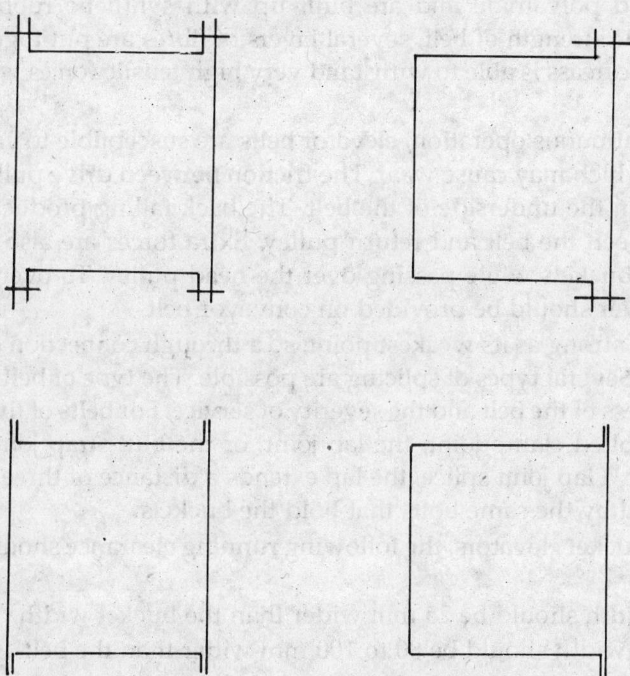

Fig. 6.14 : Cross-section of few types of elevator lags

therefore, service and inspection openings are needed. To accomplish this, the elevator legs should have service access for the installation of new belt, for inspection of the belt and for replacement of damaged buckets and shortening of belt should be provided at one or two levels. All the service opening should be sufficiently large enough and dust tight. The elevator casing near the drive and bottom pulleys should be such that when required the whole pulley can be dismantled and can be taken out.

Elevator Belts

In a normal operation of the bucket elevator, the loads exerted by the elevator height, product weight, weight of bucket belt and idle tension and the digging resistance are taken by the belt. The bucket elevator belt has no support between the drive and the return pulleys, therefore, cross stiffness of belt is very important. The stretch limitation of the conveyor belts is also very important. The total stretch of the belt under maximum load should not exceed 1–2% of the belt length. The requirements of a bucket elevator's belt are similar to those of conveyor belt. Four types of belts are used for bucket elevators and belt conveyors; (1) duck, (2) batala, (3) stitched canvas and (4) solid woven cotton. The conveyor belt should be covered with natural or synthetic rubber to overcome wear during operation.

Majority of the conveyor and elevator belts are made up of folded-ply construction. The pull resistance of the belt depends upon the applied fibre and the number of plies. Some of the belts are made by building up layers of plies that are cut or woven to the width desired. Most conveyor belts consist of synthetic fibres like polyesters and polyamide and are built up with synthetic rubber or PVC. To increase tensile strength of belt, several layers of fibres are put together to build a carcass. Such carcass is able to withstand very high tensile forces with a minimum of stretch.

During continuous operation, elevator belts are susceptible to various mechanical stresses which may cause wear. The friction between drive pulley and the belt causes wear in the underside of the belt. The back falling product is caught and crushed between the belt and return pulley. Extra forces are also exerted on the belt by rigid buckets while passing over the head pulley. To overcome all these, protective cover should be provided on conveyor belt.

A belt is as strong as its weakest point, so a through connection of the belt ends is important. Several types of splicing are possible. The type of belt splice depends on the thickness of the belt and the severity of service. For belts of five ply thickness or less, the bolted clamp joint, the lap joint, or the butt-strap joint may be used (Fig. 6.15). On a lap joint splice, the lap extends a distance of three to five buckets and is secured by the same bolts that hold the buckets.

For large bucket elevators, the following running clearance should be taken into account.

1. Belt width should be 25 mm wider than the bucket width.
2. Pulley width should be 80 to 100 mm wider than the belt.
3. The inside casing width should be 150 to 200 mm wider than the belt.
4. Bucket lip clearance should be 100 to 130 mm.

For smaller bucket elevators, lower clearances than the above can be taken.

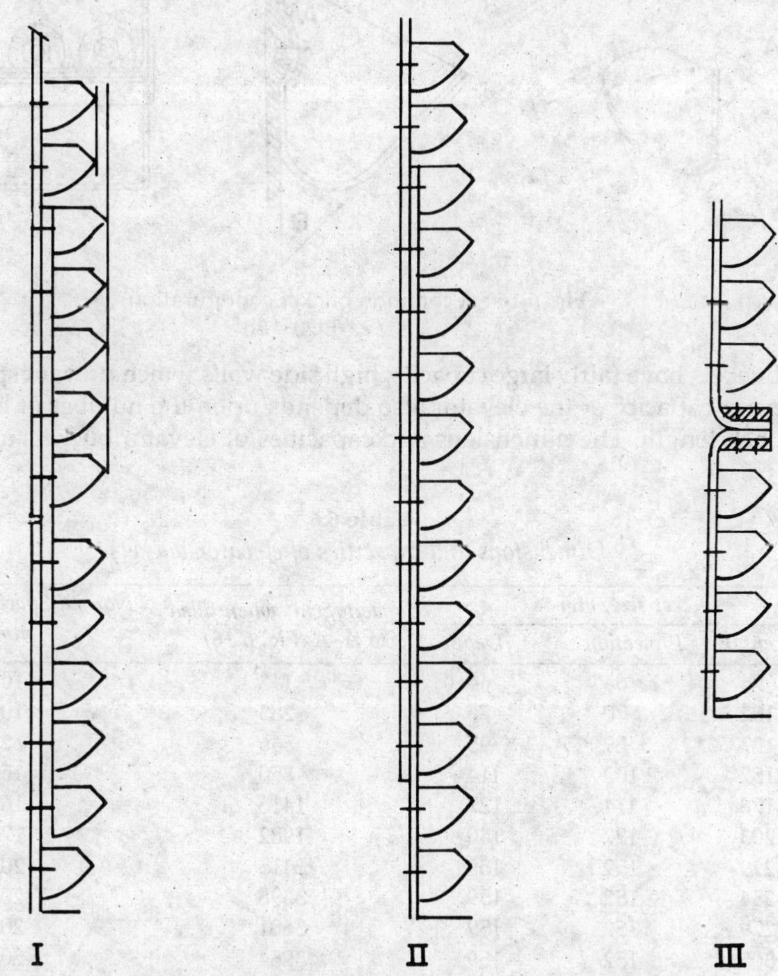

Fig. 6.15 : Belt splices
I. butt strap joint II. lap joint III. clamp joint

Buckets

As per the requirements, buckets are made of different materials and come in various shapes and sizes. The shape of the bucket is very important for filling and discharge. Digging in of buckets in the elevator boot and the centrifugal discharge at the elevator head influence the shape of buckets.

For centrifugal discharge the resultant of product weight and the centrifugal force should preferably be directed towards the lip of the bucket. The buckets should have a wide open mouth for digging and discharging the product. But at the same time the lip of bucket must not be too low and the bottom angle should not be too big. This will cause spillage of product. So there is a compromise between the above two factors. The common shape of bucket is shown in Fig. 6.16 The top angle is generally taken as 80° while the bottom angle is between 20–30°.

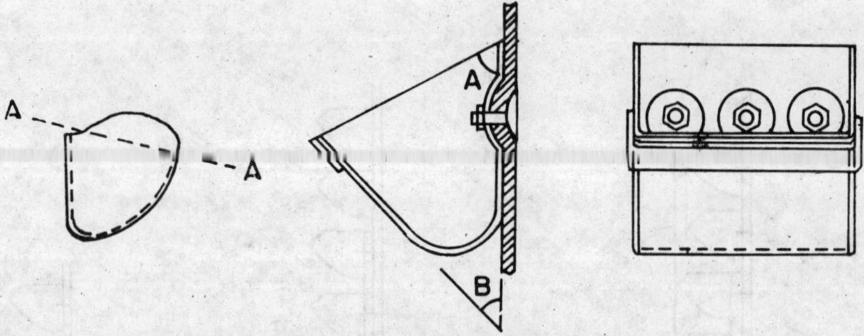

Fig. 6.16 : A common bucket configuration
$\angle A = 80^\circ$ $\angle B$ 20 – 30°

Such buckets have fairly large capacity, high side walls which prevent spilling. The conveying capacity of the elevator also depends upon the number of buckets per metre belt length. The dimensions and capacities of elevator buckets are given in Table 6.6.

Table 6.6
Dimensions and capacities of elevator buckets

Bucket size, mm			*Capacity, cm^3 when filled to A–A (Fig. 6.16)*	*Normal Spacing on belt, mm*
Length	*Projection*	*Depth*		
76	64	64	142	102
102	70	76	283	102
127	89	95	566	127
152	102	114	850	152
178	114	127	1416	165
203	127	140	1982	178
229	152	159	3115	203
254	152	159	3398	203
279	152	159	3681	203
305	152	159	3964	203
305	178	184	5380	229

Source : Wimberly, 1983

The capacity and discharge of each bucket is influenced by the previous bucket, hence the distance between two successive buckets is important. Therefore, compromise between the following factors is required, (1) the design and content of the buckets, (2) the shortest distance between successive buckets without any mutual influence, (3) for centrifugal discharge the appropriate belt speed and diameter of drive pulley. In general, the spacing would be from 2.0 to 3.0 times the projected width of bucket.

The buckets are mounted on the belt mainly by means of special bucket bolts. The flat disc head is pressed through holes in the elevator belt as shown in Fig. 6.16.

Drive mechanism

The drive mechanism of a bucket elevator is located near the elevator head. At the elevator head, the belt is turned around the drive pulley. Drive motor with gear box and couplings are mounted on a rigid and separate frame. For serving the elevator head section, the drive mechanism, a working platform is provided. Usually a ladder is provided for access to this platform.

The theoretical horsepower requirement for the bucket elevator can be calculated by the following equation.

$$hp = \frac{QHF}{4562} \qquad \ldots 6.11$$

where, Q = capacity of bucket elevator, kg/min

H = lift of elevator, m

F = factor; 1.5 for elevators loaded on the up side and 1.2 for elevators loaded on the bottom side

The theoretical horsepower should be increased 10–15% to provide for friction and power requirements for loading, power transmission and drive losses.

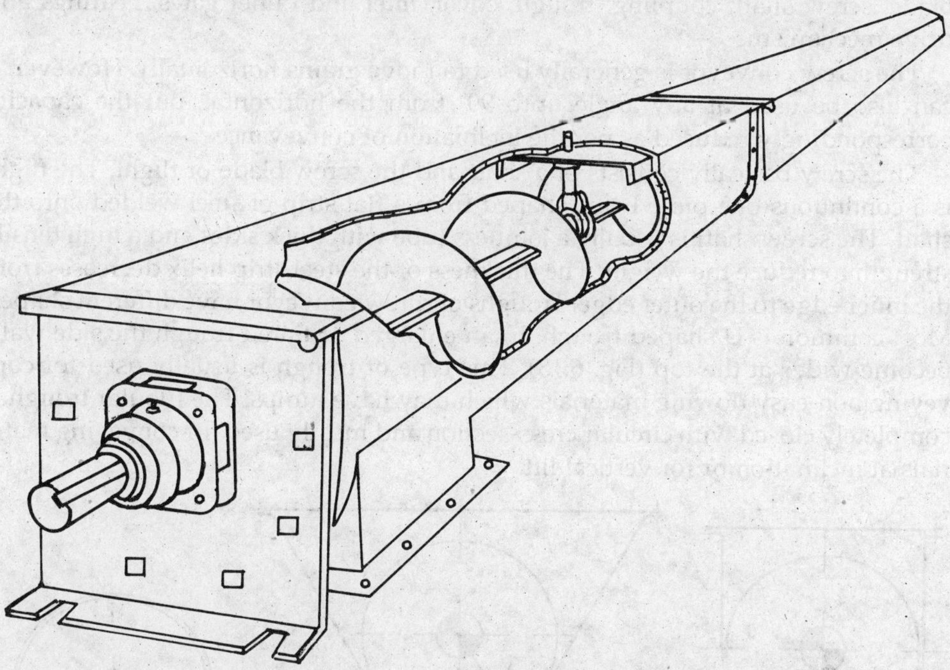

Fig. 6.17A : Diagram of a screw conveyor

Screw conveyor

The screw conveyor consists of a tubular or U-shaped trough in which a shaft with spiral screw revolves. The screw shaft is supported by end and hanger bearings. The rotation of screw pushes the grain along the trough. A typical screw conveyor is shown in Figs. 6.17 A & B. The screw conveyor is used in grain handling facilities, animal feed industries and other installations for conveying of products generally for short distances. Screw conveyor requires relatively high power and is more susceptible to wear than other types of conveyors. The pitch of a standard screw which is the distance from the centre of one thread to the centre of the next thread, is equal to its diameter. For example a 10 cm diameter screw has a pitch of 10 cm.

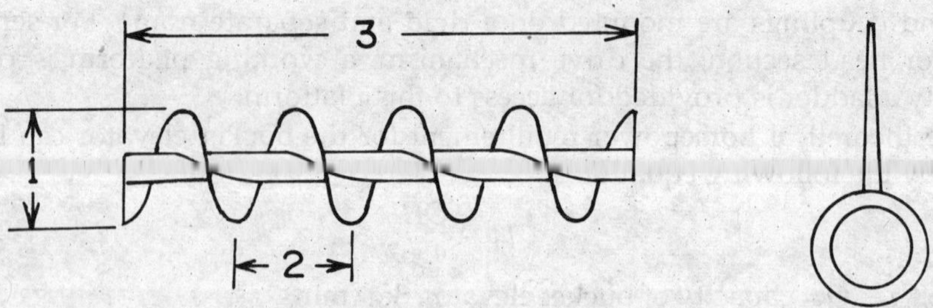

Fig. 6.17B : Screw conveyor details
1. screw diameter 2. pitch of screw 3. screw length

As the screw conveyor's driving mechanism is simpler, and no tensioning device is required, the initial cost of the conveyor is lower than any other conveyor with the same length and capacity. The main parts of a screw conveyor are, screw blade, screw shaft, coupling, trough, cover, inlet and outlet gates, bearings and drive mechanism.

The screw conveyor is generally used to move grains horizontally. However, it can also be used at any angle upto 90° from the horizontal, but the capacity correspondingly reduced as per the inclination of conveyance.

The screw basically consists of a shaft and the screw blade or flight. The flight is a continuous one piece helix, shaped from a flat strip of steel welded onto the shaft. The screw shaft is usually a jointless tube with thick sides and a high tensile strength to reduce the weight. The thickness of the steel strip helix decreases from the inner edge to the outer edge. Troughs of screw conveyor have different shapes. Most common is U shaped trough. In an enlarged or flared trough the side walls become wider at the top (Fig. 6.18). This type of trough is usually used for conveying non-easy flowing materials which may have lumps. The tubular trough is completely closed with circular cross-section and mostly used for conveying materials at inclination or for vertical lift.

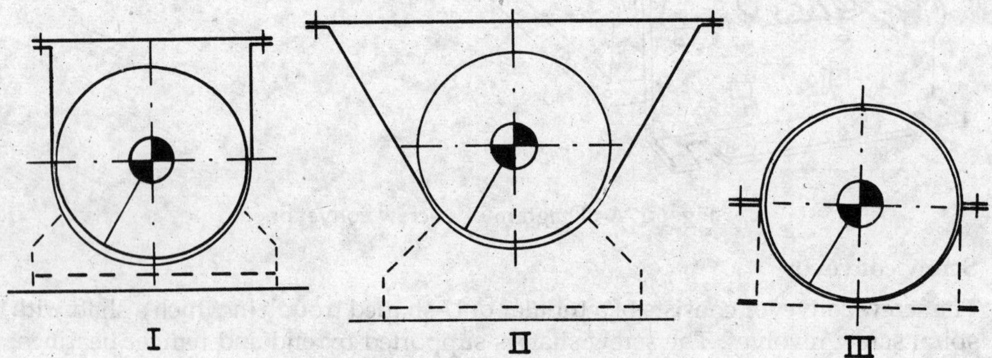

Fig. 6.18 : Various shapes of screw conveyor trough
I. U-trough II. flared trough III. tubular trough

For operational reasons, some gap is provided between the edge of the screw blade and the trough walls. Due to this gap it is not possible to completely empty the trough of a horizontal screw conveyor. If the screw conveyor is used to convey different materials, mixing of products is possible. Also when the kernels are pressed between the screw edge and trough walls, they can be damaged. During conveyance the kernels are also subjected to continuous friction with the trough walls. Screw conveyor may be designed for clockwise or counterclockwise rotation. The change in direction of rotation does not affect the capacity.

The capacity of screw conveyor is influenced by the screw diameter, inclination of the screw blade, speed of the blade, shaft diameter and cross-section of loading. The theoretical conveyance capacity of the screw conveyor can be given by the following equation.

$$\text{Capacity } Q,\ \text{m}^3/\text{hr} = 47.2\ (D^2 - d^2) \times p \times n \qquad \ldots 6.12$$

where, D = screw diameter, m

d = shaft diameter, m

p = pitch, m

n = rpm

The theoretical capacity is more than the actual capacity because of screw housing clearance and the loading factor (Fig. 6.19).

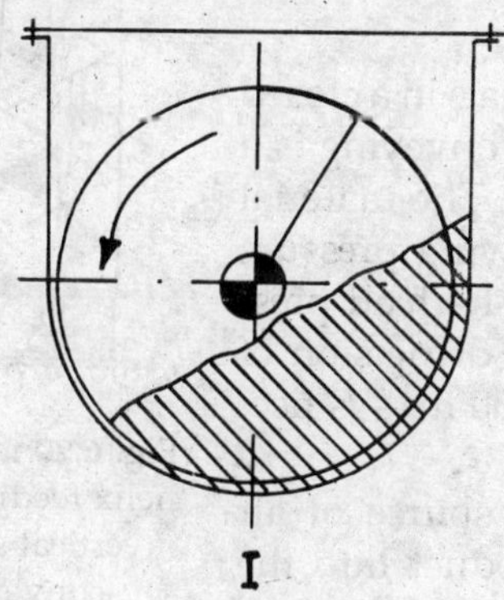

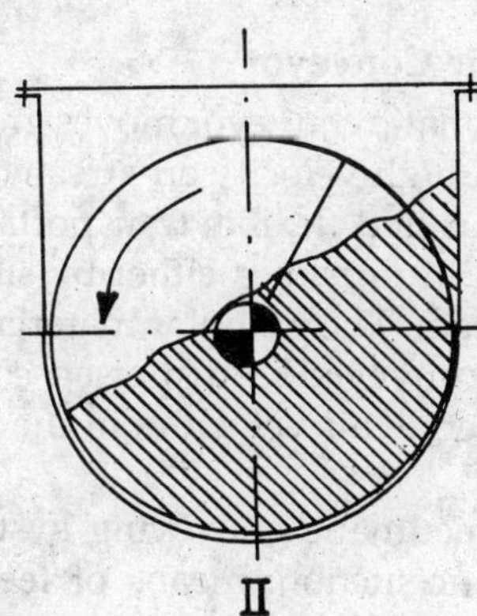

Fig. 6.19 : Screw conveyor loading factor
I. 30% loading II. 55% loading

The power requirement of screw conveyors for horizontal operation may be determined by the following equation.

$$\text{Horse power} = \frac{Q\,L\,W\,F}{4560} \qquad \ldots 6.13$$

where, Q = conveyor capacity, m^3/hr

L = conveyor length, m

W = bulk material weight, kg/m^3

F = material factor (for paddy 0.4)

If the calculated horsepower is less than 1.0, double the value.

horsepower	=	1 to 2	multiply the value by 1.5
horsepower	=	2 to 4	multiply the value by 1.25
horsepower	=	4 to 5	multiply the value by 1.1

For horsepower values of more than 5, no correction is required.

Screw conveyors can be operated in an inclined position. In this case, the material will be conveyed upward. The capacity of inclined screw conveyor decreases than the horizontal operation. A screw conveyor inclined 15 degrees will carry about 75% of the rated horizontal capacity. At an inclination of 25 degrees, it will carry about 50% of the rated horizontal capacity.

Loading and discharge of a screw conveyor can take place at several places. The loading of a horizontal or inclined screw conveyor is generally accomplished through a supply chute connected at the flat trough cover. A special feeding device as shown in Fig. 6.20 is provided for loading of material in a vertical and tubular screw conveyor.

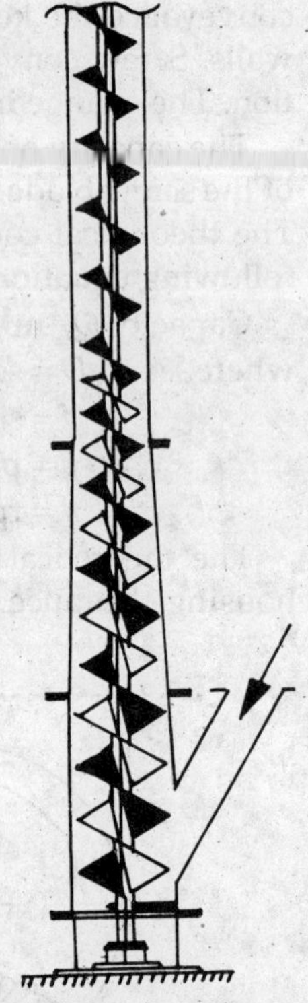

Fig. 6.20 : Double helix feeding in a vertical screw conveyor

The product supply should be regular to avoid overfilling and congestion in a screw conveyor. To regularise product flow an adjustable opening at the feeding point should be provided. The product can be discharged either at the end of the screw or the intermediate discharge can be achieved through an opening in the bottom of the trough.

Pneumatic Conveyor

The pneumatic conveyor moves granular materials in a closed duct by a high velocity air stream. Pneumatic conveying is a continuous and flexible transportation method. The material is carried in pipelines either by suction or blowing pressure of air stream. The granular materials because of high air pressure are conveyed in dispersed condition. For dispersion of bulk material, air velocities in the range of 15–30 m/s is necessary.

The pneumatic conveying system needs a source of air blowing or suction, means of feeding the product into the conveyor, ducts and a cyclone or receiving hopper for collection of product. There are three basic systems of pneumatic conveying. These are pressure or blowing system, suction or vacuum system, and combined push-pull or suck-blow system.

In blowing or positive pressure systems, the product is conveyed by using air pressures greater than the atmospheric pressure. This system consists of a fan or blower, an air-lock feeder for introducing the product into the system, ducts and suitable air and product separating device. The product is fed into the pneumatic conveying system from the bottom of a hopper.

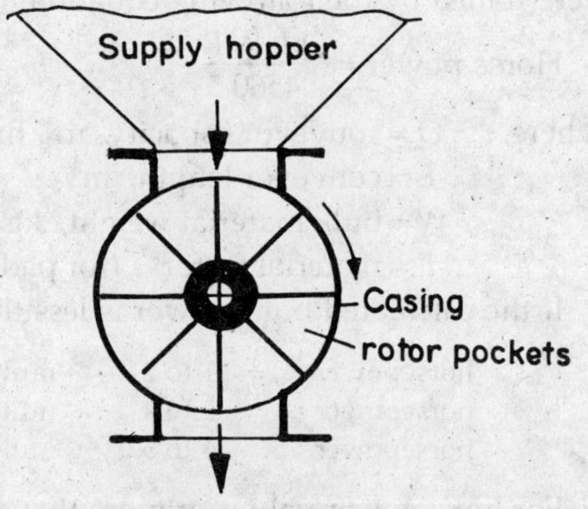

Fig. 6.21 : Rotary feeder for pneumatic conveyor

The feeder should be able to feed product at a specified rate in a pneumatic conveying system from the supply hopper at one pressure to the conveying pipeline at another pressure. The most common type of feeder is rotary feeder (Fig. 6.21). It consists of a bladed rotor with pockets at the inlet port. When the rotor moves, the products are dropped to outlet port and to the conveying pipeline. The advantage of rotary feeder is that it meters the supply of products into the conveying pipeline. It also affects the air lock which is necessary for the operation of the system.

The selection of air mover is the most important aspect of the design of a pneumatic conveying system. In design, the two factors, (1) supply air pressure and (2) the volumetric flow rate of air should be considered.

The supply pressure's value depends upon the working pressure drop required for the length of conveying line and across the filter. The magnitude of air pressure depends on the conveying length and the properties of the product to be conveyed. The volumetric flow rate of air depends on the necessary air velocity and pipe or duct size used in the system. In pneumatic conveying systems, fans and blowers with high volumetric flow rates and low pressures to positive displacement compressors producing high pressures are used.

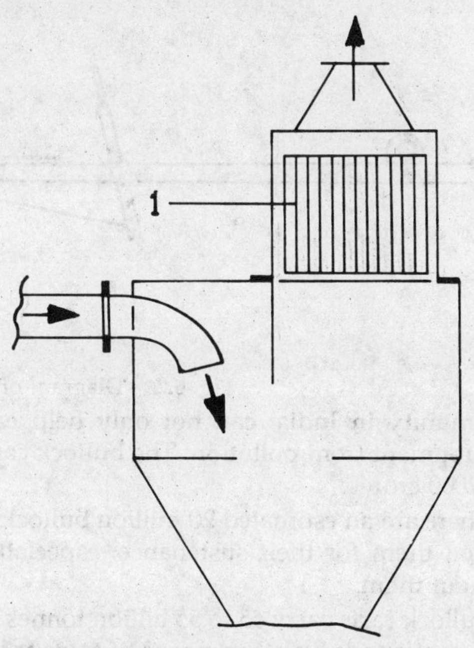

Fig. 6.22 : Separation of product particles from air by means of a fabric filter

For separation of product particles from air, air-product separators are used. Cyclones are mostly used to collect the particles. Cyclone is a device which removes the bulk of the product particle from the conveying air stream by centrifugal force. In some cyclone, a fabric filter is attached to remove residual dust and fine product particles from the air stream (Fig. 6.22). The conveying air is first passed through the cyclone and then it goes to the fabric filter for secondary separation of finer particles.

Limitations of Pneumatic Conveying

1. Erosion of solid surfaces and equipment surfaces by solid particles with conveying air stream. The rate of erosion of solid surfaces increased remarkably with conveying of abrasive products.
2. In case of bends or misaligned sections, the erosion problem becomes severe. In industrial installations, the erosion of duct system poses major problem in the operation of pneumatic conveying system.
3. In a pneumatic conveying system, chances of repeated impacts between the particles and the solid surfaces are high. Due to such impacts, product degradation results, because of this changes take place in the product size distribution and consequently the market value diminishes.

Rural Transportation

The various animals used for field operations or transportation purposes are bullocks, horses, mules, camels and buffaloes. In India about 50% of farm power is provided by bullocks. According to various expert estimates, about 10 million kilometer freight is carried by carts all over the country and nearly 70% of the arrivals of agricultural products at mandies or markets, are carried by bullock carts. As a result, there is heavy investment in animal drawn transportation system which is estimated to be around Rs. 3000 crores. The bullock cart transportation system is an extremely important part of India's rural transportation. The clumsy looking, slow moving bullock carts, (Fig. 6.23), the traditional lifeline of the rural

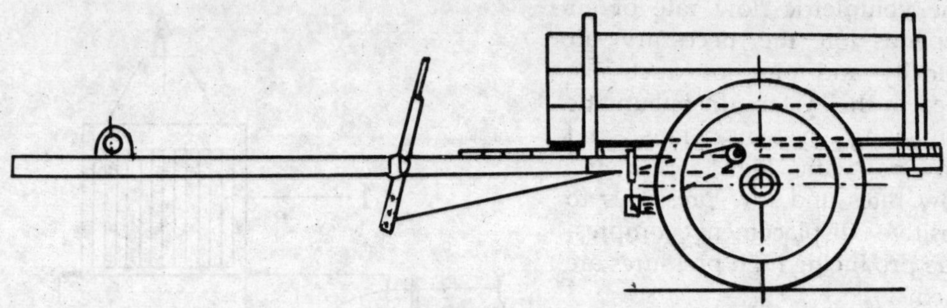

Fig. 6.23 : Diagram of a bullock cart

community in India, can not only help ease the fuel crisis but also save the environment from pollution. The bullock carts still generate an annual earning of Rs. 2000 crores.

There are an estimated 20 million bullock carts with 25 million people depending on them for their sustenance, especially the farmers, who simply can't do without them.

Bullock carts carry 45 to 55 billion tonnes of goods annually against 200 million tonnes of goods by other means of transport in other metropolitan cities. Bullock cart is the cheapest form of transport, but also the only form of transport available in some remote corners of the country. Therefore, improvement in designs of bullock carts can enhance the earnings. The design aspects of bullock cart's components are discussed below.

Bullock cart's wheel : Most of the bullock carts have wooden wheel (Fig. 6.24) constructed by the hub, the arms or spokes and falleoes separately and finally assembled together to form wheel. There are two types of wheel namely arm type wheels and spoke type wheels. Rigid wheels while moving on a deformable surface such as soft soil sinks into the soil after overcoming the bearing resistance as well as the resistance offered by the soil underneath the wheel. Thus the total rolling resistance (R) experienced by the wheels may be considered as contributed by axle bearing friction (R_b) and soil related factors (R_s).

$$R = R_b + R_s \qquad \ldots 6.14$$

The resistance caused by the bearing friction can be calculated by the following relationship

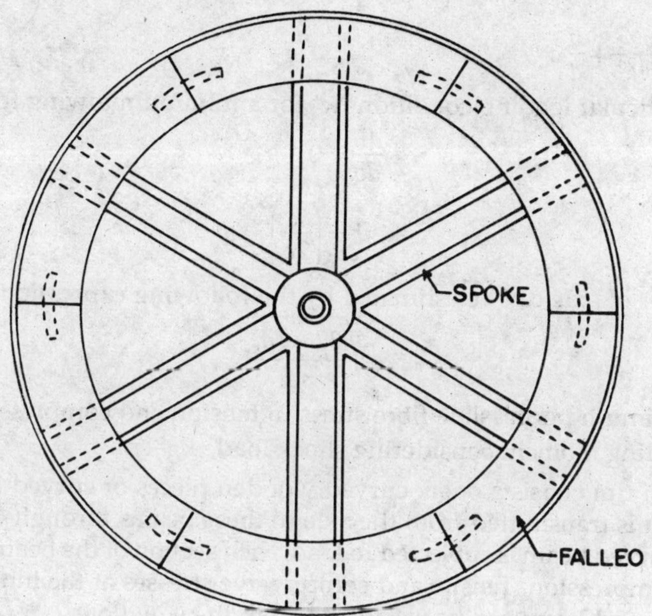

Fig. 6.24 : Diagram of a bullock cart wheel

$$R_b = W_b f \frac{D_a}{D_w} \qquad \ldots 6.15$$

where, W_b = load on the bearing

$f = \frac{2T}{W_b D_a}$, a non–dimensional coefficient

T = resistance torque in the bearing

D_a = diameter of the axle

D_w = diameter of the wheel

The towing force or pull required by a pair of bullocks under any particular loading condition depends on wheel diameter, rim width and friction at the hub and axle. It can be given as

$$F = (w, d, b, s_i) \qquad \ldots 6.16$$

where, F = towing force

d = diameter of wheel

w = weight of wheel

b = width of wheel

s_i = initial soil condition

$$\frac{Fd}{b} = K\left(\frac{w}{b}\right)^n \qquad \ldots 6.17$$

The graphical representation of above equation yields,

$$\frac{Fd}{b} = K\left(\frac{w}{b}\right) + c$$

Experimentally, it is found that the value of 'n' is generally 3/2 in varied conditions.

or, $F = \frac{k}{d}\frac{w^{3/2}}{b^{1/2}} + \frac{b}{d}c$

Under a particular loading condition 'w' for a minimum towing force,

$$\frac{dF}{db} = 0$$

or, $$b = \left(\frac{k}{2c}\right)^{2/3} w \quad \text{...6.18}$$

The diameter of axle can be estimated by the following expression,

$$D_a = \sqrt[3]{\frac{32 \times M_s}{\pi f}} \quad \text{...6.19}$$

where, f = maximum permissible fibre stress in tension and compression, kg/cm²

M_s = bending moment considering shock load, kg–cm

Falleo : Wooden rim consists of six curved wooden pieces or curved beams is the falleo. The load is transmitted from the axle to these beams through 4 spokes at a time when a curved beam is subjected to load, each section of the beam under goes tension and compression. Tensile and compressive stresses at the inner and outer fibres of the curved beam can be calculated from the equations.

$$S_{i\,(max)} = \frac{M_b \cdot h_i}{A \cdot e \cdot r_i} \quad \text{...6.20}$$

and $$S_{o\,(max)} = \frac{M_b\, h_o}{A \cdot e \cdot r_o} \quad \text{...6.21}$$

where, S_i = inner fibre stress, kg/cm²

S_o = outer fibre stress, kg/cm²

M_b = bending moment, kg–cm

h_i = distance between the neutral axis and the innermost fibre of the beam, cm

h_o = distance between the neutral axis and the outermost fibre of the beam, cm

A = cross–sectional area of beam, cm²

e = distance between neutral axis and the centroidal axis of the beam, cm

r_i = inner radius of curved beam, cm

r_o = outer radius of curved beam, cm

$$M_b = P_f \cdot R$$

P_f = load transmitted to one falleo, kg

R = radius of centroidal axis of falleo, cm

Under the no-load condition, the neutral axis and centroidal axis are coinciding each other, but when Subjected to load, the neutral axis deviates from the centroidal axis. Radius of neutral axis for loaded condition can be calculated from the following equation.

$$r_n = \frac{t}{ln\left(\frac{r_o}{r_i}\right)} \quad \text{...6.22}$$

where r_n = neutral axis of falleo, cm

t = thickness of the curved falleo, cm

r_o = outer radius of falleo, cm

r_i = inner radius of falleo, cm

$e = r_n - R$

Spoke : The bending stress in spoke can be calculated by the following equation,

$$S_b = \frac{M_t\ (D - d_o)}{n \cdot Z \cdot D} \qquad ...6.23$$

where, S_b = bending stress in spoke, kg/cm²

M_t = transmitted torque, kg–cm

D = rim diameter, cm

n = number of spokes

Z = sectional modulus of the spoke, cm

When a spoke is subjected to an axial load, the transmitted torque can be calculated from following equation,

$M_t = P_s \cdot r_i$

where, P_s = load transmitted through one spoke, kg

r_i = inner radius of wooden rim, cm

Hub : The inner diameter of hub of a wooden wheel with bush bearing is calculated by the following equation.

$$d_i = d_a + 2t + 2c \qquad ...6.24$$

where, d_i = inner diameter of hub, cm

d_a = diameter of axle, cm

t = thickness of bush bearing, cm

c = clearance between the bearing and the axle, cm

The outer diameter of hub:-

$$d_o = \frac{(B_s + 3.5)\ n}{\pi} \qquad ...6.25$$

where, B_s = peripheral width of spoke, cm

n = number of spoke

Yoke : The design of yoke should be based on the comfort and low strain on the part of bullocks.

(*i*) Design of yoke should facilitate the use of hump muscle for pulling the cart.

(*ii*) The contact area between the yoke and the hump muscle should be as large as possible

(*iii*) The bearing pressure over the contact area should be well distributed. There should not be concentration of pressure at the hump.

(*iv*) Yoke should be designed such that the curved contact area should fit the hump contour of bullock.

Soft rubber padding may be used over the curved area.

Design criteria for cart frame : The cart body area in front of the axle should be approximately equal to that in the rear of the axle so as to get stability during hauling operation.

The cart frame design should suit the size of bullocks, the pay load to be carried and the kind of road.

Bullock cart body : The total length of the cart is fixed on the basis of the diameter of the bullock-cart wheel. The total length of the bullock cart should be around three times the diameter of the wheel. The overall width of the cart may be limited to 185 cm. The width may be based on the width of streets existing in the rural areas.

Transportation of grains

Transportation of grains is an important activity at least for two reasons. First it presents logistic problems to the farmers to ensure that the grain is taken out immediately after bagging to minimise losses due to deterioration in grain quality. It may happen especially during the off-season harvest. Secondly, transportation costs are important components of the total cost of grain production.

Grain crops after harvesting and threshing are bagged. Then, these may be transported to the procurement centres, directly to the milling complexes or to any licensed or private buyers. Transportation of grains are mainly carried by bullock or animal carts, trucks, tractor-trolleys etc. At the procurement centres, selling and buying of grains are concluded. After buying, it becomes the responsibility of the buyers to store the grains and finally transport the grains to the milling complexes.

Many a times in the transportation system, vehicles are required to queue at several places. It may happen at the procurement centres while the grains are graded or transaction concluded. At the milling complexes, vehicles are required to queue for weighing purposes or for delivery at the receiving point. Little is known about the factors affecting queuing costs.

Demand for transportation services in shipping agricultural products has remarkably increased. Increase in transportation has been mainly contributed by inter-regional agricultural products, growth in population and increase in demand for food.

Exercise

1. The capacity of a toughened belt conveyor is 55 m^3/hr. With the help of Table 6.1 calculate the belt width and belt speed. If the length of belt is 50 m, estimate the horse power required for operation of the belt conveyor. Values of A and B may be taken from Table 6.2.
2. The diameter of discharge wheel of a bucker elevator is 150 mm and the projection of the bucket is 114 mm. For optimum centrifugal discharge of material, calculate the revolution per minute of operation of the elevator. Also calculate the elevator's capacity in tonnes per hour. Take material density as 740 kg/m^3.

Bibliography

Bisen, HS, 1977. Design, fabrication and testing of an improved bullock cart. Unpublished M. Tech. Thesis, IIT, Kharagpur.

Bolz, HA and GE Hgemann (Eds.) 1958. Material handling handbook. The Ronald Press Company, New York.

Boumans, G, 1985. Grain handling and storage. Elsevier Science Publishers, Amsterdam.

Deshpande, SD and TP Ojha, 1984. Theory and design of animal drawn vehicles. Technical Bulletin No. CIAE/85/46, CIAE, Bhopal.

Henderson, SM and RL Perry, 1976. Agricultural process engineering. AVI Publishing Company, Westport, Connecticut.

Michael, AM and TP Ojha, 1985. Principles of agricultural engineering. Jain Brothers, New Delhi.

Mishra, N, SH Nagabhushaniah, GC Panda and PR Das, 1979. Rolling friction at varying conditions of wheel on road surfaces. Technical papers of the National Seminar on improved bullock cart transport for rural roads in India. Dept of Applied Mechanics and Hydraulics, Regional Eng. College, Rourkela : C81–C86.

New man, CJE, 1987. Storage and ancillary equipment for bulk handling of grain in "Bulk handling and storage of grain in the humid tropics" (Eds Champ and Hinghely) ACICR Proceedings No. 22:69–100.

Ojha, TP and AC Pandaya, 1970. Performance of full size wheels predicted by model wheels. I.E. Journal 51(2) : 33–35.

Wimberley, JE, 1983. Paddy rice post harvest industry in developing countries. IRRI Los Banos, Philippines.

Chapter 7

Elementary Food Engineering

Food engineering operations are employed in food industries for production of good quality palatable and stable foods. These scientific approaches are safe and reliable and can control the spoilage of packaged foods. The knowledge of food engineering, specially the heat treatment and heat transfer have provided necessary impetus to the development of large food and canning industries. In this chapter, some of the basic food engineering operations have been described.

Thermal treatments

The thermal treatment serves to reduce the concentration of micro organisms in the foods when they are preserved by heat. The treatment may also inactivate the existing enzymes in foods.

The heat treatment process can be divided into three categories, (1) heating the foods below 100°C temperature, (2) heating the foods to a temperature of 100°C, and (3) heating the foods above 100°C temperature.

Pasteurisation is usually done below 100°C temperature to kill all pathogenic organisms and also to kill some of the spoilage organisms present in the food. Pasteurisation of milk is a common practice to preserve it. A pasteurisation process can destroy pathogens in the milk. The pasteurisation process does not effect the organoleptic properties of milk.

Pasteurisation process was first developed in 1864 by Pasteur to prevent abnormal fermentations in wine. In present days, pasteurisation is applied to a wide range of foods, like; milk, canned and bottled fruits, cream, ice-cream, fruit juices, pickles, beers etc. Pasteurisation of milk is performed by heating every particle of milk to at least 61°C and holding it at that temperature for 30 minutes, or heating the milk to 71°C and holding it at 71°C for at least 15 seconds. Pasteurisation destroys most of the bacteria present in milk thus improves the keeping quality of milk. The taste of milk is not affected by pasteurisation.

Since the temperature of pasteurisation is below 100°C, food preserved by this method is subjected to less damage than high temperature sterilisation. Owing to this, after pasteurisation the foods may be left with residual enzyme activity. This may cause spoilage of food during storage.

There are two general types of pasteurisers, (1) the batch type, and (2) the continuous type or high temperature short time.

Batch Pasteurisation : Bulk foods such as milk and fruit juices may be pasteurised in a barrel type heat exchanger to 61°C for 30 minutes. The barrel also has stirring arrangement and may be used for both heating with steam or hot water and cooling by chilled water. Foods sealed in containers may be pasteurised in batches. In case of milk pasteurisation in batch pasteurisers, after the milk is removed, the barrel should be given a pre-rinse with 49°C water to melt the milk fat. Then, gives it is brushed and given a hot water rinse.

Continuous Pasteurisation : Bulk liquids are generally handled by high temperature short time (HTST) pasteurisers. Almost all the modern dairies are equipped with continuous pasteurisers. In this method, the foods are pasteurised through plate heat exchangers which usually comprise four stages, (1) pre heating or regeneration, (2) heating, (3) holding, and (4) cooling. In plate type pasteurisers, milk is heated to 71°C and is maintained at that temperature for 15 seconds in holding tubes.

In a continuous pasteuriser, milk at 5°C is fed into the regeneration unit by a pump. The cold milk is heated in the regeneration section by outgoing pasteurised milk. In this section, the cold incoming milk is heated to 55°C from 5°C by pasteurised milk. From the regeneration unit the warm milk is pumped to plate heater to heat the milk at pasteurisation temperature of 71°C. The milk is then passed to the holding tube. Then, the milk comes to regeneration unit where it is cooled to 18°C by incoming cold milk at 5°C. The pasteurised milk is then pumped directly into the cooling section where it is cooled to 4°C for final packaging (Fig. 7.1). Generally 1°C ice-water or brine is circulated in the cooling section.

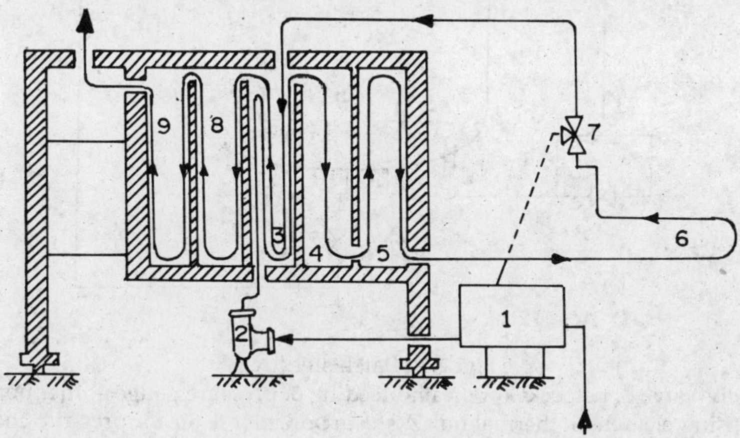

Fig. 7.1 : High temperature short time continuous milk pasteuriser
1. balance tank 2. milk pump 3. regenerator 4. alter 5. heater
6. holding pipe 7. flow diversion 8. water cooler 9. chilled water cooler

Sterilisation

Sterilisation of foods is carried at temperatures above 100°C. At such temperatures, pressures above atmosphere are generated. The two common methods of sterilisation are, (1) sterilising in containers, and (2) sterilisation of food before placing in the container.

Sterilisation of food in containers can be done by following methods.

1. Indirect heating by saturated steam
2. Forced convection of hot air
3. By direct flame contact

Sterilisation by saturated steam is most common method. Low-acid foods are sterilised at temperatures above 100°C with the help of pressurised sterilisers. With steam steriliser, it is necessary to ensure adequate venting of air from the retort and from the container surfaces.

Batch sterilisers

In batch and non-agitating type of sterilisers the labour requirement is high but still they are quite popular. The batch sterilisers (Fig. 7.2) are flexible and can accommodate containers of different sizes. The batch sterilisers are also suitable

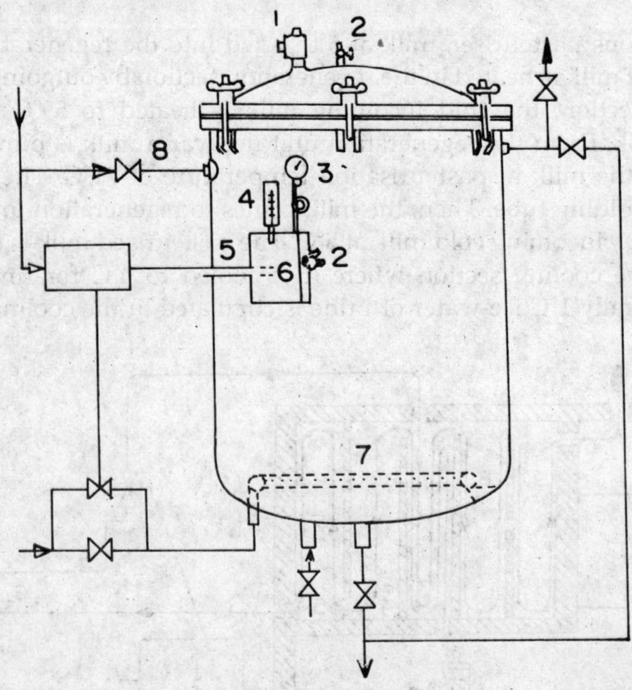

Fig. 7.2 : Batch steriliser

1. safety valve 2. petcocks for steam bleeding 3. pressure gauge 4. thermometer 5. sensing element 6. thermal box 7. steam spreader 8. air for pressure cooling

for different types of processes. But in comparison to continuous sterilisers the steam and water consumption are higher. Horizontal and side loading type batch sterilisers occupy more space than top loading and vertical types. But the loading and discharge of containers are easier in case of horizontal types. In batch sterilisers the steam pressures are in the range of 350–415 KN/m^2.

Continuous pressure steriliser

There are three types of continuous sterilisers.

1. The continuous pressure cooker-cooler

2. The continuous rotary sterilisers
3. The hydrostatic steriliser

The continuous pressure cooker-cooler sterilisers are of non- agitating types. The containers which are to be sterilised are carried on a roller track or a chain conveyer through the preheating, sterilising and cooling sections.

In the rotary sterilisers there are three cylinders. The inner walls of these cylinders have helical track (Fig. 7.3). The containers are carried in the track by the flanges at the periphery of cylinder. The helical track causes agitation of containers by the combined rolling and sliding action. Rotary sterilisers have large capacity.

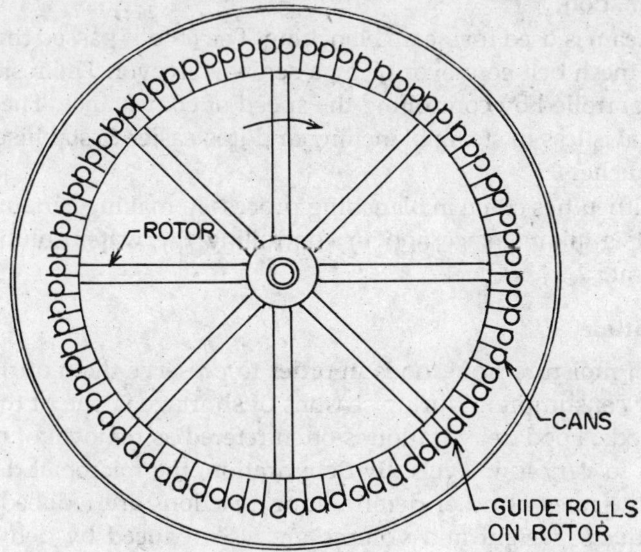

Fig. 7.3 : Inside view of a continuous rotary steriliser

In the hydrostatic sterilisers, hot water columns are used to generate pressure. The temperature of water in the preheater zone and air cooling section is between 107 to 118°C. The water seal temperatures in the steam chamber are between 114 to 127°C. The containers are carried on a twin roller-chain conveyor operating at slow speed of around 2 m/min. The containers are usually carried with their long horizontal axes. This helps the convective heat transfer within the container.

Blanching

Blanching is an important heat process in the preparation of vegetables and fruits before dehydration and canning/packaging. Blanching is done to inactivate enzymes and to destroy the peroxides. Blanching is the process of heating the food rapidly to a predetermined temperature, holding it at that temperature for a definite period and then either cooling the material or passing it to subsequent processing without delay.

Besides destroying enzymes, following changes in food may also be brought about by blanching.

1. Cleaning of raw material and reduction in bacterial load.
2. Softening and shrinking of foods which help to fill the containers.
3. The texture of dehydrated foods is improved by blanching.

4. Heat sensible vitamins are lost.
5. Leaching of water soluble nutrients.

Blanching can be done by two methods, (1) immersion blanching, with the help of hot water, and (2) steam blanching. Immersion blanching is carried out by two ways, first, passing the foods at a controlled rate through a perforated drum rotating in a tank of water. The hot water is thermostatically controlled at temperatures of 75–100°C. In the other method, food is suspended in water and heated to blanching temperature. Remarkable loss of water soluble materials takes place in immersion blanching process. It is also necessary to maintain plant hygiene to avoid contamination.

Saturated steam is used for steam blanching. The food is passed through a steam chamber on a mesh belt conveyor or by a screw conveyor. The residence time of blanching is controlled by controlling the speed of conveyance. The loss of water soluble material is less in steam blanching and it is easier to stabilise the blancher than water blanchers.

The main difficulties come in blanching process in making certain that uniform heat-treatment is taking place and in controlling the water soluble losses and effluent disposal.

Food Dehydration

The removal of moisture from foods in order to preserve them during seasons of abundance for consumption during seasons of shortage is one of the oldest preservation methods. Food dehydration is often refered as removal of moisture from food products to very low levels. By dehydration, the microbial deterioration is arrested and the rates of other deteriorating reactions are reduced significantly. The food products' weight and volume are also reduced by dehydration. This results in improvement of efficiency of product transportation and storage.

Dehydration of fruits and vegetables during seasons of abundance is a common practice for their preservation. Dehydration of horticultural produce represents one of the most effective low cost methods of food preservation. One of the most important parameters in food dehydration is the equilibrium condition of food that establishes a limit to the process.

The components of foods include proteins, fats, carbohydrates, vitamins, enzymes and inorganic salts. The water present in food products is not in pure form. It may be in the form of solutions of solids, gel or bound in numerous ways with the solid material.

During the dehydration process, movement of soluble solids takes place. This movement may be restricted by cell walls which act as semi-permeable membrane. Due to shrinkage of products this can cause solids migration. Movement of solids towards centre of the food pieces may also happen. Animal and vegetable products show some degree of shrinkage upon dehydration. It has been reported that the bulk density and porosity of dried vegetable pieces depend to a large extent on the drying conditions.

It has been reported by many investigators that during the dehydration of food products a hard impermeable layer is formed at the surface. This impermeable layer reduces the drying rate. This phenomenon is known as "case hardening" and may be a result of migration of soluble solids to the surface and surface's higher temperature.

Drying time calculation

For calculation or prediction of constant rate drying period, the following general equation holds good.

$$\text{Constant rate drying, } R_c = \frac{dm}{dt} = \frac{m_0 - m_c}{t_c} \qquad \text{...7.1}$$

where, m_0 = initial moisture content, kg water/kg dry solids

m_c = critical moisture content, kg water/kg dry solids

t_c = time for constant rate drying

During falling rate drying, the following analysis holds good.

$$-\frac{dm}{dt} = \frac{R_c}{m_c}(m) \qquad \text{...7.2}$$

or,

$$\frac{m_c}{R_c}\int_{m_c}^{m}\frac{dm}{m} = \int_{t_c}^{t} dt \qquad \text{...7.3}$$

The limits of integration in dehydration process would be between the critical moisture content m_c and the predesired final moisture content m.

Thus

$$t - t_c = \frac{m_c}{R_c}\, ln\left(\frac{m_c}{m}\right) \qquad \text{...7.4}$$

The falling rate drying time can be given by,

$$t_F = \frac{m_c}{R_c}\, ln\left(\frac{m_c}{m}\right) \qquad \text{...7.5}$$

The total drying time is given by the following expression.

$$t = t_c + t_F = \frac{m_0 - m_c}{R_c} + \frac{m_c}{R_c}\, ln\left(\frac{m_c}{m}\right) \qquad \text{...7.6}$$

The equation 7.6 do not represent the complete dehydration time as there may be more than one falling rate period in case of some food products.

Problem : Cut and blanched pieces of cauliflower are dehydrated in a cabinet dryer. The initial moisture content of cauliflower was 80% (wb) and it is to be dried to 4% moisture content (wb). The critical moisture content is 25% (wb). Constant rate drying continues for 6 minutes, estimate the total drying time for the product.

Solution :

$$\text{The initial moisture content, } m_0 = \frac{0.8}{0.2} = 4.0\,\frac{\text{kg water}}{\text{kg solids}}$$

$$\text{Critical moisture content, } m_c = \frac{0.25}{0.75} = 0.333\,\frac{\text{kg water}}{\text{kg solids}}$$

$$\text{Final moisture content, } m = \frac{0.04}{0.96} = 0.0416\,\frac{\text{kg water}}{\text{kg solids}}$$

Constant rate drying period = 6 min.

$$R_c = \frac{m_0 - m_c}{t_c} = \frac{4.0 - 0.333}{6}$$

$$= 0.611 \text{ kg water/kg solids/min}$$

$$\text{drying time, } t = \frac{m_0 - m_c}{R_s} + \frac{m_c}{R_c}\ ln\left(\frac{m_c}{m}\right)$$
$$= \frac{3.667}{0.611} + \frac{0.333}{0.611}\ ln\left(\frac{0.333}{0.0416}\right)$$
$$= 6.00 + 0.545\ \ ln\ 8.0$$
$$= 7.133 \text{ minutes}$$

Dehydration Equipment

The dehydration methods and equipment can be grouped in the following four categories.

1. *Dehydration by heated air* : In this method, the food product comes in contact with a moving stream of air and the transfer of heat takes place by convection.

2. *Dehydration by direct contact with a heated surface* : In this method, dehydration takes place when the food product comes in contact with a heated surface. The transfer of heat is due to conduction.

3. Dehydration of food products by a radiating or dielectric source.

4. *Dehydration of food by freeze drying* : In this method, the food products are frozen and then sublimated to vapour. The heat transfer takes place under low pressure condition.

Tray or compartment dryers

These types of drying equipment provide means of dehydration of horticultural crops in pieces. The system has trays or similar product holders to expose the product to heated air in an enclose space.

It consists of an insulated cabinet (Fig. 3.36) in which trays are placed one over other with gap in between. The heated air can be circulated inside the cabinet with a fan either horizontally between the trays of food product or vertically through the trays and food. Air movement over the product surface is at relatively high velocities to achieve efficient heat and mass transfer. In cross-flow systems air velocities of 2–5 m/s are used. The chief advantage of tray or compartment dryers are, (1) relatively cheap to build and maintain, and (2) flexibility to accommodate various food products. With slight variation in the tray dryer a vacuum can be incorporated within the chamber. The vacuum makes the possibility of maintaining lowest possible pressure in the space around the food product. By reducing the pressure, the temperature for moisture evaporation can be brought down. This results in product quality improvement. The disadvantages of the tray dryer are that these are batch system and non-uniform drying of product at different locations within the chamber can take place. To overcome non-uniform drying and to improve uniformity of drying, rotation of the product trays is the normal practice.

Drum dryer

The drum dryers are consist of one or more hollow metal cylinders. Based on the number and mode of operation, the drum dryers are classified as single drum, double drum and twin drum as shown in Fig. 7.4. The drums are hollow and revolve on horizontal axes. These drums are heated by circulating steam, hot water or other high temperature heat transfer liquid through the drums. The drums are

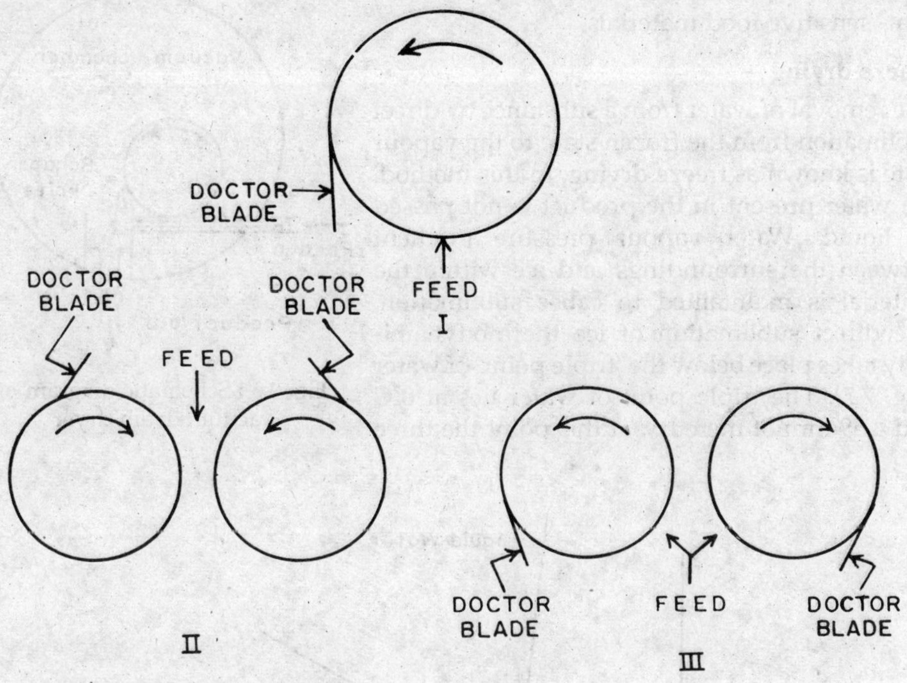

Fig. 7.4 : Various types of drum dryers
I. single drum II. double drum III. twin drum

usually rotated with the help of variable speed motors or other power source. A film of the wet material of uniform thickness is applied to the drum surface. As the drum rotates, the wet material get dried in contact with the hot drum surface. The dried material is removed from the drum surface by a scraping device as shown in Fig. 7.5. The scraper should touch the drum surface in uniform manner and should be sharp enough to take out the dried product from the drum surface.

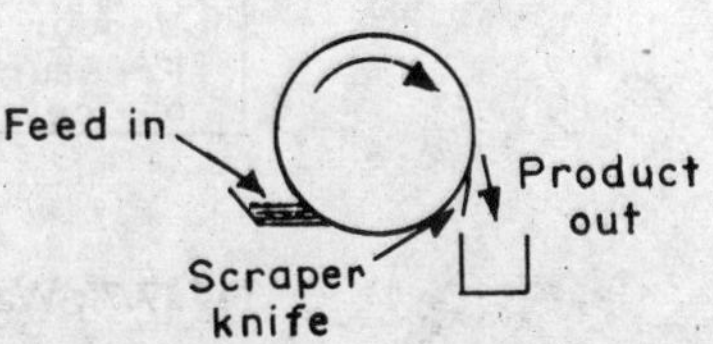

Fig. 7.5 : Schematic diagram of a single drum dryer with scraper knife

A single drum dryer consists of only one drum, while the double drum dryer has two drums or rolls which rotate toward each other at the feeding end. The clearance between the two rolls depends upon the thickness of the layer of feed material applied to the drum for drying. The twin drum looks like a double drum dryer but the operation is different. As illustrated in the Fig. 7.4, in a twin drum dryer, the drums rotate in opposite directions at the top and feeding is done at the bottom.

For dehydration of heat-sensitive materials the drying drum is enclosed in a vacuum tight chamber as shown in Fig. 7.6. The drying takes place at pressure lower than atmospheric. Such dryers are quite expensive and normally used only for very heat- sensitive food materials.

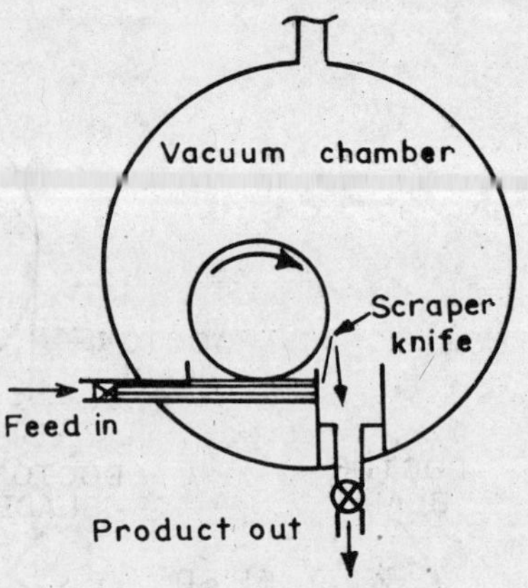

Fig. 7.6 : Schematic diagram of vacuum drum dryer

Freeze drying

The removal of water from a substance by direct sublimation from the frozen state to the vapour state is known as freeze drying. In this method, the water present in the product is not passed as liquid. Water vapour pressure gradient between the surroundings and ice within the material is maintained to cause sublimation. The direct sublimation of ice thermodynamically takes place below the 'triple point' of water (Fig. 7.7). The triple point of water lies at 0°C and 4.59 mm of mercury, at this point the three

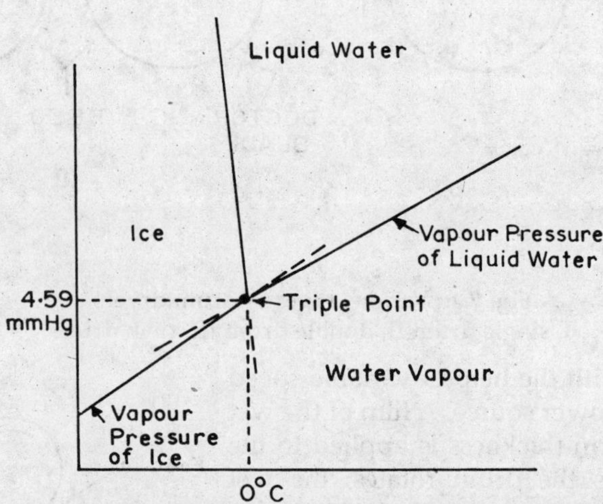

Fig. 7.7 : Water-liquid-solid phase diagram

phases of water viz. solid ice, liquid water and water vapour coexist at equilibrium. Below this point, the water vapour pressure of liquid water is more than ice. This results in direct sublimation.

In a freeze drying system, dehydration takes place in three stages. First, the water of product is removed by formation of ice crystals. These ice crystals later on removed from the body of material by sublimation. After removal of all the ice, the food may contain small amount of water which may be removed by evaporation in the freeze dryer. In freeze drying system, the size and shape of original food product can be retained and the shrinkage, which is a problem with other drying system, is negligible in freeze drying. Also, the migration of soluble solids is limited. The losses due to heat damage is minimised in freeze drying system. Freeze drying usually provides good retention of trace volatile flavour and aroma compounds present in the food.

The operation of freeze drying is carried out at low pressure, usually at 50 N/m^2. The heat is supplied either by radiation or conduction from hot surfaces. The sublimated moisture condenses on the refrigerated coil located at other end of dryer.

In conventional freeze dryers, condenser is provided to remove the water vapour formed during the process and a heating system to provide the necessary latent heat of sublimation to the frozen product (Fig. 7.8).

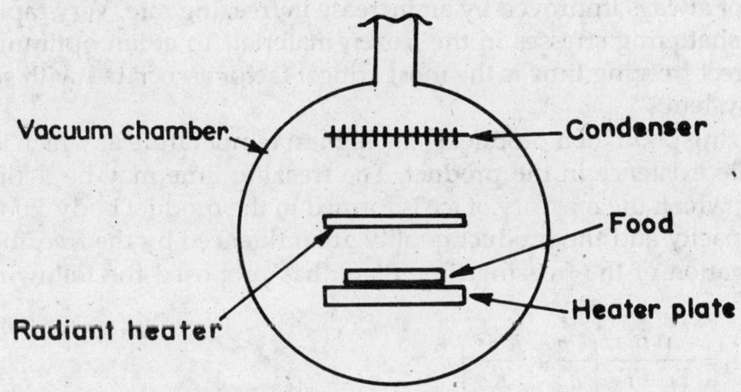

Fig. 7.8 : Schematic diagram of a vacuum freeze drying system

High capacity tunnel freeze dryers have been developed for dehydration of food. The tunnel dryer consists of a cylindrical tunnel of about 2 m diameter. The length of tunnel depends upon the desired drying capacity.

Freezing

From ancient times, food has been preserved by freezing, where sufficiently low temperatures exist. By reducing the temperature of food material below 0°C, the growth rates of micro organisms is significantly reduced. The microbial activities responsible for deterioration of food are also arrested. When the food is kept below freezing point, formation of ice crystals within the product takes place, due to this phenomenon the availability of free water for reaction is converted to the solid state, less water is available for deteriorative reactions.

The sensible heat and latent heat of fusion from the product are removed in food freezing process. For this, the product must be exposed to low-temperature medium for sufficient time. When sensible heat and latent heat of fusion are removed, the product temperature is lowered and the inside water transformed into ice crystals. But, around 25% product water remains in liquid state in frozen food during storage.

The frozen food can be preserved for longer duration and the product quality can be maintained. But the quality of product depends on the freezing process and the frozen-storage conditions. The quality of frozen product is mainly dependent upon the time of freezing *i.e.* the time required to bring down the temperature of product to below the initial freezing temperature. Some food products require rapid cooling so that small ice crystals may be formed within the product and the texture of product is not affected. Rapid cooling is cost intensive, therefore, for some products which are not influenced by structural changes, is generally not followed. In order to maintain the product quality, it is necessary to maintain the proper storage temperature of the frozen foods. Higher storage temperatures impaire the quality of frozen foods.

Freezing time

As stated earlier the rate of freezing affects the quality of food product. Normally, rapid freezing results in high quality product. But, it has been reported that food quality is not always improved by an increase in freezing rate. Very rapid freezing may cause shattering stresses in the frozen material. To attain optimum product quality, correct freezing time is the most critical factor associated with selection of a freezing system.

The 'freezing point' of a product is the highest temperature at which ice crystals have a stable existence in the product. The freezing time may be defined as the time during which the majority of ice is formed in the product body. Food freezing system's capacity and the product quality are influenced by the freezing time.

For estimation of the freezing time Plank has proposed the following expression.

$$t_f = \frac{\rho L_f}{T_F - T_\infty}\left(\frac{Pa}{h_c} + \frac{R a^2}{K}\right) \qquad \ldots 7.7$$

where, t_f = freezing time

ρ = product density

L_f = latent heat of fusion

T_F = initial freezing temperature

T_∞ = temperature of cooling medium

a = product size (diameter)

h_c = convective heat transfer coefficient

K = product's thermal conductivity

P and R = constants dependent on product shape

Constant	*Infinite plate*	*Infinite cylinder*	*Sphere*
P	$\frac{1}{2}$	$\frac{1}{4}$	$\frac{1}{6}$
R	$\frac{1}{8}$	$\frac{1}{16}$	$\frac{1}{24}$

Freezing technology and equipment

There are various techniques by which foods can be frozen. The common methods of food freezing are given below.

1. Freezing foods by immersion in cold brine
2. Freezing the material between two cold plates
3. Freezing the food in a blast of cold air
4. Freezing on a flat metal plate underneath which brine flows
5. Freezing product under fluidized condition by a upward blast of cold air. The freezers are developed based on the followings, (1) freezing by indirect contact with a refrigerant, (2) freezing by immersion in a refrigerating medium, and (3) freezing in a blast of cold air.

Freezing by indirect contact of refrigerant

The plate contact freezer is widely used indirect contact system of freezing. The product is kept in between two refrigerated plates. The plates are pressed against product (Fig. 7.9). The product is frozen while held between the cold plates. The

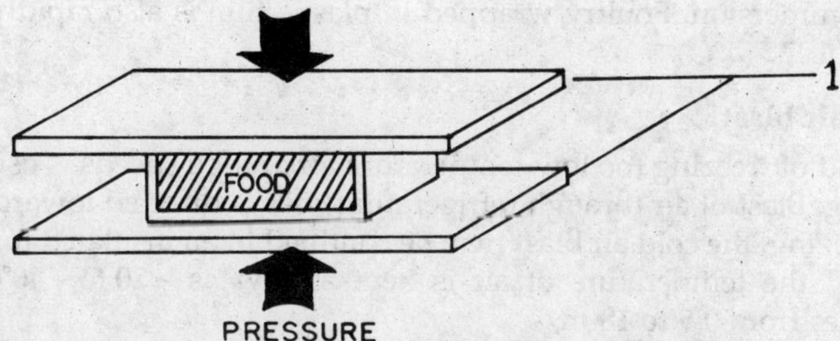

Fig. 7.9 : Schematic diagram of a plate freezer
1. hollow refrigerated freezer plates

plates are mounted parallel to each other and can be arranged horizontally or vertically. The product to be frozen is kept in packages. Before the product is placed in the freezer, the plates are cooled to the desired low temperature. After loading, the product is squeezed between the plates by application of pressure to enhance the heat transfer through plates and package. The cabinet of the freezer is closed and the product is left until its temperature reaches about – 18°C. Thickness of food package influences the freezing time. Plate contact freezers are available as batch type or continuous type. The batch freezers have flexibility to accommodate various product types and sizes.

Freezing by direct contact with a cooled refrigerant

In this method of freezing, food products are frozen in direct contact with refrigerant (Fig. 7.10). The product is frozen by direct immersion in low temperature

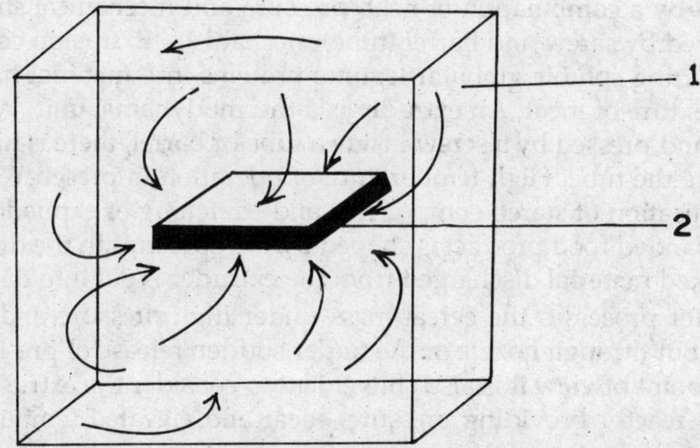

Fig. 7.10 : Schematic diagram of direct-contact freezing of foods
1. refrigerant 2. food product

brine. As there is no barriers between the refrigerant and the product, the system operates more efficiently. Also the liquids are good heat conductors, a product can be frozen quickly by direct immersion in low temperature liquids like sodium chloride brine and sugar solutions.

A refrigerating medium should be edible and remains unfrozen at – 18°C, is suitable for immersion freezing. Food freezing with direct contact can also be achieved by low temperature air at high speeds. Canned citrus juice concentrate is frozen by immersion. Poultry wrapped in plastic film is also rapidly frozen in brine or glycol.

Freezing by air blast

In this method of freezing too low-temperature air at high speeds is used to affect quick freezing. Blast of air through refrigerating coils is directed towards product. For rapid freezing, the cold air blast may be confined in an insulated tunnel. In air blast freezers, the temperature of air is kept as low as – 20 to – 40°C and the velocity ranges from 0.5 to 18 m/s.

The products to be frozen are placed on trays. They may be kept in packages or in loose condition. These trays are placed on freezing coils in a low-temperature room in which cold air is blowing over the product.

Tunnel freezing is normally adopted for freezing with air blast. The product is loaded on trolley mounted trays. The tray racks are passed into the tunnel in counter direction to the movement of product. In tunnel freezing, dehydration may take place if the product is kept unpackaged.

With necessary changes, the air blast freezing system can be converted into a fluidised bed freezing system. In such system a powerful stream of air is forced through a layer of the food being carried on a mesh conveyor. The product is frozen in suspension stage. Freezing of peas is quite common by fluidised-bed freezer. Product in fluidised state freeze in 3 to 5 minutes. Fruits and vegetables are extensively frozen by fluidised-bed freezers.

Extrusion Cooking

Extrusion cooking is the process of cooking moistened starchy and/or proteinous food material by a combination of heat, pressure and mechanical shear. The process is achieved by screw and barrel (tube) mechanism. Extrusion cooking is capable of converting soluble globular legume proteins into material having fibrous and chewy texture of meat. An extruder is a thermodynamic unit. When material is conveyed and pressed by a screw inside a tube or barrel, there is high temperature rise inside the tube. High temperature of operation in presence of water promotes gelatinisation of starch components and stretching of expnadable components. The expanded food product is shaped by the openings in the die. The expanded and cooked material discharged from the extruder is cut into desired lengths.

An extruder processes the cereal mass under high pressure and temperature and forces it out through nozzle or die under sudden release of pressure. From an engineering point of view it is also convenient to consider the extrusion-cooker as a continuous reactor providing pressure, shear and elevated temperature (Linko et al 1982). A combination of the amount of pressure and the shape of the expulsion nozzle determines the shape of the product. The size and shape obtainable from extruder is limitless. Extruded products in a wide variety of shapes like rings, stars, curls or balls can be formed. The transformations of product during extrusion are irreversible.

The extrusion cooking process is high temperature short time process in which moist, soft grain is fed into the extruder where the desired temperature and pressure are obtained over the required period of residence time. For cooking of the product generally external heat is not supplied, heat for cooking is achieved through shear and friction in the extruder. Feed rate, design and arrangement of screws, barrel and die, speed of operation and the properties of raw materials effect the shear and friction conditions in the extruder and the amount of heat generated inside the tube.

An increase in extrusion temperature results in a higher degree of gelatinisation. Moisture content increase has positive effect at high temperatures. When the screw speed and nozzle diameter are increased, the gelatinisation of starch reduces. Complete gelatinisation can be obtained at 110°C and 24 to 27% moisture content. It is found that in extrusion cooking even at low moisture content gelatinisation is usually possible.

The basic elements of proper extrusion-cooking are the following.

(*i*) Feeding of granular or milled material in the extruder should be continuous at desired feed rates.

(*ii*) The material should be preconditioned with steam at controlled temperature of 82 to 99°C at atmospheric pressure.

(*iii*) The moisture should be uniformly applied to the product.

(*iv*) The equipment may transform the granular or processed floury material into a dough at 82–110°C.

(*v*) The temperature of dough should be elevated to 115–200°C during the last 10 to 30 second in the extruder to cook and expand the product.

(*vi*) Formation of desired shape and size of the product by a nozzle or die at the end of the process. The extrudate is cut into desired lengths.

(*vii*) Drying and cooling of the extrudate.

Snacks, break fast cereals, bread crumbs, pet foods and many other starch-containing products for human and for animals are produced by extrusion-cooking.

An extruder has three basic components viz. screw, barrel or tube and die. The screw of an extruder may be divided into two zones, (1) compression zone and (2) metering zone as shown in Fig. 7.11.

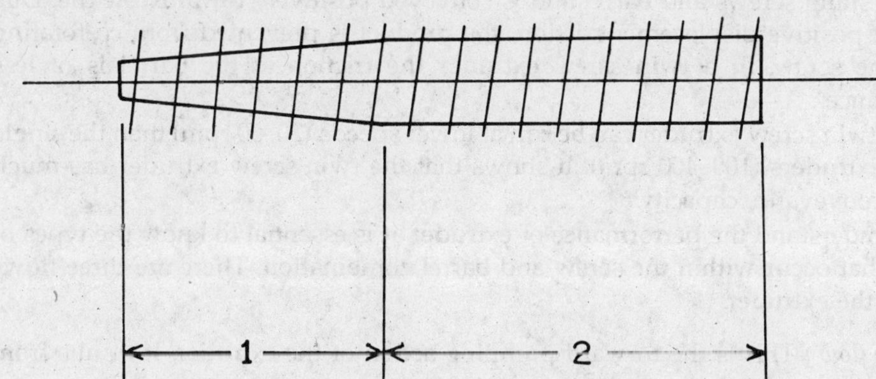

Fig. 7.11 : Schematic diagram of an extruder screw
1. compression zone 2. metering zone

The extruders are mainly of two types, single screw extruder and twin screw extruder. In a single screw extruder, a single screw operates inside the tube, while in twin screw extruder there are two intermeshing screws. These are either counter-rotating or co-rotating. The barrel of extruder may be cylindrical or conical (Fig 7.12). Some extruders may have conical barrel for increasing pressure as the material advances inside the tube.

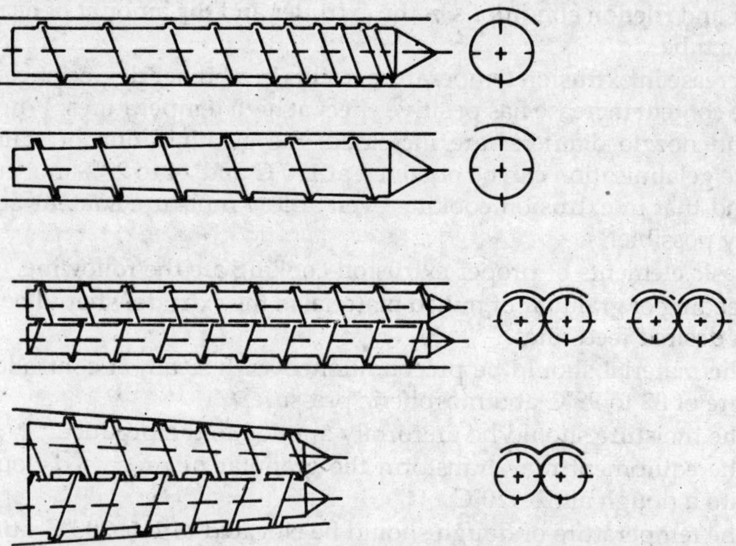

Fig. 7.12 : Schematic diagrams of some of the commercial food extruder-cooker designs

The main difference between the single screw extruder and twin screw extruder is in their conveying machanism. In single screw extruder, the conveying action is the result of friction effects, (1) the friction between screw and product, and (2) the friction between barrel and product. The single screw extruder needs the barrel wall for a good conveying action. In single screw extruders, product may co-rotate with the screw. In a twin screw extruder, the product is enclosed between the intermeshing screws and barrel and is conveyed positively towards the die. Due to such positive displacement action, the product is prevented from co-rotating with the screw. In a twin screw extruder the friction at the barrel is of less importance.

The twin screw extruder can be run at lower speeds (20–60 rpm) than the single screw extruders (100–400 rpm). It shows that the twin screw extruder has much better conveyance capacity.

To understand the performance of extruder, it is essential to know the types of flows that occur within the screw and barrel combination. There are three flows within the extruder.

1. *Drag flow :* This is the forward pumping action of the extruder. It results from the relative motion between the screw and barrel.

2. *Pressure flow :* This is the flow that occurs due to the presence of pressure gradient between the feed and the discharge screw. This flow occurs in the con-

tinuous helical channels formed by the flights and root of the screw and barrel. During normal extruder operation, the direction of this flow is from the discharge end of the screw to the feed end of the screw.

3. *Leakage flow* : It is similar to pressure flow and is driven by a pressure gradient. This flow occurs in the clearance between the screw flights and the barrel and within any slots in the barrel wall or surface. Leakage flow reduces the extruder output.

The net output of an extruder can be given by the following expression.

$$Q_n = Q_d - Q_p - Q_l \qquad \text{...7.8}$$

where, Q_n = net output

Q_d = drag flow

Q_p = pressure flow

Q_l = leakage flow

The net output can also be given by

$$Q_n = \frac{VWH}{2} - \frac{WH^3}{12\,\mu}\left(\frac{P}{L}\right) \qquad \text{...7.9}$$

where, V = peripheral screw speed

W = screw pitch

H = screw depth

P = discharge pressure

L = helical length of screw

μ = viscosity of material

Every extruder has a forming nozzle or die at its discharge end. This die puts resistance to the flow of material. It is this resistance of flow that determines the pressure in an extruder. For Newtonian materials, the characteristics of a die can be given by,

$$Q = K_d\,(P) \qquad \text{...7.10}$$

where Q = flow through the die

K_d = function of the die geometry and fluid viscosity

P = pressure drop

Evaporation

Evaporation is an unit operation to concentrate a solution with a volatile solvent. It can also be defined as an unit operation in which water or volatile solvents are removed to obtain concentrated liquid products from dilute liquid foods. Almost all the liquid foods are evaporated to remove a portion of water by vaporising it. Evaporation also provides microbiological stability and reduces the transportation and storage costs. Evaporation is different from dehydration in the sense that at the end of the process we get thick liquor and the food product remains in liquid state. Evaporation of liquid foods is also carried out to pre-concentrate a liquid food for further processing like spray drying or drum drying, crystallisation.

Evaporation is carried out in evaporators, which consists of a heat exchanger enclosed in a chamber. The heat exchanger supplies sensible heat and latent heat of evaporation to the feed. Normally saturated steam is used as the heating medium in food industries. The product inside the evaporator chamber is kept under vacuum, this increases the temperature difference between steam and the product. Thus the product boils at lower temperatures. The other important parts of an evaporator is a separator to separate vapour from the concentrated liquid and a condenser to cause condensation of the vapour and its removal from the system.

In a single effect evaporator (Fig. 7.13), there are heat exchanger, separator and condenser while in multieffect evaporator there are more than one heat exchanger

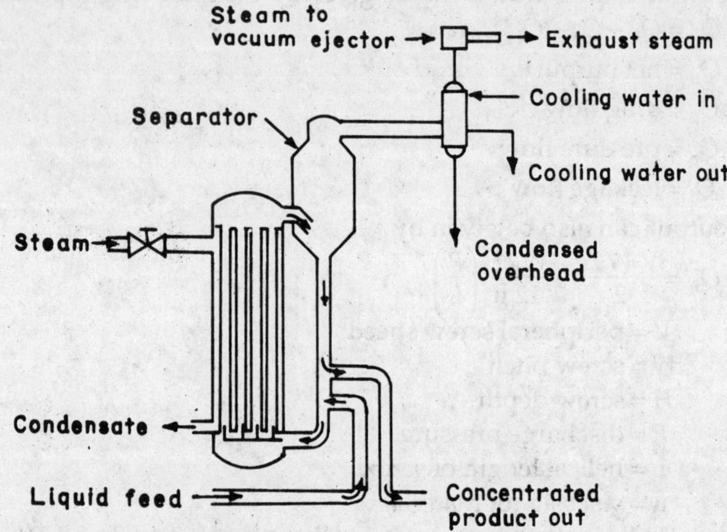

Fig. 7.13 : Diagram of single effect evaporator

and separator. In a multieffect evaporator, the vapour produced from first heat exchanger is used as heating medium in subsequent heat exchanger (Fig. 7.14). The steam is used in first heat exchanger to evaporate liquid while vapour is used

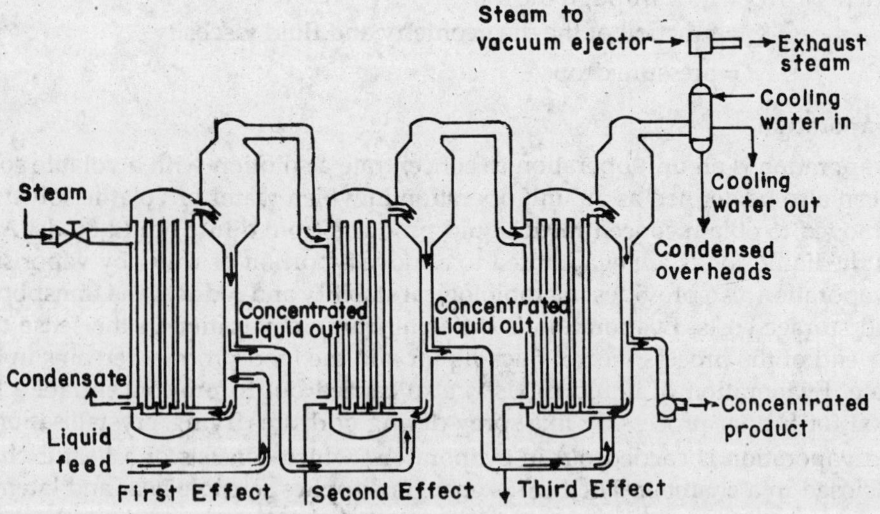

Fig. 7.14 : Multi effect evaporator

in wet exchanger to evaporate water of partially concentrated liquid. Such system gives higher energy use efficiency.

The heat load of a single effect evaporator

The overall rate of heat transfer 'q' through the heating surface of an evaporator is the product of three factors as given below

$$q = U\,A\,\Delta T$$

where, U = overall heat transfer coefficient

A = area of the heat transfer surface

ΔT = overall temperature drop

The overall rate of heat transfer is equal to the product of mass flow rate of the heating medium usually steam and the latent heat of condensation of inlet steam.

$$q = U\,A\,\Delta T = \dot{m}_s\,\lambda_s \qquad \ldots 7.11$$

where, $\dot{m}_s$ = mass flow rate of steam, kg/s or kg/hr

λ_s = latent heat of condensation of inlet steam, k Cal/kg

In calculations, the heat losses from the system by convection and radiation from hot surfaces are neglected. If the heat losses are neglected, then all the heat given up by the condensing steam is transferred for evaporation of water from the liquid food. The heat given up by the steam is equal to the sum of sensible heat gained by liquor and the latent heat of vapour produced during the course of evaporation.

$$\dot{m}_s\,\lambda_s = \text{sensible heat gained by liquor} + \text{latent heat of vaporisation of water}$$

$$= (\dot{m}_f - \dot{m}_p)\,\lambda + \dot{m}_f\,C_p\,(T_p - T_f) \qquad \ldots 7.12$$

where, $\dot{m}_f$ = mass flow rate of feed, kg/s or kg/hr

$\dot{m}_p$ = mass flow rate of thick liquor or product, kg/s or kg/hr

λ = latent heat of vaporisation, k cal/kg (at the temperature of product)

c_p = Specific heat of feed, k cal/kg°C

T_p = temperature of thick liquor, °C

T_f = temperature of feed, °C

The temperature, T_p is the boiling point of the concentrated liquid which increases with increasing concentration. The increase in boiling point of solution over that of water is known as the "boiling point elevation" of the solution. Estimation of boiling point elevation can be done by the Dühring rule. According to this rule linear relationship exists between the boiling point temperature of solution and the boiling point temperature of water at the same pressure. But the linear relationship do not exist over a wide range of temperatures.

The boiling point elevation should be substracted from the temperature drop predicted from the steam tables.

Material balance

The overall material balance of evaporator can be presented by the following equation

$$\dot{m}_f = \dot{m} + V$$

where, V = mass of solvent vaporised in the evaporator, kg/s or kg/hr.

The material balance on the liquid

$$\dot{m}_f\,(1 - x_f) = \dot{m}\,(1 - x_p) + V \qquad \ldots 7.13$$

where, x_f = mass flow rate of solid fraction in feed, kg/s or kg/hr

x_p = mass flow rate of solid fraction in product or thick liquor, kg/s or kg/hr

$$\text{The economy of evaporation} = \frac{\text{quantity evaporated/hr.}}{\text{steam required/hr.}}$$

The enthlpy balance of the system can be given as per following expression

$$q = \dot{m}_s \lambda_s = (\dot{m}_f - \dot{m}_p) H_v - \dot{m}_f H_f + \dot{m}_p H \quad ...7.14$$

where, H_u = Specific enthalpy of vapour

H_f = specific enthalpy of thin liquor

H = specific enthalpy of thick liquor

Problem : In a single effect evaporator, orange juice is being concentrated at steady-state condition. The feeding rate of orange juice is 2160 kg/hr with 10% solids. The juice has to be concentrated to 74% total solids. The inlet feed temperature is 42°C while the product boils at 61.5°C inside the evaporator. The specific heats of dilute and thick orange juice are 0.93 and 0.55 k cal/kg°C respectively. The overall heat transfer coefficient is 808 k cal/m^2–hr–°C. Assuming negligible boiling point elevation, calculate the mass flow rate of thick juice, steam required, economy of operation and the total heat transfer area. The temperature of saturated steam entering the evaporator is 130°C.

Solution :

(*i*) Mass flow rate of feed in terms of total solids

$$= (0.10)\,(2160) = 216 \text{ kg/hr}$$

Therefore, mass flow rate of product (thick juice)

$$0.74 \times \dot{m}_p = 216$$

or, $$\dot{m}_p = 291.9 \text{ kg/hr}$$

Mass flow rate of vapour, $\dot{m}_v = \dot{m}_f - \dot{m}_p$

$$= 2160 - 291.9$$

$$= 1868.1 \text{ kg/hr}$$

(*ii*) Using heat balance

$$\dot{m}_s \lambda_s = (\dot{m}_f - \dot{m}_p)\, \lambda + \dot{m}_f C_p (T_p - T_f)$$

Latent heat of vaporisation at 61.5°C = 625.1 k cal/kg

Latent heat of condensation of steam at 130°C = 651.01 k cal/kg

$$\dot{m}_s \lambda_s = 1868.1 \times 625.1 + 2160 \times 0.93\,(61.5 - 42)$$

$$= 1167749.3 + 39171.6$$

or, $$\dot{m}_s = \frac{1206920.9}{651.01} = 1853.92 \text{ kg/hr}$$

(*iii*) $$\text{Economy of operation} = \frac{\text{quantity evaporated/hr}}{\text{steam required/hr}}$$

$$= \frac{1868.1}{1853.92} = 1.007 \text{ kg water evaporated/kg steam}$$

(*iv*) Total heat transfer area

$$q = U\,A\,\Delta T = \dot{m}_s \lambda_s$$

or, $$A \times 808 \times (130 - 61.5) = \dot{m}_s \lambda_s$$

or, $$A = \frac{1853.92 \times 651.01}{808 \times 68.5}$$

$$= 21.8 \text{ m}^2$$

Heat Exchanger

Transfer of heat energy is utilised to a great extent in many food processing operations. Two most common modes of heat transfer are conduction and convection. The conduction and convection mechanisms are also termed as molecular heat transfer and turbulent heat transfer where heat exchange takes place by molecular action and by mixing process respectively. The above two modes of heat transfer can occur separately or simultaneously.

In food processing operations the transfer of heat energy is mostly achieved by heat exchangers. Heat exchangers are the devices in which heat is transformed from a hot fluid to a cold fluid. The transfer of heat between the two fluids may be carried out either by direct contact or by transmission through a separating wall. The transfer of heat takes place by conduction, convection and in case of large temperature differences even by radiation. With the availability of suitable materials and with support of good research work done in this area, the design of heat exchangers have been greatly perfected. As a result the flow of warm and cold fluids are oriented through the heat exchangers in such a manner that the maximum possible heat transfer between the two streams take place. The common types of heat exchangers are :

1. Plate heat exchanger
2. Double pipe heat exchanger
3. Shell and tube heat exchanger
4. Shell and tube heat exchanger with baffles
5. Plate fin type heat exchanger

Plate heat exchanger : Plate heat exchangers are primarily used for low viscosity fluids like milk. In this exchanger, warm and cold fluids flow through alternate passages between vertical plates (Fig. 7.15). The plate heat exchanger is commonly used in HTST (high temperature short time) units for heating the milk

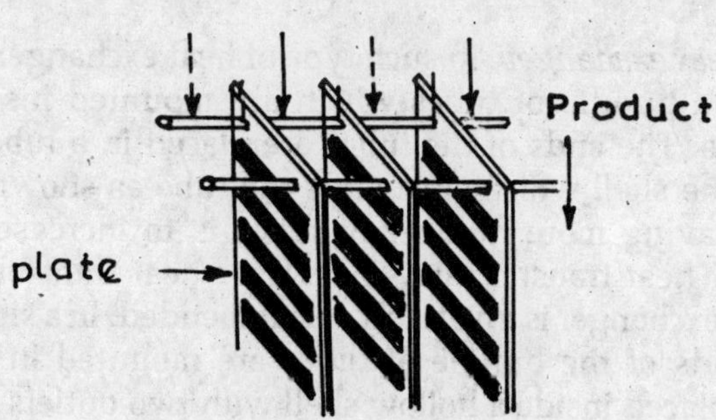

Fig. 7.15 : Plate heat exchanger

well below its boiling point. The plate exchangers are quite efficient in heat transfer, occupy less space and are compact. The construction is simple and easy to clean after use. The plates are designed to provide uniform and less turbulent flow of fluids with a high heat transfer rate. The plate thickness may vary from 1.25 to 3.00 mm and the spacing between the plates is 1.25 to 7.75 mm. The corrugated or channeled surface of plates impart turbulence to flowing liquid which results increased heat transfer rates. The corrugations also add strength to the plates.

Double pipe heat exchanger : It is perhaps the simplest type of heat exchanger. A schematic diagram of a double pipe heat exchanger is shown in Fig. 7.16. The double pipe heat exchanger has two concentric pipes or tubes with one fluid flowing through the inner control pipe and the other fluid flowing either parallely or counter currently in the larger tube along the available annular space between the two tubes. The heat transfer rate between two streams depends upon velocities or mass flow rates of the two fluids. The double pipe heat exchanger may be used for transfer of heat between liquid to liquid, liquid to gas or gas to gas. The selection of material of construction of pipes depends upon the type of fluids being handled in the exchanger. In the double pipe heat exchangers product to product regeneration is not possible and it requires higher pressure than the plate heat exchanger of same capacity. It also occupies more space than the plate type for the same capacity. It is generally used for heating of food products of moderate viscosity.

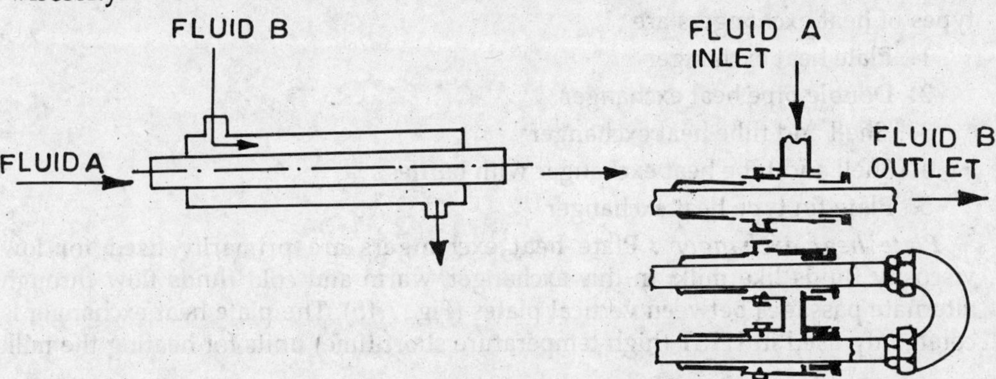

Fig. 7.16 : Double pipe heat exchanger

Shell and tube heat exchanger : In such type of heat exchangers one of the fluids is carried through a bundle of tubes which are mounted inside a shell which surrounds the tubes. The ends of the tubes are placed in a tube sheet. The other fluid is passed in the shell which surrounds the tube as shown in Fig. 7.17. The main purpose of having more number of tubes is to increase the surface area available for higher heat transfer rate. For a large heat transfer surface area, the shell and tube heat exchanger is invariably recommended. In a single shell and tube exchanger, both ends of the bundle of tubes are mounted in tube sheets. This bundle of tubes is placed inside a hollow shell with two outlets at top and a drain at the bottom. The shell and tube heat exchanger can be used as a heater or cooler

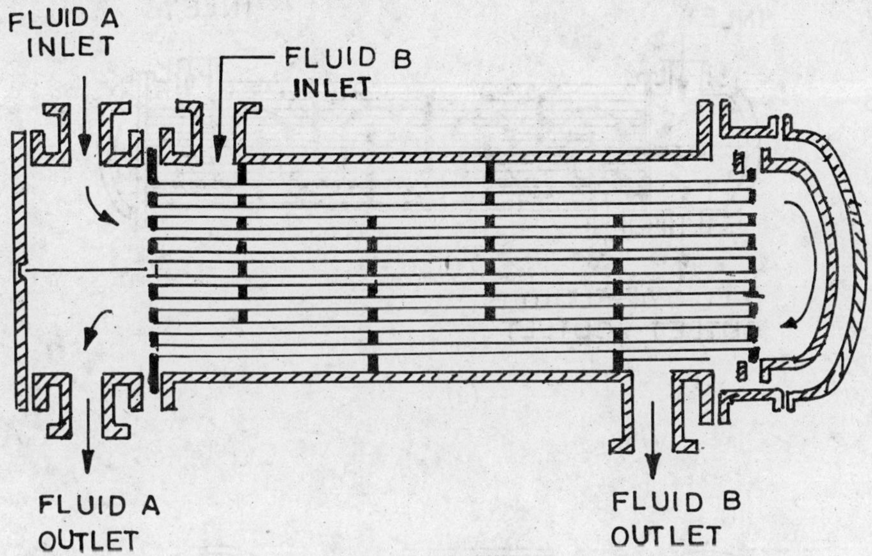

Fig. 7.17 : Shell and tube heat exchanger

as desired. In shell and tube heat exchanger both parallel and counter flow arrangements are possible.

Shell and tube heat exchanger with baffles : In a shell and tube heat exchanger the heat transfer coefficients are of comparable importance, and both should be large if a satisfactory overall coefficient is to be obtained. The velocity and turbulence of the inside shell are as important as those of the tube side fluid. In shell and tube heat exchanger for improving the heat transfer rate, baffles are introduced in heat exchangers in order to allow the fluid surrounding the tubes to be in close contact with the surface of the tubes. The arrangement of baffles may be disc and doughnut type or segment type as shown in Fig. 7.18.

Plate fin type heat exchanger : A plate fin type heat exchanger is a cross flow type arrangement. It is used to transfer high amount of heat as compared to any other cross flow type exchanger. A general plate fin heat exchanger is shown in Fig. 7.19. In the tubes, the fluid to be cooled or heated is passed while gas flows across the tubes along the plates.

Heat exchangers are also classified on the basis of the relative direction of motion of two fluids. The classification is as follows :

(i) Parallel flow

(ii) Counter flow, and

(iii) Cross flow

Parallel flow heat exchangers are those in which the hot and cold fluids flow in the same direction, as shown in Fig.7.20. The net result of this type of variation in Δt is that the exchanger is much more effective for a unit area of heat transfer surface at the entrance than it is near the exit.

In counter flow arrangement, the two fluids exchange heat with each other in opposite directions, as shown in Fig. 7.21. In this case the driving force is nearly

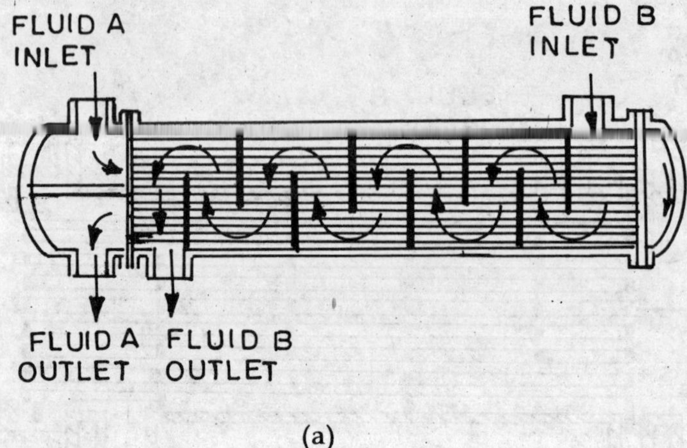

(a)

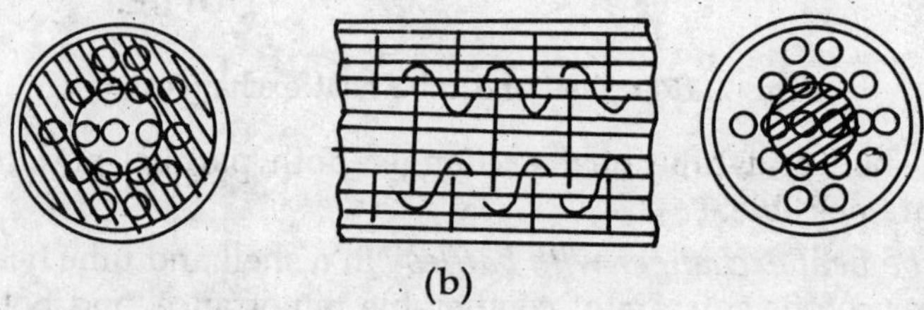

(b)

Fig. 7.18 : (a) Shell and tube type heat exchanger with baffles
(b) Disc and doughnut baffles

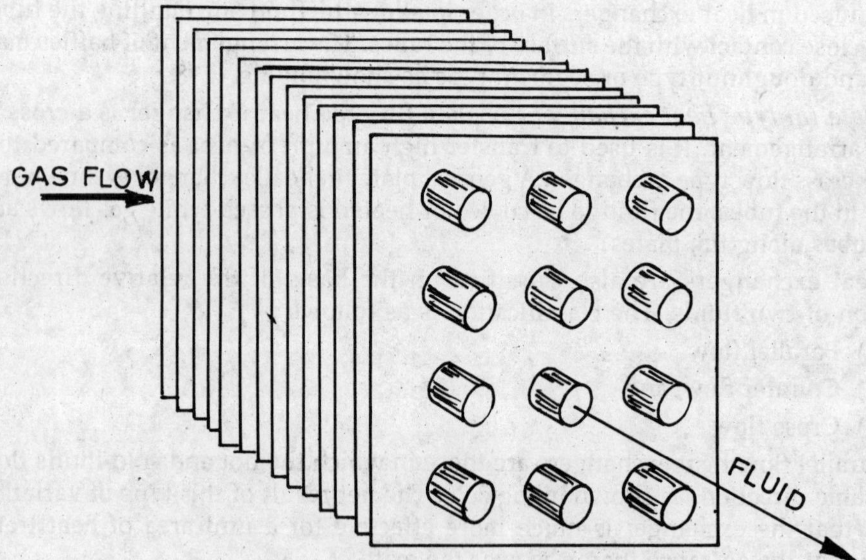

Fig. 7.19 : Plate fin type heat exchanger

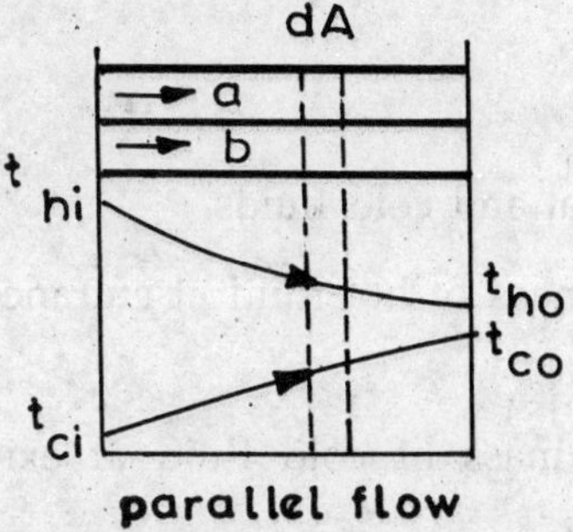

Fig. 7.20 : Parallel flow heat exchanger

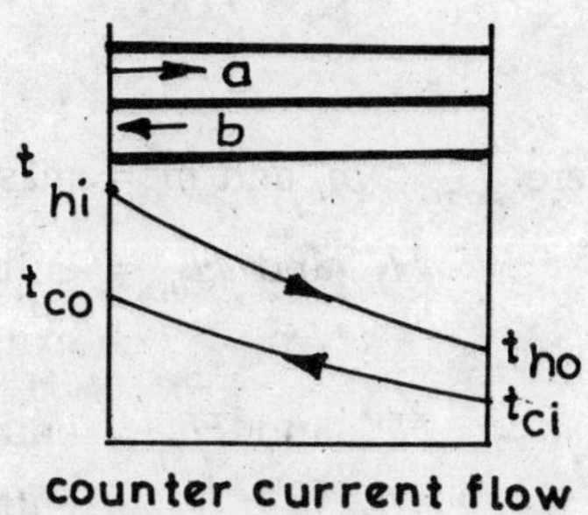

Fig. 7.21 : Counter-current flow heat exchanger

constant through out the length of the exchanger. More heat is exchanged in this case and the temperature of the cold fluid may be raised to nearly the initial temperature of the hot fluid. For the same amount of heat exchanged, a counter flow heat exchanger is usually smaller than the parallel flow heat exchanger.

In cross flow heat exchangers the direction of flow of one fluid is perpendicular to that of the other fluid. Automobile radiator is an example of such type of heat exchanger (Fig. 7.22)

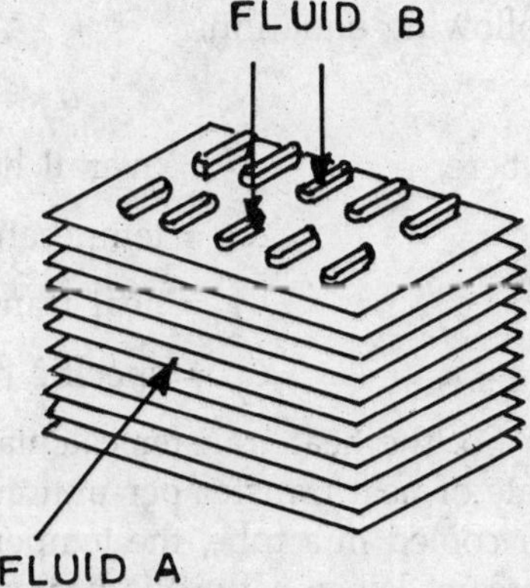

Fig. 7.22 : Cross flow heat exchanger

Energy balances : The problems on heat transfer by heat exchangers are solved on the basis of energy balances and the estimations of rates of heat transfer. The heat gained by the cold fluid is equal to that by the hot fluid, Thus,

$$q_c = -q_h \quad \text{... 7.15}$$

where, q_c = heat gained by cold fluid

q_h = heat lost by hot fluid

The rate of heat transfer of one fluid stream through the exchanger may be given as:

$$m(H_2 - H_1) = q \quad \text{... 7.16}$$

where, m = mass flow rate of fluid

H_1 & H_2 = enthalpies per unit mass of stream at entrance and exit, respectively

When we neglect the comparatively very small amount of heat transfer between the cold, and hot fluids and the ambient to that of heat transfer through walls of heat exchanger, therefore,

$$m_h\left(H_{h_2} - H_{h_1}\right) = q_h \quad \text{... 7.17}$$

$$\text{and } m_c\left(H_{c_2} - H_{c_1}\right) = q_c \quad \text{... 7.18}$$

where, m_h and m_c = mass flow rates of hot and cold fluids

H_{h_2} and H_{h_1} = enthalpies per unit mass of hot fluid at entrance and exit, respectively

H_{c_2} and H_{c_1} = enthalpies per unit mass of cold fluid at exit and entrance, respectively

or, $$m_h (H_{h_2} - H_{h_1}) = m_c(H_{c_2} - H_{c_1}) \quad \text{... 7.19}$$

The rate of heat transfer in any heat exchanger may be expressed by the following equation,

$$d_q = - U \Delta t \, dA = mC_p \, \Delta t \quad \text{... 7.20}$$

where, U = overall heat transfer coefficient

Δt = temperature difference

A = heat transfer surface area

C_p = specific heat of fluid

So, the heat transfer calculations are based on the area of heating surface. The rate of heat transfer per unit area is called as heat flux. When a fluid being heated or cooled in a tube, the temperature will vary through out the cross-section of the stream. When a fluid is heated in a tube, the temperature of fluid near to wall of tube is higher and the temperature of fluid gradually lowers towards centre and vice versa. Due to such temperature gradient, an average temperature of fluid is given. As shown in Figs. 7.20 and 7.21 the temperature difference (Δt) keeps on varying from point to point along the tube. Since the heat flux is proportional to the Δt, the flux also varies with the tube length.

$$\frac{dq}{dA} = U\Delta t \quad \text{... 7.21}$$

For calculation of the total heat transfer on the basis of total heat exchanger area, the above equation must be integrated.

Let us consider an incremental length of the heat exchanger (surface area dA) as shown in Fig. 7.20. For integration and calculation of heat exchanger surface requirements and overall heat transfer coefficients, certain assumptions are taken into account and these are accepted, (1) the specific heat of fluid and overall thermal conductance do not change with temperature, (2) the overall coefficient U is constant, (3) the temperature of fluid at any point is uniform, (4) heat exchange with the ambient is negligible, (5) the fluid gains or looses no heat, except through transfer surface and (6) the flow is steady.

$$dq = \text{U} \, \Delta t \, dA \quad \text{... 7.22}$$

$$\frac{d(\Delta t)}{dq} = \frac{\Delta t_2 - \Delta t_1}{q_t} \quad \text{... 7.23}$$

where, Δt_2 = temperature difference at entrance

Δt_1 = temperature difference at exit

q_t = total heat

$$\frac{d\Delta t}{U\Delta t dA} = \frac{\Delta t_2 - \Delta t_1}{q_t} \quad \text{... 7.24}$$

$$\int_{\Delta t_1}^{\Delta t_2} \frac{d\Delta t}{\Delta t} = \frac{U(\Delta t_2 - \Delta t_1)}{q_t} \int_0^A dA$$

$$\ln \frac{\Delta t_2}{\Delta t_1} = \frac{U(\Delta t_2 - \Delta t_1)}{q_t} A$$

$$q_t = UA \frac{\Delta t_2 - \Delta t_1}{\ln \frac{\Delta t_2}{\Delta t_1}} = UA\, \Delta t_m$$

here $\Delta t_2 = \theta_i$

$\Delta t_1 = \theta_o$

Δt_m = log mean temperature difference (LMTD)

$$\therefore \quad \Delta t_m = \frac{\theta_i - \theta_o}{\ln \frac{\theta_i}{\theta_o}} \quad \text{... 7.25}$$

and

$$q = m_h C_{p_h} (t_{h_i} - t_{h_o}) = m_c C_{p_c} (t_{c_o} - t_{c_i}) \quad \text{... 7.26}$$

where, m_h and m_c = mass flow rates of hot and cold fluids, respectively

C_{p_h} and C_{p_c} = specific heats of hot and cold fluids, respectively

t_{h_i} and t_{h_o} = temperature of hot fluid at entrance and exit, respectively

t_{c_i} and t_{c_o} = temperatures of cold fluid at entrance and exit, respectively

The above approach is useful for the analysis of heat exchanger when the inlet and outlet temperatures of the fluids are known. But for a particular type of heat exchanger the inlet and outlet temperatures are to be determined, the analysis can be done on the basis of effectiveness of the exchanger. The effectiveness of an exchanger is the change in temperature of fluid in proportion to the maximum possible change. The effectiveness is defined as the ratio of actual heat transfer to the maximum possible heat transfer. The maximum amount of heat transfer could be attained if the temperature difference of one of the fluids is equal to the maximum temperature difference present in the exchanger ie $t_{h_i} - t_{c_i}$.

Effectiveness of exchanger, $E = \dfrac{\text{actual heat transfer}}{\text{maximum possible heat transfer}}$

$$= \frac{t_{h_i} - t_{h_o}}{t_{h_i} - t_c}$$

Since $t_{h_i} - t_{h_o} = \left(t_{h_i} - t_c\right) - \left(t_{h_o} - t_c\right)$ ... 7.27

$$E = \frac{\left(t_{h_i} - t_c\right) - \left(t_{h_o} - t_c\right)}{t_{h_i} - t_c}$$

$$= \frac{t_{h_i} - t_c}{t_{h_i} - t_c} = \frac{t_{h_o} - t_c}{t_{h_i} - t_c}$$

$$= 1 - \frac{t_{h_o} - t_c}{t_{h_i} - t_c}$$

$$\frac{t_{h_o} - t_c}{t_{h_i} - t_c} = e^{-\left[\frac{UA}{e_{Ph}\, m_h}\right]} \quad \text{...7.28}$$

Therefore, $E = 1 - e^{-\left[\frac{UA}{e_{Ph}\, m_h}\right]}$ ... 7.29

For same amount of heat transfer to attain maximum temperature difference the value of the product of mass flow rate of fluid and its specific heat should be minimum

or, $q_{max} = m_{c_{pmin}}\left(t_{h_i} - t_{c_i}\right)$ ... 7.30

$= C_{min}\left(t_{h_i} - t_{c_i}\right)$ ... 7.31

The effectiveness of parallel heat exchanger

$$E_p = \frac{c_{Ph}\left(t_{h_i} - t_{h_o}\right)}{c_m\left(t_{h_i} - t_{c_i}\right)} \quad \text{... 7.32}$$

$$= \frac{C_{P_c}(t_{c_o} - t_{c_i})}{C_{min}(t_{h_i} - t_{c_i})} \quad \text{... 7.33}$$

$$= \frac{1 - e^{UA\left(\frac{1}{C_{P_c}} + \frac{1}{C_{Ph}}\right)}}{C_{min}\left(\frac{1}{C_{P_c}} + \frac{1}{C_{P_h}}\right)} \quad \text{... 7.34}$$

The effectiveness of counterflow heat exchanger, E_c

$$E_c = \frac{c_{p_c}(t_{c_o} - t_{c_i})}{c_{\min}(t_{h_i} - t_{c_i})} \quad \text{... 7.35}$$

$$= \frac{c_{p_h}(t_{h_i} - t_{h_o})}{c_{\min}(t_{h_i} - t_{c_i})} \quad \text{... 7.36}$$

$$= \frac{1 - e^{-UA\left(\frac{1}{C_{Pc}} + \frac{1}{C_{Ph}}\right)}}{c_{\min}\left[\frac{1}{c_{Pc}} - \frac{1}{c_{Ph}} e^{-UA\left(\frac{1}{C_{Pc}} + \frac{1}{C_{Ph}}\right)}\right]} \quad \text{... 7.37}$$

In a shell and tube heat exchanger the shell side and tube side heat transfer coefficients are of comparable importance, and both must be large if satisfactory overall coefficient is to be attained. The velocity and turbulence of the shell side fluid are as important as those of the tube side fluid. The heat transfer coefficient for the tube side fluid (h_t) in a shell and tube exchanger can be calculated from the following equation.

$$\frac{h_t}{c_p v}\left(\frac{c_p \mu}{k}\right)^{2.3}\left(\frac{\mu_w}{\mu}\right)^{0.14} = \frac{0.023\left[1 + \left(\frac{D}{L}\right)^{0.7}\right]}{\left(\frac{Dv}{\mu}\right)^{0.2}} \quad \text{... 7.38}$$

Where, C_p = specific heat of fluid at constant pressure, k cal/kg –°C

v = fluid mass velocity, kg/m^2 – s

μ = absolute viscosity, kg/m – s

μ_w = absolute viscosity of fluid at wall temperature, kg/m – s

k = thermal conductivity, k cal/h_r – m–°C

D = tube diameter, m

L = length of tube, m

The coefficient for the shell side h_s cannot be calculated by above equation. In the shell the direction of flow of fluid is partly parallel to the tubes and partly across the tubes because the cross-sectional area of the stream and the mass velocity of the stream vary as the fluid crosses the tube bundle across the shell. The "Donohue" equation provides an equation for calculating the shell side heat transfer coefficient. This equation takes into account the weighted average mass velocity v_c of the fluid flowing parallel with the tubes and that flowing across the tubes.

The mass velocity v_p, parallel with the tube is the mass flow rate divided by the free area for flow in the baffle window S_b.

$$S_b = f_b \frac{\pi D_i}{4} - N_b \frac{\pi D_0^2}{4} \quad \text{... 7.39}$$

Where, f_b = generally 0.1955

D_i = inside diameter of shell, m

N_b = number of tubes

D_o = outside diameter of tube, m

The area for transverse flow between the tubes in the row closest to the centreline of the exchanger

$$S_a = PD_i\left(1 - \frac{D_0}{d_c}\right) \quad \text{... 7.40}$$

where, P = baffle pitch

d_c = centre to centre distance between tubes

Therefore, the mass velocities

$$v_p = \frac{m}{S_p} \text{ and } v_a = \frac{m}{S_a}$$

$$v_c = \sqrt{v_p v_a}$$

The 'Donohue' equation is

$$\frac{h_s D_o}{k} = 0.2\left(\frac{D_o v_c}{\mu}\right)^{0.6}\left(\frac{c_p \mu}{k}\right)^{0.33}\left(\frac{\mu}{\mu_w}\right)^{0.14} \quad \text{...7.41}$$

where, h_s = heat transfer coefficient

Problem : The structure of a tubular heat exchanger is given below,

Inside diameter of exchanger = 600 mm

tube outside diameter = 15 mm

number of tubes = 121

length of tubes = 1800 mm

pitch of tubes = 25 mm

Standard 25% battles are spaced at 300 mm. Liquid milk at an average temperature of 20°C is being heated in the shell side of the heat exchanger at a rate of 3000 kg/hr. If the outside tube surface is at 70°C, calculate the heat transfer coefficient of milk.

Solution : $D_o = \frac{15}{1000} = 0.015$ m

$$D_i = 600 \text{ mm} = 0.6 \text{ m}$$

$$d_c = \frac{25}{1000} = 0.025 \text{ m}$$

$$P = \frac{300}{1000} = 0.3 \text{ m}$$

$$S_a = 0.6 \times 0.3 \left(1 - \frac{0.015}{0.025}\right)$$

$$= 0.072 \text{ m}^2$$

For 25% baffles, factor $f_b = 0.1955$

$$N_b = 0.1955 \times 121 = 23.64 \text{ say 24 tubes}$$

Therefore,

$$S_p = 0.1955 \frac{3.1416 \times 0.6 \times 0.6}{4} - 24 \frac{3.1416 \times 0.015 \times 0.015}{4}$$

$$= 0.051 \text{ m}^2$$

$$v_a = \frac{3000}{0.072} = 41666.7 \text{ kg/m}^2 - \text{hr}$$

and

$$v_p = \frac{3000}{0.051} = 58823.5 \text{ kg/m}^2 - \text{hr}$$

$$v_c = \sqrt{v_a v_p}$$

$$= 49507.4 \text{ kg/m}^2 - \text{hr}$$

The properties of milk are

$$\mu \text{ and } \mu_w = 2.12 \ Cp = \frac{2.12}{1000} \text{ kg/m} - \text{s}$$

$$C_p = 0.93 \ k\text{cal/kg} - °\text{C}$$

$$k = 0.43 \ k\text{cal/hr} - \text{m} - °\text{C}$$

$$\frac{h_s D_o}{k} = 0.2 \left(\frac{0.015 \times 49507.4}{0.00212}\right)^{0.6} \left(\frac{0.93 \times 0.00212}{0.43}\right)^{0.23}$$

$$\left(\frac{0.93}{0.93}\right)^{0.14} = 165.1$$

or,

$$h_s = \frac{165.1 \times 0.43}{0.015}$$

$$= 4732.8 \ k\text{cal/m}^2 - \text{hr} - °\text{C}$$

Problem : In a counter flow type exchanger milk is being cooled by chilled water at 1°C. The milk is flowing at a rate of 250 kg/hr and cooled from 35°C to 4°C. The flow rate of water is four times the flow rate of milk. If the overall thermal conductance of the exchange is 510 *k* cal/hr – m² – °C and the specific heat of milk is 0.93 *k* cal/kg – °C, estimate the cooler surface area.

Solution : In the above problem it is given that

$$t_{h_i} = 35°C,\ t_{h_o} = 4°C$$

and $t_{c_i} = 1°C$, we have to determine the temperature of cooled water

Since $$q = m_h\, c_{p_h}\left(t_{h_i} - t_{h_o}\right) = m_c\, c_{p_c}\left(t_{c_o} - t_{c_i}\right)$$

Therefore, $250 \times 0.93\ (35 - 4) = 250 \times 4 \times 1\ (t_{c_o} - 1)$

or, $$t_{c_o} - 1 = \frac{250 \times 0.93 \times 31}{250 \times 4}$$

or, $$t_{c_o} = 8.2°C$$

For counter flow exchanger

$$\theta_i = 35 - 8.2 = 26.8°C$$

$$\theta_o = 4 - 1 = 3°C$$

$$\Delta t_m = \frac{\theta_i - \theta_o}{\ln\frac{\theta_i}{\theta_o}} = \frac{26.8 - 3}{\ln\frac{26.8}{3}}$$

$$= 10.86°C$$

$$q = UA\,\Delta t_m = 250 \times 0.93\ (35 - 4)$$

or $$A = \frac{250 \times 0.93 \times 31}{510 \times 10.86}$$

$$= 1.3\ m^2$$

Problem : In a parallel flow heat exchanger milk with a flow rate of 500 kg/hr and at temperature of 40°C enters the exchanger. The milk is being cooled by chilled water at 1°C and flow rate of 2000 kg/hr. The effective heat transfer area is 3.0m² and the overall heat transfer coefficient is 980 *k* cal/m² – hr – °C. The specific heat of milk is 0.93 *k* cal/kg – °C, determine the out going temperature of milk and water and the effectiveness of the exchanger.

Solution :

$$m_h\, c_{p_h} = 500 \times 0.93 = 465$$

$$m_c\, c_{p_c} = 2000 \times 1 = 2000$$

heat loss by milk = heat gained by water

$$\therefore \quad 500 \times 0.93\ (40 - t_{h_o}) = 2000\ (t_{c_o} - 1)$$

$$\frac{t_{c_o} - 1}{40 - t_{h_o}} = \frac{465}{2000} = 0.2325$$

or,
$$t_{c_o} - 1 = 9.3 - 0.2325\ t_{h_o}$$

or,
$$t_{c_o} + 0.2325\ t_{h_o} = 10.3$$

$$465\ (40 - t_{h_o}) = q = 980 \times 3 \times \Delta t_m$$

Since
$$\Delta t_m = \frac{\theta_i - \theta_o}{\ln \dfrac{\theta_i}{\theta_o}}$$

for counter flow exchanger

$$\theta_i = t_{h_i} - t_{c_i} = 40 - 1 = 39°\text{C}$$

$$\theta_o = t_{h_o} - t_{c_o}$$

$$465\ (40 - t_{h_o}) = 980 \times 3 \left[\frac{39 - (t_{h_o} - t_{c_o})}{\ln \dfrac{39}{(t_{h_o} - t_{c_o})}} \right]$$

or,
$$\ln \frac{39}{(t_{h_o} - t_{c_o})} = 6.32 \left[\frac{39 - (t_{h_o} - t_{c_o})}{(40 - t_{h_o})} \right]$$

$$= 6.32 \left[\frac{(40 - 1) - (t_{h_o} - t_{c_o})}{(40 - t_{h_o})} \right]$$

$$= 6.32 \left[\frac{(40 - t_{h_o}) - (t_{c_o} - 1)}{40 - t_{h_o}} \right]$$

or,
$$\ln \frac{39}{t_{h_o} - t_{c_o}} = 6.32 \left[1 + \frac{t_{c_o} - 1}{40 - t_{h_o}} \right]$$

$$= 6.32\ (1 + 0.2325) = 7.789$$

or,
$$\frac{39}{t_{h_o} - t_{c_o}} = e^{7.789} = 2413.9$$

$$t_{h_o} - t_{c_o} = \frac{39}{2413.9} = 0.016$$

$$t_{h_o} = t_{c_o} + 0.016$$

$$= 10.3 - 0.2325\ t_{h_o} + 0.016$$

or, $t_{h_o} = 8.36°C$

and $t_{c_o} = 8.35°C$

Effectiveness of the heat exchanger (counter flow)

$$E_C = \frac{c_{p_h}(t_{h_i} - t_{h_o})}{c_{\min}(t_{h_i} - t_{c_i})}$$

Here $c_{\min} = c_{p_h}$

$$\therefore \quad E_C = 1 \times \frac{t_{h_i} - t_{h_o}}{t_{h_i} - t_{c_i}}$$

$$= \frac{40 - 8.36}{40 - 1}$$

$$= 0.811$$

Problem : Calculate the required heat exchanger area for parallel as well as counter flow arrangements, if 600 kg/hr of water is cooled from 80°C to 50°C by means of flow rate of cooled water at 10°C and at 900 kg/hr. Assume the overall heat transfer coefficient to be 500 k cal/m^2 – hr – °C and specific heat of water 1 k cal/kg °C

Solution : The rate of heat transfer in terms of hot water

$$q = 600 \times 1\ (80 - 50)$$

$$= 18{,}000\ k\,\text{cal/hr}$$

since heat lost = heat gained

$$18{,}000 = 900 \times 1\ (t_{c_o} - 10)$$

or, $t_{c_o} = 30\ °C$

$$q = UA\,\Delta t_m$$

Therefore, the required heat transfer area $= \dfrac{q}{U\,\Delta t_m}$

(i) For parallel flow arrangement

$$\Delta t_m = \frac{\theta_i - \theta_o}{\ln \frac{\theta_i}{\theta_o}}$$

$$\theta_i = t_{h_i} - t_{c_i} = 80 - 10 = 70\ °C$$

$$\theta_o \lambda = t_{h_o} - t_{c_o} = 50 - 30 = 20°C$$

$$\Delta t_m = \frac{70 - 20}{\ln \frac{70}{20}} = 39.9°C$$

therefore, $$A = \frac{q}{U\,\Delta t_m} = \frac{18000}{500 \times 39.9}$$

$$= 0.9 m^2$$

(ii) For counter flow arrangement

$$\Delta t_m = \frac{\theta_i - \theta_o}{\ln \frac{\theta_i}{\theta_o}}$$

$$\theta_i = t_{h_i} - t_{c_o} = 80 - 30 = 50°C$$

$$\theta_o = t_{h_o} - t_{c_o} = 50 - 10 = 40°C$$

$$\Delta t_m = \frac{50 - 40}{\ln \frac{50}{40}}$$

$$= 44.8°C$$

$\therefore$ $$A = \frac{q}{U\,\Delta t_m} = \frac{18000}{500 \times 44.8}$$

$$= 0.8\ m^2$$

Variation in overall heat transfer coefficient : One of the assumptions in the analysis of exchanger was that the overall heat transfer coefficient remains constant, but infact the value of the overall heat transfer coefficient varies from one end to the other. In case the variation in overall coefficient of heat transfer is calculated as

$$U = a + b\,\Delta t$$

where, a and b = constants

The equation 7.24 becomes,

$$\frac{d(\Delta t)}{\Delta t\,(a + b\,\Delta t)} = \frac{dA(\Delta t_2 - \Delta t_1)}{q_t}$$

Integrating

$$(\Delta t_2 - \Delta t_1) \cdot \frac{A}{q_t} = \frac{1}{a} \ln \left[\frac{\Delta t}{a + b\,\Delta t} \right]_{\Delta t_1}^{\Delta t_2}$$

$$= \frac{1}{a} \ln \left[\frac{\Delta t_2 \, (a + b\Delta t_1)}{\Delta t_1 \, (a + b\Delta t_2)} \right]$$

Putting, $a + b\,\Delta t_1 = U_1$ and $a + b\,\Delta t_2 = U_2$

we have
$$a = A\,\frac{U_1\,\Delta t_2 - U_2\,\Delta t_1}{\Delta t_2 - \Delta t_1}$$

and
$$q = A\,\frac{U_2\,\Delta t_1 - U_1\,\Delta t_2}{\ln\left(\dfrac{U_2\,\Delta t_1}{U_1\,\Delta t_2}\right)}$$

In case of double pipe heat exchanger, for increasing the heat transfer rate it is common practice to increase the effectiveness by providing a U bend to the pipe. If such an arrangement is made in a single shell is known as one shell pass, two tube heat exchanger. For the arrangement the analysis for heat exchanger can be done by incorporating a correction factor applied to the LMTD. The values of correction factor may be obtained from curves in Fig. 7.23. The heat transfer can be given by

$$q = UAF\,\Delta tm$$

where, F = correction factor

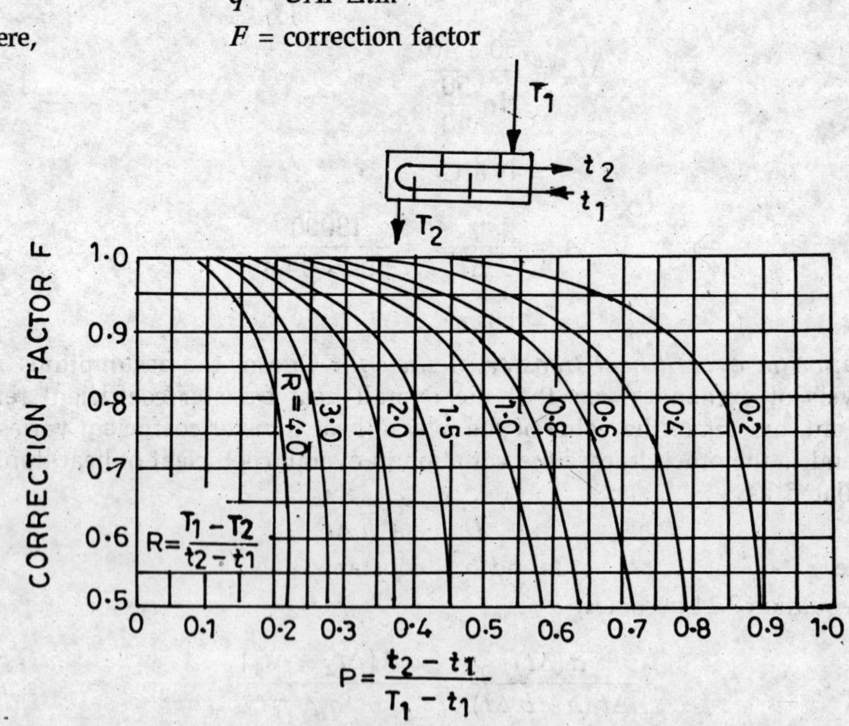

Fig. 7.23 : Correction factors for one shell pass multiple passes of tubes

Problem : In a single shell exchanger with 8 tubes HOT fluid flows through the shell side and oil flows through the tubes. The hot fluid enters at 60°C and leaves the exchanger at 40°C, while the oil enters at 10°C and leaves at 25°C. If the overall coefficient of heat transfer is 200 k cal/m^2–hr–°C and heat transfer rate is 40,000 k cal/hr, find the total surface area required for heating.

Solution : In this case the correction factor is to be determined for single shell pass

$$R = \frac{T_1 - T_2}{t_2 - t_1} = \frac{60 - 40}{25 - 10} = 1.33$$

$$P = \frac{t_2 - t_1}{T_1 - t_1} = \frac{25 - 10}{60 - 10} = 0.3$$

From Fig. 7.23 the correction factor F is 0.95

$$\theta_i = t_{h_i} - t_{c_o} = 60 - 25 = 35$$

$$\theta_o = t_{h_o} - t_{c_i} = 40 - 10 = 30$$

$$\Delta t_m = \frac{5}{\ln \frac{35}{30}}$$

$$= 32.4°\text{C}$$

$$q = UA\,\Delta t_m\,F$$

$$A = \frac{q}{U \Delta t_m\,F}$$

$$= \frac{40{,}000}{200 \times 32.4 \times 0.95}$$

$$= 6.49 \text{ m}^2$$

Exercise

1. Sliced and blanched potato chips are dehydrated in a tray dryer. The initial moisture content of chips was 85% (wb) and it is to be dried to 5% moisture content (wb). The critical moisture content is 24% (wb). The constant rate drying continues for 7 minutes, calculate the total drying time for the product.
2. In a single effect evaporator, fruit juice is being concentrated at steady state condition. The feeding rate of juice is 1582 kg/hr with 12% solids. The juice is to be concentrated to 72% total solids. The inlet feed temperature is 35°C while the product boils at 62°C inside the evaporator. The specific heats of dilute and thick fruit juice are 0.91 and 0.49 k cal/kg °C respectively. The overall heat transfer coefficient is 785 k cal/m^2 - hr – °C. Assuming the boiling point elevation as negligible, calculate the mass flow rate of thicker

juice, steam required, economy of operation and the total transfer area. The temperature of saturated steam entering the evaporator is 125°C.

3. In a counter flow type heat exchanger milk is being cooled by chilled water at 1°C. The milk is flowing at a rate of 300 kg/hr and cooled from 30°C to 4°C. The flow rate of water is 4.5 times the flow rate of milk. If the overall thermal conductance of the exchanger is 520 *k* cal/m^2 – hr –°C and specific heat of milk is 0.9 *k* cal/kg – °C, calculate the cooler surface.
4. In a parallel flow heat exchanger milk with a flow rate of 450 kg/hr and at temperature of 35°C enters the exchanger. The milk is being cooled by chilled water at 1°C and flow rate of 1800 kg/hr. The effective heat transfer area is 2.8 m^2 and the overall heat transfer coefficient is 975 *k* cal/m^{2}hr °C. The specific heat of milk is 0.93 k cal/kg °C calculate the outgoing temperatures of milk and water and the effectiveness of exchanger.
5. Calculate the required heat exchanger areas for parallel as well as counter flow arrangements if 500 kg/hr of orange juice is cooled from 40°C to 10°C by cooled water at 5°C. The flow rate of cooled water is 800 kg/hr. Assume the overall heat transfer coefficient to be 510 *k* cal/m^2 hr °C, the specific heats of juice and water are 0.85 *k* cal/kg °C and 1.0 *k* cal/kg °C respectively.

Bibliography

Ahmad Tufail, 1985. Dairy Plant Systems Engineering. Kitab Mahal, Allahabad.

Brennan, JG, JR Butters, ND Cowell and AEV Lilly, 1976. Food Engineering Operations. Applied Science Publishers Limited, London.

Clark, JP, 1978. Dough rheology in extrusion cooking. Food Tech. 32 : 76.

Cleland, AC and RL Earle, 1977. A comparison of analytical and numerical methods of predicting the freezing times of foods. J. Food Science 42 : 1390-1395.

Donohue, D.A. 1949. Ind. Eng. Chem. 41 : 2499.

Fennema, O and WD Powrie, 1964. Fundamentals of low temperature food preservation. Advances in Food Res. 13 : 220.

Foust, A.S., L.A. Wenzel, C.W. Clump, L Maus and L.B. Andersen, 1980. Principles of unit operations, 2nd Edition, John Wiley and Sons, New York.

Frazier, PJ, A Crawshaw, NWR Daniels and PW Russell Eggitt. 1982. Optimisation of process variables in extrusion texturing of soya. Paper presented at the "Extrusion Cooking Symposium" 7th World cereal and bread congress, Prague.

Heldman, DR, 1988. Factors influencing food freezing rates. Food Tech. 37(4) : p 103-109.

King, CJ and JP Clark, 1968. Convective heat transfer for freeze drying of foods. Food Tech. 22 : 1235-1239.

Linko, P, YY Linko and J Olkku, 1982. Extrusion cooking and bioconversions. Paper presented at "Extrusion Cooking Symposium", 7th World cereal and bread congress, Prague.

Levine, L., 1988. Understanding extruder performance. Cereal Foods World Vol. 33 No. 12 : 963-970.

McCabe, WL, JC Smith and P Harriott, 1985. Unit operation of chemical engineering, 4th ed. McGraw Hill, New York.

Michael, AM and TP Ojha, 1987. Principles of agricultural engineering, 2nd ed. Vol. I. Jain Brothers, New Delhi.

Mustakas, GG, WJ Albrecht, GN Bock Walter, JE McGhee, WF Kwolek and EL Griffin, 1970. Extruder processing to improve nutritional quality, flavour and keeping quality of full fat soy flour. Food Tech. 24 : 1290-1296.

Singh RP and CY Wang, 1977. Quality of frozen foods. A review J. Food Process Eng. 1(2) : 97.

Van Arsdel, WB, MJ Copley and AJ Morgan Jr. (Eds.), 1973. Food Dehydration. Second Edition Vol. I AVI Publishing Co. Westport Connecticut.

Van-Zuilichem, DI, W Stolp and LPBM Janssen, 1983. Engineering aspects of single and twin-screw extrusion-cooking of biopolymers in "Extrusion Cooking Technology" Ronald Jowitt (Ed.) Elsevier Applied Science Publishers London and New York.

Woolrich, WR and AF Novak, 1977. Refrigeration Technology in 'Fundamentals of Food Engineering' pp 23-79. AVI Publishing Co. Westport Connecticut.

Chapter 8

Agricultural Wastes and By-Products Utilisation

Agricultural wastes are excesses of agricultural production that have not been effectively utilised. With remarkable increase in agricultural production in India, the availability of different types of agricultural by-products, wastes and residues have also increased. These include crop residues, processing by-products, animal refuse and forest biomass. It is estimated that about 350 million tonnes of such products are annually produced in India in the forms of straw, stalks, husk, cobs, shells, sticks, peels etc. But bulk of such materials remain unutilised and their disposal becomes a problem and also becomes a source of environmental pollution. This happens due to their scattered and seasonal availability, bulky nature, low density and physio-chemical characteristics. The annual availability of agricultural wastes which could be used for energy production is given in Table 8.1 and the thermal capacity of agricultural wastes is given in Table 8.2.

Table 8.1

The total annual organic matter production of different crops and matter available for energy uses (million tonnes)

Crop	*Total organic matter production*	*Organic matter available for energy uses*
Cereals	275	70
Pulses	35	20
Oilseeds	25	10
Sugar cane	100	50
Cotton	5	5
Others	5	5

Table 8.2

Agricultural wastes and their thermal capacity

Crop wastes	*Moisture Content, %*	*Thermal capacity,, k. cal/kg.*
Paddy	10.6	3200
Groundnut	12.0	4200
Bagasse	15.0	3800
Cotton	12.0	3350
Maize	8.0	3500

Crop wastes	*Moisture Content, %*	*Thermal capacity,, k. cal/kg.*
Pearl millet	11.2	3850
Lentil	10.1	3810
Coconut	6.0	4350
Bamboo	10.5	3925
Fuel wood	7.0	3500

Most waste management approaches are methods of concentration and /or relocation of wastes. Recycling, reprocessing and utilisation of the wastes in a positive manner offer the possibility of returning the excesses to beneficial use. Most of the agricultural wastes are source of number of beneficial uses like building materials, feeds, fuels, paper and board, chemical, medicines etc. as given in Table 8.3.

Table 8.3

Economic products from agricultural wastes /by-products

Economic product	*Agricultural wastes/by-products*
1. Fuel	Animal dung, community wastes, crop wastes such as rice and wheat straw, maize and sorghum cobs/stalks, rice husk, bagasse etc.
2. Protein food	Oil cakes, meat wastes, okara etc.
3. Oils and fats	Rice bran, tea., coffee, cotton, mahua seeds, animal fats, non-edible oil seed such as neem, sal, karanja etc.
4. Cattle and poultry feed	Straw, husk, acquatic weeds, leaves, forest grasses, mango stone, wastes from banana, pea, tomato and potato, oil cakes, bagasse, meat and fish wastes etc.
5. Pulp, paper and boards	Bamboo, bagasse, sticks, stalks, husk, straws, shell and leaves of certain crops, cotton and jute wastes, paper cuttings.
6. Furfural	Corn cobs, coconut shells, bagasse, rice husk, sorghum stalk
7. Activated carbon	rice husk, coconut shell, groundnut shell, jute stalk, bamboo, bagasse
8. Alcohal	Molasses, saw dust
9. Starch	banana pseudostem, mangokernel
10. Organic manure	Crop wastes, weeds, cattle wastes, oilcakes, sewage etc.
11. Monosodium gluconate	Bagasse.

Fuels and Plant nutrients

Various agricultural wastes have been traditionally used as fuel which could be used as solid, liquid or gaseous product. The processes for converting such organic materials into gaseous or liquid fuels include both dry (non-biological) and wet (biological) processes. The processes are based on thermal, chemical and enzymatic conversion methods.

The dry process is burning of waste materials in excess of air with near complete combustion of the material. Commercial applications of this process are mainly for drying, steam production or electrical generation. Another dry process is combustion of wastes in a controlled atmosphere, which is known as gasification process. It involves burning of tightly packed organic matter with a limited air supply at temperature above 1100°C. This gives producer gas which has an energy content of about one sixth of that of petroleum gas.

Agricultural wastes and by-products such as rice husk, ground nut shell, cotton sticks, vegetable wastes can be dry burned to power IC engines. In another method the agricultural wastes are dry heated in the absence of oxygen which results in destructive distillation. The chief product of this process is charcoal. The process is also known as pyrolysis which is a refined method of destructive distillation. In involves heating of organic matter at high temperature (550–1100°C) in the absence of oxygen for several hours to produce various proportions of gas,, oil and charcoal. Charcoal may be used directly or as briquetted fuel.

The production of charcoal in the developing countries is widespread and long established craft. It provides a valuable product for both domestic and industrial markets. In all charcoal manufacturing processes, one of the most important factors affecting the yield is the moisture content of raw material used. Efforts to season the wood to a moisture content of below 35–40% (wb) are generally worth while. On an average 0.4 kg of charcoal is obtainable from 1.0 kg of wood.

Another dry process is hydrocarbonisation or liquification. In this process high temperature and pressure in combined way is applied to wastes to produce oil/gas. But the operating cost of hydrocarbonisation process is high.

Composting is one of the methods to recover and reuse a portion of the nutrients and organic fraction remain in agricultural wastes. The high carbon content agricultural wastes are frequently combined with animal wastes for composting. Composting is a microbiological process and it is a natural process. The composting can be done under controlled and uncontrolled biological systems. The system for treatment of liquid or solid wastes may be aerobic or anaerobic.

Aerobic process is the biological treatment of wastes in which dissolved oxygen is present. The oxidation of organic matter is caused by molecular oxygen. It yields chemical energy to most microorganisms in the process.

There are some microorganisms which can function without dissolved oxygen. These microganisms are known as anaerobic micro organisms or anaerobes. The anaerobic microorganisms utilise compounds which do not have dissolved oxygen as oxidising agents eg. carbon dioxide, partially oxidised organic compounds, sulfate and nitrate.

The major objectives in composting are to stabilise putrescible organic matter, to conserve the crop nutrients and organic matter as much as possible, and to produce a uniform, relatively dry product. This is used as a soil conditioner and garden supplement.

Building material

Various types of industrial wastes are also available which could be used to produce building materials. Almost 30 million tonnes of rice husk (from rice milling industries), 1.5 million tonnes of red mud (from aluminium industry) and fly ash (from thermal power plants) are available in India which can be suitably processed to make inexpensive cement like binders. The process of making binder for masonry and flooring/non-structural concrete work involves, admixing rice husk with red mud and fly ash and local clay soil. This mixture is moistured with water and 50 mm balls are made. The balls are dried and fired in an open grate or kiln. The resultant ash is ground with lime in 2:1 ratio to a fineness passing through IS sieve 15. The binder thus obtained can be used for structural works.

The process of making rice husk ash bricks involves burning of the rice husk, grinding the resultant ash and mixing it with 10% lime. This mixture is moistured with water and the bricks are casted in suitable moulds are cured.

Feeds

In addition to the use of fruit and vegetable processing solids for animal feeds, animal wastes have been considered as a portion of cattle and poultry feeds. Animal wastes contain considerable amount of energy and nutritive value. Farm wastes such as poultry droppings, cattle dung and other fine textured cellulosic materials have been identified as potential alternative feed resources especially for goats and ruminants. These feed stuffs can be fed in favourable combinations with cake, wheat bran, mineral mixture and salt. Molasses and urea can also be mixed depending upon the initial nutrients availability and the energy value required for the final product. Ingredients of new proportion of feed and conventional feed is given in Table 8.4.

Table 8.4
Ingredients of conventional and new ration for animals

Ingredients	*Conventional feed*	*New pelleted feed*
Maize grain, %	37	37
Groundnut cake (deoiled), %	30	12
Wheat bran, %	30	18
Salt, %	1	1
Mineral mixture, %	2	2
Dried poultry excreta, %	—	30

The utilisation of dried chicken as part of feed for poultry and ruminants as well as the utilisation of dried ruminant excreta for ruminants has been suggested by research scientists. Wastes from ruminants are composed of a variety of nitrogen containing compounds, vitamins and undigested feed products. These contain cellulose, hemicellulose and lignin. Where bedding is used, barn wastes also contain low quality roughage such as straw, wood chip and saw dust.

Due to population growth the availability of the worlds raw material is decreasing. This phenomenon poses great threat to global food shortages. This necessiated the need for recycling of organic matters. The livestock scientists have discovered that poultry and other animal wastes could be supplemented in animal and poultry feed. The possibility of recycling of poultry and animal wastes may prevent animal husbandry to compete with man to some extent for grains. Ruminants have special digestive capacity and can take animal wastes. Properly processed animal wastes are wholesome in appearance, taste and smell. These wastes are almost 50% higher in crude protein than the conventional feeding ration. The animal wastes also contain crude fibre, calcium, phosphorus, other minerals and trace elements, vitamins etc.

The feed costs is usually 60–70% of the total livestock production cost, and since animal wastes are cheap they lead to a significant reduction in the cost of meat, milk and other animal products.

Biogas production

The biogas technology is based on the phenomenon of microbial decomposition of organic mater in the absence of air to yield a gas, mainly consisting of methane and carbon-dioxide. In this process different groups of microbes act upon complex organic matter in four stages. In the first stage, the hydrolytic bacterias degrade the high molecular weight substances like cellulose, starch, proteins, fats etc. present in the organic matter into small molecular weight compounds like fatty acids, amino acids, carbon-dioxide and hydrogen. In the second stage the acetogenic bacterias act upon the end products of the first stage and convert them into acetates and hydrogen. Simultaneously a third group of bacterial called homoacetogens convert the hydrogen and simple carbon compounds produced in first and second stages into acetates. In the fourth stage the methanogens convert the acetates and some other simple compounds like formate, carbon-dioxide and hydrogen into methane. This methane, can be collected and used as a source of energy. The stabilised organic product after digestion contains many useful substances and can be used as a source of humus and plant nutrients.

The anaerobic digestion of wastes, unlike other fermentation process not require sterilisation and subsequent innoculation. This technology is able to deal with combination of several wastes. Due to stabilisation of wastes the technology can be considered as one of the profitable ways to tackle the problem of organic waste disposal.

India, having the largest cattle population in the world offers tremendous potential for the development of biogas technology. The combined population of cattle and buffaloes in over 240 million which yield about 1000 million tonnes of cattle wastes annually. If 50% of the cattle waste could be made available for biogas generation, there is potential for setting up 12–18 million small biogas plants in the country. Apart from offering benefits of energy and fertilisers, introduction of this technology in rural areas is likely to bring about indirect benefit of improved sanitation.

Common factors affecting biogas production are type of basic organic material, temperature, pH of media, solid concentration, agitation etc.

The materials having high lignin content are difficult to decompose and therefore, and not preferred for biogasification. However, materials having a non-lignin C:N 20 to 25:1 are good for biogasification. Therefore, with a C:N ratio of 20:1 animal dung is considered to be a suitable material for biogas production.

The biogas production with naturally occuring microbes takes place within the temperature range of 10–40°C. However, at temperatures of 33–35°C maximum amount of biogas can be produced per unit weight of dry organic materials under atmospheric conditions.

The acid forming microbes can function well within the pH range of 5.5 to 7.5, whereas the methane generating microbes function well within the pH range of 6.5 to 8.5, therefore, the pit range of 6.5–7.5 is optimum.

The optimum solid content for biogas production is 8–12%. The 1:1 input of water and dung with 18–20% solid makes the solid contents to 9–10% and the same is recommended for biogas production.

Slow agitation/mixing of slurry helps in improving the biogas production.

Fibre

Fibre is one of the basic needs of human beings, Fibres are used for making clothes, ropes, mats, net, cushions and several other useful articles. The fibres may be classified into two groups; man-made fibres and natural fibres. Cotton, coconut, jute, sunhemp, flax are the main plants for the raw material for the fibre industries in India. However, there are several other plants like munj, bhabbar, aak etc. which also have good quality fibre.

Fibre is contained in the outer part of stem of plants like, jute, sunhemp, linseed, khimp etc. While husk of the coconut fruit contain major portion of fibres. The coconut fibre ropes are called "coir". The basic processes involved in the separation of fibre from the plant's component may be physical or chemical. As far as possible physical processes should be preferred over chemical process at cottage level. The physical process of fibre separation involves:

Stripping/separation of fibrous part of plant

Retting followed by extraction mannually or mechanically, cleaning and drying

Beating followed by cleaning

The operation of stripping involves the separation of fibrous component of plant by a stripper or hands. Retting is a process in which stem, leaves or shells are soaked in water for certain period depending upon the bondage of fibres, and climate. The retting operation is biological. The process is usually anaerobic, but it can also be carried out under aerobic conditions.

After retting, the fibres are extracted either manually or mechanically. The capacity of extraction depends upon the condition of retted material, types of fibres and skill of operator. In mechanical extraction, the stalk undergoes crushing, the pith fails laterally and the fibres get separated from the woody matter. The mechanical scutching machine consists of inter-meshed fluted rollers. These primary and secondary rollers crush the stalk and leaves and separate the pith. The machine can be driven manually or by some power source. The effective removal of fibre from the entire length of stalk without significant loss of fibre is, therefore, governed by a suitable combination of number of pairs of rollers and designs of breaking and crushing rollers. The extraction ability of machine also depends upon the normal load, stalk moisture, dimensions and peripheral velocity of rollers, gap between rollers and pitch, height, profile and thickness of tooth.

Floss available from aak balls and bast fibre can be used to make cloth. Aak floss is first cleaned by removing seeds. Later on tapes are made which are used to prepare silvers. Then the yarn is prepared by blending cotton in the ratio of 1:1, 2:1 or 3:1. The yarns of different blends are used to prepare cloth on handloom.

Rice by-products utilisation

In rice mills during processing of paddy-rice, the by-products generated are, (1) rice husk 20–24%, (2) rice bran 4–6%, (3) rice germ 1–2% and (4) small broken 2–3%. The rest are large brokens and head rice.

For modernisation of rice mills it is also necessary that rice by-products may profitably be used so as to generate extra income to mill owners. Economic utilisation of milling by-products is a great incentive to millers. The research and developmental works in this area is of particular importance to keep the pace of modernisation of rice mills.

Husk consists of one fifth of paddy by weight and makes up the largest milling by-product of rice milling. In India about 16 million tonnes of rice husk is generated annually. The husk is fragile, porous and has low bulk density (90 to 160 kg/m^3). It contains about 15–18% silica and is low in nutritional value and adds very little to the soil as fertiliser.

The fuel value of the husk has been reported as 3000–3500 k cal per kg, which is nearly one half of soft coal. Therefore, husk can be used for production of hot gases, steam, and running of boilers and steam engines. In Table 8.5 the calorific values of some of the Indian varieties of paddy hulls are given.

Table 8.5

Calorific values of hulls obtained from different varieties of paddy

Paddy variety	*Higher Calorific Values, k cal/kg*	*Lower Calorific Value, k cal/kg*	*Moisture Content, %*	*Calorific value at no moisture*
IR-8	2848.6	2802.2	8.67	3090
Patnai	3215.6	3106.5	10.38	3560
Padma	2902.5	2867.4	6.62	3065
Dular	3252.6	3196.6	10.50	3560
Basmati	2831.6	2798.2	6.16	2980

The percentage of volatile matter, fixed carbon and ash can be obtained by proximate analysis. The ash is the non-combustible residue left after complete combustion of the husk. Proximate analysis of some paddy varieties is given in Table 8.6.

Table 8.6

Proximate analysis of paddy husk

Paddy Variety	*Volatile matter, %*	*Fixed Carbon, %*	*Ash, %*
IR-8	68.6	14.0	17.4
Patnai	69.3	14.9	15.8
Padma	68.7	12.7	18.6
Dular	67.6	15.9	16.5
Basmati	66.6	16.9	16.5

Paddy husk ash has several uses. The husk ash can be used for the followings.

1. Manure
2. Active carbon
3. Production of sodium silicate
4. Production of cement
5. Ceramic industry
6. Production of furfural
7. Silicon tetra chloride
8. Calcium silicide

Rice bran and germ

Apart from paddy husk, the other by-products of rice mills are rise bran, polish and rice germ. Rice bran is a source of proteins, oil and nutrients. Although the rice bran is very useful by-product but it is underutilised and frequently wasted.

Rice bran is a good source of oil and high grade edible oil is extracted from raw bran, followed by refining. It is rich in vitamin E and other nutrients and has advantageous effects in lowering blood cholestrol level.

Rice germ is richer in protein and fat, but lower in fibre than bran. Flour is prepared from rice germ which is used as cattle and poultry feed. The proximate composition and vitamin contents of rice bran, rice polish and rice germ is given in Table 8.7.

The composition of rice bran depends on a variety of factors associated with the rice grain itself and milling process. Major factors associated with the rice grain are varietal and environmental variability which affects, (1) average chemical composition, (2) distribution of chemical constituents, (3) thickness of outer layers, (4) size and shape of grains and (5) resistance of grains to breakage and abrasion.

Major factors associated with the milling process are, (1) processing methods and machines, (2) milling conditions.

Table 8.7

Proximate composition and vitamin contents of rice bran, rice polish and rice germ

Constituents	*Bran*	*Polish*	*Germ*
Protein, %	14.1	10.8	24.1
Fat, %	18.8	9.1	19.3
Crude fibre, %	5.1	1.1	—
Ash, %	11.2	5.8	6.2
Sugars, %	8.7	2.6	7.4
Vitamins, μg/g			
Thiamine	24.0	22.0	65.0
Riboflavine	2.0	2.2	5.0
Niacin	336.0	330.0	83.0
Pyridoxine	25.0	20.0	16.0
Plantothenate	27.7	33.3	30.0
Insitol	4627.0	4536.0	3725.0
Cholime	1700.0	1020.0	3000.0

Rice bran oil

The bran received from whiteners is cleaned to separate broken rice, germ etc. This is usually done by screening. The next unit operation is thermal treatment. Rice bran contains an enzyme which is called 'Lipaze'. Due to presence of this enzyme the rice bran oil quality deteriorates after its separation from the brown rice. Lipaze acts on the oil and hydrolyses it, as a result after passage of time free fatty acids (FFA) increase. Increase of FFA makes the rice bran oil inedible and, therefore, the oil is used for industrial uses like soap making etc. Studies undertaken to stabilize the bran reveals that if the fresh bran is thermally treated action of lipaze can be arrested. The activity of lipaze may be retarded by thermal treatment at 100–105°C and holding the bran at this temperature for about 5 to 10 minutes. Thermally treated bran is known as stabilised bran. Oil extracted from such stabilised bran is fit for edible purpose. The popular way of extracting bran oil is by solvent extraction method using 'Hexane'. But the extracted oil needs further purification or refining.

Refining of bran oil for edible purpose is generally carried by the following processes which are listed below.

(1) Dewaxing process to remove mainly wax, organic impurities and moisture.
(2) Degumming process to remove mainly gums and mucilages.
(3) Neutralisation process to remove mainly free fatty acids.
(4) Decolourisation process to remove mainly colouring matters, metallic ingredients.
(5) Deodorisation process to remove odourous matters
(6) Winterisation process to remove mainly saturated glycerides.

Crude rice bran oil of high acid value is usually used for making solidified oil, stearic and oleic acid, glycerine and soap. Powder soap is also produced from bran oil.

Crude oil can also be combined with glycerine to make neutral oil. It can be combined with a drying agent to make oil for anticorrosive uses. Oil can also be sulfonated to make lubricants. Such lubricants are used in textile or leather finishing. The gummy waste from rice oil mixed with defatted bran or wheat bran for making cattle feed.

Bibliography

Bohra, CP, 1990. Technologies and cost economics of biogas plant. Lecture delivered in Summer Institute at CIAE, Bhopal.

Chakravorty. J, S, Prasad and MS Maity, 1976. Utilisation of paddy husk for making cement. RPEC Reporter 2 (1) : 52.

Indian Council of Agricultural Research, 1974. Fuel gas and manure by anaerobic fermentation of organic materials ICAR Tech. Bul. (Agric) 46, New Delhi.

Kaushik, SK and RM Vasan, 1985. Investigations of concerete mixes and mortars using industrial wastes. Paper presented at the National Seminar on "Role of building material industries in conservation of wastes into wealth", New Delhi March 19–21.

Loehr, Rc, 1974. Agricultural wastes management. Problems, processes and approaches. Academic press. New York.

Muller, Zo, 1984. Feed from animal wastes; feeding manual. FAO animal production and health. Paper 28 FAO, Rome.

Pryor, WJ and JK Connor, 1964. A note on the utilisation of chickens of energy from feces. Poultry Science 43:833–834.

Singh Jai, 1990. Forage conservation technology for rural employment and security. Lecture delivered in Summer Institute at CIAE, Bhopal.

Srivastava, PK and RP Kachru, 1988. Agro industrial residues: Valuable source for building materials. Productivity. 29(3) : 247–255.

Varshney, Ac and K. L. Bhoi, 1986. Development of equipment for processing munj (saccharum munja) fibre. Agric. waste. 17 : 1–6.

Varshney Ac and K.L. Bhoi, 1988. Cloth from bast fibre of the calotropis procera (Aak) plant. Biological waste, 26: 229–232.

Appendices

Appendix I

Conversion Factors

Length

1 m	= 3.2808 ft = 39.37 in
1 ft	= 0.3048 m = 30.48 cm
1 km	= 0.621 mile
1 μ	= 10^{-6} m = 10^{-3} mm

Mass

1 kg	= 2.205 lbm
1 lbm	= 0.453 kg
1 g	= 0.0022 lbm
1 quintal	= 100 kg
1 tonne	= 1000 kg

Area

1 sq cm	= 6.155 sqm
1 sq in	= 6.4515 sqcm
1 sq m	= 10.769 sqft
1 acre	= 0.4047 hectare

Volume

1 cucm	= 0.061 cu in
1 cu in	= 16.389 cucm
1 cuft	= 0.0283 cum
1 litre	= 1000 cc = 61.024 cu in

Density

1 g/cc	= 62.43 lbm/cft
1 lbm/cft	= 16 kg/cum

Temperature

K	= 273.16 + °C

°C	$= \frac{°F - 32}{1.8}$
R	= 459.69 + °F

Force

1 N (Newton)	= 10^5 g cm s^{-2} (dyne)
	= 1 kg m s^{-2} = 0.12 kgf
	= 0.2481 lbf
1kgf	= 2.205 lbf
	= 9.806 N
1 lbf	= 0.453 kgf
	= 4.4482 N
1 dyne	= 1 g cm s^{-2} = 10^{-5} kg m s^{-2}

Pressure

1 bar	= 750.06 mm Hg
	= 401.85 water
	= 0.987 atmosphere
	= 10^5 N/m^2
1 N/m^2	= 1 pascal
	= 10^{-5} bar
	= 10^{-2} kg/m s^2
1lbf/ft^2	= 4.788 × 10^2 dyne/cm^2 = 47.88 N/m^2
1 atm	= 1.0133 × 10^6 g/cm s^2
	= 760 mm Hg at 0°C
	= 14.696 blf/in^2 = 1.03 kgf/cm^2
	= 1.0132 bar
	= 1.0132 × 10^5 N/m^2

Work or Energy

1 erg	= 10^{-7} Joule
	= 2.39 × 10^{-8} Cal
1 Joule	= 10^7 erg
	= 1 Nm
	= 0.239 Cal
	= 3.7251 × 10^{-7} hp hr = 2.778 × 10^{-7} kW hr
1 BTU	= 1.055 × 10^{10} erg
	= 1.055 × 10^3 Joule
	= 251.16 Cal
	= 2.9307 × 10^{-4} kW hr
1 Cal	= 4.184 kJ
	= 3.9657 BTU
	= 1.1622 × 10^{-3} kW hr

Power

1 kW	= 1 Joule/s
	= 0.8 k Cal/hr = 14.34 Cal/min

1 hp	= 550 ft lbf/s
	= 746 Watt
	= 596.8 k Cal/hr
	= 0.7068 BTU/s
1 BTU/hr	= 0.293 W
1 J/s	= 1 W
Energy flow rate	
1 W	= 0.86 kCal/hr
	= 3.464 BTU/hr
1 BTU/hr	= 0.252 kCal/hr
	= 0.293 Watt
1 ton refrigeration	= 200 BTU/min
	= 50 kCal/min
	= 3.517 kW
Enthalpy	
1 BUT/lb	= 0.5556 kCal/kg
	= 2.325 kJ/kg = 2.325 J/g
Viscosity	
1 g/cm s (poise)	= 10^{-1} kg/m s = 100 centipoise
	= 1 dyne · s/cm^2
1 lbm/ft hr	= 0.4134 cp (centipoise)
1 kg/m s	= 1 Pa S
1 N s/m^2	= 1 Pa S
1 kg/m hr	= 0.6719 lbm/ft hr
Thermal conductivity	
1 watt/m °K	= 2.3901 × 10^{-3} Cal/sm °K
	= 0.8598 kCal/hr m °C
1 Cal/hr m°C	= 0.672 BTU/hr ft °F
	= 1.1623 watt/m°K
Specific heat	
1 BTU/lbm°F	= 4.1865 J/g°C
	= 1 Cal/g°K

Appendix II

Physical properties of dry air at atmospheric pressure

Temperature, °C	Density, kg/m^3	Specific heat, kCal/kg.°C	Viscosity, (times 10^6 kg/m.s)	Thermal Conductivity, k Cal/hr. m°C
0	1.252	0.2414	17.45	0.020
10	1.206	0.2412	17.85	0.021
20	1.164	0.2417	18.24	0.021
30	1.127	0.2419	18.68	0.022
40	1.092	0.2422	19.12	0.023
50	1.057	0.2426	19.51	0.023
60	1.025	0.2429	19.90	0.024
70	0.996	0.2431	20.39	0.024
80	0.968	0.2434	20.79	0.025
90	0.942	0.2438	21.23	0.026
100	0.916	0.2441	21.67	0.026

Appendix III

Thermal Velocities and drag coefficients of food grains

Grain	Terminal Velocity, m/s	Drag Coefficient
Rice	7.11	0.776
Wheat	9.82	0.602
Sorghum	10.29	0.442
Maize	13.85	0.552
Bengalgram	11.81	0.641
Pigeon-pea	13.48	0.425
Black gram	11.07	0.464
Green gram	10.70	0.483
Pea	14.63	0.452
Cowpea	12.48	0.493
Groundnut kernel	13.23	0.586
Soybean	13.40	0.472

Appendix IV

Frictional Coefficients of grains with respect to different surfaces

Grain	Moisture Content, %	Coefficient of internal friction	Coefficient of static friction			
			Polyethylene	Wood	Steel	R.C.C
Rice	11.8	0.571	0.525	0.427	0.467	0.490
Wheat	13.3	0.496	0.326	0.354	0.377	0.456
Soybean	12.3	0.464	0.285	0.314	0.336	0.345
Pigeon-pea	10.8	0.344	0.266	0.275	0.292	0.307

Appendix V

Thermal properties of maize, paddy and what

Grain	*Moisture Content range, % wb*	*Thermal conductivity, Cal/cms.°C × 10^5*	*Thermal diffusivity, $cm^2/s \times 10^3$*	*Specific heat, Cal/g.°C*
Wheat				
W G–357	8–24	29.202 + 0.152 M	1.011 – 0.0080 M	0.326 + 0.0082 M*
Kalyan–227	8–24	29.756 + 0.240 M	1.099 – 0.0080 M	0.334 + 0.0087 M
Maize				
Vijay	10–24	35.275 + 0.280 M	1.065 – 0.0060 M	0361 + 0.0092 M
Local	10–24	30.424 + 0.490 M	1.167 – 0.0050 M	0.291 + 0.0104 M
Paddy				
Jaya	10–24	24.485 + 0.350 M	1.158 – 0.0219 M	0.244 + 0.0114 M
Joshna–349	10–24	23.550 + 0.330 M	1.703 – 0.0259 M	0.218 + 0.014 M

* M = Moisture Content

Appendix VI

Latent heat of vaporisation for maize, wheat and free water, k Cal/kg

Moisture Content, % wb	*Temperature of evaporation, °C*				
	0	*10*	*21*	*38*	*65*
Maize					
12	653	647	642	631	614
13	636	631	625	614	594
14	622	617	611	600	583
15	608	603	597	586	569
Wheat					
12	667	661	653	642	622
13	644	639	633	622	606
14	628	622	617	606	589
15	617	611	606	594	578
Free water	597	592	586	576	560

Appendix VII

Design of Solar cabinet dryer's trays and collector's area.

(*i*) Design of trays: For design considerations of trays, same loading density as generally used for open sun-drying may be put on each tray.

$$\text{Area of tray} = \frac{\text{Total quantity be dried}}{\text{Quantity per sq.m. in open drying}}$$

From the total tray area, number of trays required are calculated by choosing required tray size.

(*ii*) The heat required for drying can be calculated by the following expression.

$$Q = \frac{M_1 - M_2}{100 - M_2} W \times 1 + w \times t \times C_p$$

Where, Q = heat required for drying, k Cal

M_1 and M_2 = initial and final moisture contents of product, %(w.b.)

W = quantity of material per batch, kg

I = latent heat of Vaporisation, 600 k Cal/kg

t = temperature difference between drying and ambient air, °C

C_p = specific heat of material, k Cal/kg

(*iii*) Efficiency of the collection system depends upon many factors such as transparent cover, type of collector surface, season, solar collector's azimuth etc. The efficiency of indirect dryers are taken to be 60% whereas for direct type dryer the same is taken to be 50%. It is assumed that solar drying time is half of the open sun time. For calculation of collector's area the following expression can be used.

(*a*) Collector area for indirect type dryers

$$\text{Area, } m^2 = \frac{Q}{T/2 \times s.d \times 0.60}$$

(*b*) Collector area for direct type dryers

$$\text{Area, } m^2 = \frac{Q}{T/2 \times s.d \times 0.50}$$

Where, T = time required in open sundrying, days

s = average sunshine hours per day

d = solar constant of location, k Cal/m^2/hr

Index

R

S